Pelican Books
An Economic History of the U.S.S.R.

Alec Nove is Professor of Economics at the
University of Glasgow. His publications include
The Soviet Economy (1961), *The Soviet Middle
East* (1965; with J. A. Newth), *Was Stalin Really
Necessary?* (1965), *Socialist Economics* (edited
with D. M. Nuti, Penguin Books 1972), *Stalinism
and After* (1975), *Planning: What, How and Why*
(1975), *The Soviet Economic System* (1977), *East–West
Trade: Problems, Prospects, Issues* (1978) and
Political Economy and Soviet Socialism (1979).
Professor Nove is married, with three sons.

ALEC NOVE *An Economic History of the U.S.S.R.*

PENGUIN BOOKS

Penguin Books Ltd, Harmondsworth, Middlesex, England
Penguin Books, 625 Madison Avenue, New York, New York 10022, U.S.A.
Penguin Books Australia Ltd, Ringwood, Victoria, Australia
Penguin Books Canada Ltd, 2801 John Street, Markham, Ontario, Canada L3R 1B4
Penguin Books (N.Z.) Ltd, 182–190 Wairau Road, Auckland 10, New Zealand

First published by Allen Lane The Penguin Press 1969
Published in Pelican Books 1972
Reprinted 1975
Reprinted with revisions 1976
Reprinted 1978, 1980
Reprinted with revisions 1982

Copyright © Alec Nove, 1969
All rights reserved

Made and printed in Great Britain by
Hazell Watson & Viney Ltd.
Aylesbury, Bucks
Set in Monotype Times

Contents

Preface

For fifty years now the Soviet regime has been established in Russia. While its leaders have been dealing with perplexing problems of many kinds, observers in the West have been writing books and articles. If these were laid end to end, they would certainly stretch rather far. Why, then, should the author, who is not even a historian anyway, have the temerity to add to their number? 'A moratorium on books about Russia!' some critics might say. Another book on this subject requires some justification.

The justification is this: there is no convenient and compact economic history of twentieth-century Russia. There has, of course, been some scholarly work in this field. E. H. Carr's monumental history has much to say about the economy. However, the Pelican edition of his volume on the economic events of just the period 1917–23 is over four hundred pages long. This is not a criticism: there are no wasted pages. There is scope for a shorter survey of the whole epoch. For the early years of Soviet power the work of Dobb is most valuable, and the late Alexander Baykov covered the period up to the war. Particulars of these and other books will be found in the Bibliography. The author has benefited from reading and re-reading them and also the quite numerous Soviet works, contemporary and recent, to which references will be found in the notes.

Anyhow, readers may find that there are some advantages in relative brevity; perhaps it will be possible to sketch in some general patterns which of necessity escape notice in a more microscopic survey.

One issue is bound to loom large: the relative importance of politics. This is, of course, an *economic* history. Yet Lenin once wrote: 'Politics cannot but have dominance over economics. To argue otherwise is to forget the ABC of Marxism.' This may surprise those who believe that the dominance of *economics* is the ABC of Marxism, but it does illustrate the proposition that Soviet politics dominated, and altered, economic relations. Out of the horse's mouth, too.

It is undeniable that politics have, in an important sense, been dominant. None the less, this must not be taken too literally or understood too superficially: it does not mean that politicians could do what they liked with the economy; it does not mean that economic issues were unreal. Politicians responded to economic problems, struggled with varying success with economic perplexities. Being for most of the period in command of the major part of economic life, the politicians were, for most of their waking hours, the board of directors of the great firm U.S.S.R. Ltd. In other words, their actions as politicians were interpenetrated by their function as super-managers. Thus, Lenin notwithstanding, it is hardly possible to draw a clear line between politics and economics.

In this book we will concentrate on *economic* policies, decisions, events, organizations, and conditions without for a moment wishing to suggest that other emphases are not perfectly legitimate too. In a general history the balance would naturally be different, even if this author were writing it. Perhaps this is most obvious in the relatively brief chapter on the war of 1941–5. This great drama deserves the closest attention from many points of view, but the purely military side is so overwhelmingly important that it seemed right, in an economic history, to give these years a somewhat cursory treatment.

All dates are given in the *present* calendar, i.e. the revolution happened on 7 November, not 25 October 1917. Quotations from Russian works are translated by the author, except where the quotation is from an edition in English.

The manuscript was read and critically reviewed by Dr Sergei Utechin and Mr Jacob Miller, and I owe a great deal to their generosity in picking holes in insufficiently closely knit arguments and drawing attention to numerous errors and omissions. Professor R. W. Davies was kind enough to show me a draft of work in progress which I drew upon in writing Chapter 4, and Dr Moshe Lewin generously allowed me to borrow from his storehouse of information on the Soviet peasantry. Some chapters are based on seminar papers delivered in various British and American universities, and have benefited from critical comments from the participants. Roger Clarke was most helpful in looking up facts and checking tables. Naturally I take full responsibility for all mistakes of fact and interpretation.

Finally, grateful acknowledgement must be made to Mrs M. Chaney and Miss E. Hunter for not complaining (very often) at the succession of semi-legible drafts which descended upon them, and turning these into a manuscript which could be sent to the printers.

1. *The Russian Empire in 1913*

INDUSTRIAL GROWTH

In the last years before it was engulfed by war and revolution, the Russian Empire had reached a level of development which, though leaving it well behind the major industrialized Western powers, was none the less appreciable. It would be quite misleading to assume that the communists took over a wholly undeveloped and illiterate country with a stagnant economy. So our first task must be to take a brief look at the progress of the empire, and at least by implication to consider whether she was well on the road to a modern economy when the process was interrupted in 1914.

Russia in 1854 faced the Western powers with an obsolete social organization and obsolete weapons. Society was still dominated by an inflexible caste system, and most of the peasants were serfs owned by the landed proprietors, the State or the Crown. Industry had languished since 1800. At this date Russian output of metal had been equal to Britain's; by 1854 she had fallen very far behind. The only railway of importance that had been completed ran from St Petersburg to Moscow, with a line to Warsaw under construction. The Russian army in the Crimea had to be supplied by horse and cart on dirt tracks; this army consisted of serfs serving virtually for life, and was poorly armed and equipped. The fleet had no steamboats and could only be sunk to block the entrance to Sevastopol. The military failure in the Crimea was a great shock to Tsar and society alike. The empire as a military power had failed to keep pace with the changing world; it had to be modernized. No doubt all this helped to precipitate the abolition of serfdom. At first the government remained strongly influenced by conservative and traditional views, as may be seen from the limitations on the mobility and enterprise of peasants which formed part of the emancipation provisions of 1861. It was only gradually that the conscious pursuit of industrialization became a major motivation of policy.

However, the need for railway building was clearly understood after the terrible lessons of the Crimea, and the building of them vitally influenced the development of the Russian economy in the second half of the nineteenth century. No one doubts that in the fifty-three years which separated the abolition of serfdom from the outbreak of the First World War there had been rapid economic growth and major social change, and it is certainly of interest to compare Russia's growth at this period with that of other countries and with the subsequent achievements of the communists. This is an extremely difficult task, especially as the statistics are often confusing and defective. In pre-revolutionary times systematic figures tend to be available only for large- and medium-scale industry, whereas handicrafts and small workshops were still extremely important. It is a task beyond the scope of the present study to attempt any recalculation of statistics and growth rates under the empire. So far as industry is concerned, Goldsmith's admirably thorough reconstructions (which also cite and develop indices calculated by the eminent Russian economist Kondratiev) give a number of different index series, depending on the weights used. For simplicity I will cite here his value-added 'imputed' weights with 1900 prices.[1]

Industrial output (manufacturing and mining) (1900 = 100)

1860	13·9	1896	72·9	1905	98·2
1870	17·1	1897	77·8	1906	111·7
1880	28·2	1898	85·5	1907	116·9
1890	50·7	1899	95·3	1908	119·5
1891	53·4	1900	100·0	1909	122·5
1892	55·7	1901	103·1	1910	141·4
1893	63·3	1902	103·8	1911	149·7
1894	63·3	1903	106·5	1912	153·2
1895	70·4	1904	109·5	1913	163·6

For the period 1888–1913 this index gives one a growth rate of just about 5 per cent per annum. This was fairly high – higher on a *per capita* basis than in either the United States or Germany. However, the much slower rate of increase in agriculture, and the high share of agriculture in Russia's employment and national income, made the overall performance appear much more modest. Rough national income estimates, made by Goldsmith, show Russian growth rates well below those of the United States and Japan, a little below that of Germany, though above Britain and France. With Russia's very rapid increase in population, the *per capita* figures were less favourable still. Goldsmith considers that Russia's real income per head was *relatively* higher in 1860, in comparison with the United States and Japan, than in 1913; in other words their growth was more rapid than Russia's. Growth, though very rapid in certain years, was exceedingly uneven. For example, in the decade 1891–1900 industrial production more than doubled, and, in particular, there was a very marked advance in heavy industry. This was the consequence of the protective tariffs introduced in 1891, and of the deliberate policy followed in subsequent years by Count Witte, who became Minister of Finance. The output of pig iron in Russia trebled during the decade, while production in Germany increased in these same years by only 1·6. Output of oil during this decade kept pace with that of the United States, and in fact in 1900 Russia's oil production was the highest in the world, being slightly ahead of America's. The same decade saw a great railway boom, with the total track mileage increasing by 73·5 per cent. However, an economic crisis led to a slowdown of growth in the years 1900–5 and again in the period 1907–9. This particularly affected manufacture of iron and steel, and it was not until 1910 that the output of pig iron surpassed the 1900 level. From then until the outbreak of war there was another sharp upswing in industrial production. A recent Soviet textbook, which is not likely to overstate the achievements of Tsarism, has put forward the following estimates: during the period 1860–1910 the world's industrial production increased by six, Great Britain's by 2·5, Germany's by six and Russia's by 10·5.[2]

The argument is advanced by Soviet historians that the economic growth of the empire was none the less much too slow and that Russia remained very far behind the more advanced countries. In percentage terms Russia's growth compared favourably with her rivals, but it was still inadequate in relation to her rich natural resources and to the great gap which separated her from western Europe and the United States. The following figures given by a present-day Soviet source, and relating to the present boundaries of Russia, compare Russia's production figures with those of the U.S.A. and the United Kingdom:

1913	Russia	U.S.A.	U.K.
Electricity (milliard Kwhs)	2·0	25·8	4·7
Coal (million tons)	29·2	517·8	292·0
Oil (million tons)	10·3	34·0	–
Pig iron (million tons)	4·2	31·5	10·4
Steel (million tons)	4·3	31·8	7·8
Cotton textiles (milliard metres)	1·9	5·7	7·4

(SOURCE: *Promyshlennost' SSSR*, 1964, pp. 112–16.)

It is interesting to note that the oil industry failed to maintain its rate of progress and in fact fell back in the first decade of the century.

An original and skilful attempt to measure the relative progress of the powers was made by P. Bairoch.[3] The result strongly supports the view that Russia, despite her very considerable growth, was not making much headway in catching up with the more advanced countries. Bairoch's calculations are based on a combination of the following: consumption of raw cotton and coal, production of pig iron, the railway network, and power generation. For Russia he uses mainly data taken from Goldsmith's study. All figures are expressed *per capita*. Given the statistical inadequacies of the nineteenth century, this method has much to commend it, though the author would be the last to claim its accuracy. Russia's industrial performance is shown in the table on page 15. This table shows that, far from overtaking even Spain, Russia in these fifty years fell behind Italy. The author comments: *'Dès la fin du XIXième siècle, c'est la Russie qui se place au dernier rang des pays européens étudiés ici.'*

Similar conclusions follow for the calculations of S. N. Prokopovich:[4]

	National income		
	1894	1913	Growth
	(Roubles *per capita*)		(*per cent*)
United Kingdom	273	463	70
France	233	355	52
Italy	104	230	121
Germany	184	292	58
Austria-Hungary	127	227	79
Russia (in Europe)	67	101	50

Relative industrial progress of world powers 1860–1910

	Raw cotton (Kg. per head)		Pig iron (Kg. per head)		Railways *		Coal (Kg. per head)		Steam power (h.p. per 1000 persons)		Ranking list	
	1860	1910	1860	1910	1860	1910	1860	1910	1860	1910†	1860	1910
Germany	1·4	6·8	14	200	21	75	400	3190	5	110/130	6	4/5
Belgium	2·9	9·4	69	250	30	102	1310	3270	21	150	2/3	3
Spain	1·4	4·4	3	21	6	58	–	330	–	4	8	8
U.S.A.	5·8	12·7	25	270	19	122	420	4580	25	150/180	2/3	1
France	2·7	6·0	25	100	18	87	390	1450	5	73	5	6
Italy	0·2	5·4	2	8	6	38	–	270	–	14/46	9/10	9
Japan	–	4·9	–	5	–	14	–	230	–	7/10	11	11
U.K.	15·1	19·8	130	210	44	69	2450	4040	24	220/240	1	2
Russia	0·5	3·0	5	31	1	24	–	300	1	?/16	9/10	10
Sweden	1·5	3·6	47	110	3	76	90	910	–	55/150	7	7
Switzerland	5·3	6·3	–	–	28	88	–	–	–	85/190	4	4/5

* Total length related to population and area.
† The higher figure includes other forms of power.
– Negligible or not available.
NOTE: Most figures represent an average over several years.

SOURCE: A. Kogan, Problemy ekonomiki, No. 1 (1939, p. 34)

The Russian engineer-economist, Professor Grinevetsky, came to the same conclusion. Quoting Russia's backwardness in metal-goods industries, he wrote:

These comparisons eloquently speak of the fact that Russia in her pre-war economic growth was not merely not catching up the younger countries with powerful capitalist development, but was in fact falling behind. This conclusion would be very sad for our social-political vanity, but it must be considered as an indubitable fact.[5]

Russia was thus the least developed European power, but a European power none the less. She was capable of overwhelming militarily and competing economically with a partly developed European state such as Austria–Hungary. But her development was exceedingly uneven both industrially and geographically. Her modern industry was very modern indeed, with a marked tendency to large and well-equipped factories using the most up-to-date Western models. These were principally in the areas of St Petersburg and Moscow, in Russian Poland and in the Ukraine. The main metallurgical centre was now in the south, using Donets Basin coal. The older Urals metallurgical centre was declining. Most of the rest of the country had very little industry other than handicrafts. Apart from the oil of Baku, the southern and eastern territories were particularly primitive.

A disproportionate share of some industries was concentrated in areas lost to Russia after the First World War and the civil war (the Baltic states, and territories which became part of Poland and Roumania). The following table illustrates this:

	1912	
	Total value of production	
	In retained territory	In lost territory
	(millions of roubles)	
All industry	6059	1384
Wool	344	297
Leather	76	44
Paper	61	33
Jute and sacks	28	14
Woodworking	163	53
Chemicals	223	64
Cotton fabrics	1389	364
Metal goods	1137	258

(SOURCE: V. Motylev, *Problemy ekonomiki*, No. 1 (1929), p. 36.)

The relative importance of small-scale (workshop and artisan) industry at this period may be illustrated by the following figures: in 1915 it employed about 67 per cent of those engaged in industry or 5·2 million persons. It produced 33 per cent of industrial output,[6] i.e. the output per head was only a quarter of that of workers engaged in large-scale industry. This shows the contrast between the modern and the old, between great industrial plant and tiny cottage-industry or workshop, which, of course, was and is to be found in other developing countries.

A similar unevenness characterized the growth of different sections of industry. Thus while there was impressive growth in the metallurgical, textile, fuel and food-processing industries, engineering lagged far behind. Most industrial equipment continued to come from abroad. This weakness was to be a principal cause of the catastrophic shortage of armaments when war broke out, not least because of the dominance of Germany as a supplier of equipment.

CAPITAL, DOMESTIC AND FOREIGN

The progress of Russian industrialization suffered from relative shortage of capital, as well as from a poorly developed banking system and a generally low standard of commercial morality. The traditional Muscovite merchants, rich and uneducated, were far from being the prototypes of a modern commercial capitalism. The situation changed towards the end of the nineteenth century, and particularly during the rapid industrialization which characterized the nineties. There was a marked growth of both Russian and foreign capital, and an equal improvement in the banking system. Russian entrepreneurs of a modern type began more and more to emerge. Under cover of the protective tariff of 1891, and with the establishment of a stabilized rouble based on the gold standard, foreign capital received every encouragement. This was particularly the work of Count Witte, who exercised a dominant influence over Russian financial and commercial policy at this time. His public statements and papers make it abundantly clear that he was pursuing deliberately a policy of industrialization, and that the dominant motive was the traditional one that a relatively backward Russia must catch up with the more developed powers, particularly in her potential to produce the means of national power, above all armaments. While anxious to obtain foreign financial help in the form of loans and investments, Witte was frankly surprised that such help should be forthcoming. When an economic conflict broke out with Germany, and Bismarck placed a ban on German credits to the Russian Empire, Witte sent the following memorandum to the Tsar:

True enough, what sense is there for foreign states to give us capital? ... Why create with their own hands an even more terrible rival? For me it is evident that, in giving us capital, foreign countries commit a political error, and my only desire is that their blindness should continue for as long as possible.[7]

A number of Russian authors have made calculations concerning the role of foreign capital in Russia's development. According to figures cited by Lyashchenko, in 1900 about 28½ per cent of the capital of private companies was foreign-owned, in 1913 about 33 per cent. During these years foreign capital invested in Russia increased by 85 per cent, while Russian capital increased by 60 per cent. While the growth of foreign investments thus somewhat exceeded that of native investments, the latter were none the less rising by a very substantial percentage at this time. Foreign capital was invested in varying degrees in the different industries; it was above all in the oil industry that the foreigners were dominant. However, according to Lyashchenko, they also provided about 42 per cent of the capital in the metal goods industries, 28 per cent in textiles, 50 per cent in chemicals, 37 per cent in woodworking. Russian banks formed close links with foreign banks, and were effective in the cartelization of Russian industry, through the creation of so-called Syndicates, which followed the depression of 1900–3.

The investment of foreign capital in railway building and industry, and also successive loans, especially from France, to the Russian government, created a major problem for the Ministry of Finance to ensure a large enough surplus in visible trade to enable the necessary repayments, profit remittances and interest charges to be met. One consequence was the constant concern of the government to increase the export surplus of agricultural produce; this led to an effort to restrict consumption by the peasants and to increase sales by levying taxes on the peasants. Yet Russian industry, particularly in consumer goods, depended to a considerable extent on the purchasing power of the peasants for its market. This dilemma is by no means peculiar to Russia, and raises an issue of great interest in the economics of development. Russian economists brought up on the principles of the Manchester school used to assert, in their discussions after the abolition of serfdom in 1861, that the poverty of the peasants and the lack of a sufficient peasant market held back the development of Russian industry, since it made industrial investment unprofitable. Yet Witte's policy of deliberately encouraging and sponsoring industrialization involved a reduction in peasant purchasing power as part of the means of compelling them to sell foodstuffs which they would rather themselves consume, in order to meet the material and financial requirements of industrialization. In the words of Gerschenkron: 'The problem of peasant demand lost its previous significance, and its relation to industrialization was thoroughly reversed. ... To reduce peasant consumption meant increasing the share of national output available for investment. It meant increased exports. ...'[8] There is more than a purely superficial similarity between the policy then pursued and that adopted by Stalin over thirty years later.

The government experienced great difficulty in raising sufficient revenue, particularly in years of war or of international tension. Despite its inadequate equipment, the Russian army was a particularly heavy drain on financial resources. It is therefore understandable that it was Witte who persuaded Tsar Nicholas II to call the first disarmament conference ever held, at the Hague in 1899.

Statistically speaking, industrial development was now proceeding at a satisfactory rate. However, the Russian Empire at the beginning of the twentieth century was beset by many dangers and was in a state of social and political disequilibrium. This arose in part out of the very fact of a rapid transformation of a formerly semi-feudal and agrarian society. Much of this instability arose from peasant attitudes, which we must now briefly examine.

AGRICULTURE AND THE PEASANTS

The abolition of serfdom in and after 1861 opened a new era in Russian social relations. Yet the settlement of that year caused a deep dissatisfaction. Rural unrest was still serious and contributed greatly to the revolutionary waves of the twentieth century. There were several reasons for this.

Firstly, under the settlement of 1861 the land was divided between the landlords and the peasants. This offended the peasants' age-long sense of fairness. Serfdom had been imposed largely for state reasons, to provide an economic basis for the service gentry to enable them to serve the Tsar in civil and military capacities. They were in effect paid by the Tsar with land, and peasants had an obligation to maintain the gentry in the service of the Tsar. When in 1762 this duty of service was finally abandoned, this removed the only possible justification for the peasants' attachment to the land, as the peasants understood it. Let me illustrate the original *raison d'être* of serfdom with a historical example. In 1571 a Tartar invasion from the Crimea caused many thousands of peasants and their families to be led away into slavery. To prevent a recurrence of these events a standing army was necessary, and a strong monarchy. Many peasants understood this. They were in a position to evade their obligations by moving eastwards and southwards across an open frontier, and in fact did so on a large scale during the wars of Ivan the Terrible, which reduced revenue, numbers of recruits and the value of land grants to the gentry. The peasants' attachment to the soil thus had a rational purpose. However, by the late eighteenth century the gentry had the right to become merely parasitic landlords (though in fact many of them did continue to serve the state in various capacities), while the majority of the peasants were reduced to a state of slavery. The belief survived among them that 'they were the lord's but the land was God's'. They disputed the legitimacy of the ownership of the land by the lords, and considered that those who cultivated the land should have the full use of it. They therefore saw the 1861 settlement as depriving them of land to which they had a legitimate right.

Secondly – adding insult to injury – they were obliged to purchase their share of the land in instalments (redemption dues).

Thirdly, the peasants did not achieve equality before the law, or real personal freedom. Their land was held not by them but by the village community. This institution was often known as the *mir* or the *obshchina*. The heads of families in the village controlled land utilization. In most parts of European Russia they periodically redistributed strips within a three-field system familiar to students of medieval farming. The peasant was not allowed to leave his village without the authority of the community, and all the households of the village were jointly liable for taxes and for redemption dues.

Fourthly, a rapid increase in population pressed upon the means of subsistence. Towns were growing, but not at a rate sufficient to prevent a rapid growth in numbers of rural inhabitants, as the following figures demonstrate:

	European Russia Urban	Rural
	(millions)	
1863	6·1	55·3
1897	12·1	82·1

This trend continued after the turn of the century; it has been estimated that there was a 20 per cent increase in the rural population between 1900 and 1914. The size of peasant holdings and the traditional methods of cultivation made it increasingly difficult to feed the larger numbers, and there were insufficient incentives to use new methods. The situation was made worse by the imposition of taxes which were part of Witte's policy, mentioned above. It is true that there were lands available east of the Urals to which some of the surplus peasants could go, but resettlement was impeded by the joint responsibility of the village for taxes and redemption dues, which naturally led to fellow-villagers refusing to allow individuals to leave.

Needless to say, the above are only broad generalizations, which did not have universal application. Thus in Siberia and in parts of northern Russia there had never been serfdom at all, and a self-reliant class of peasants developed there long before 1861 and continued thereafter. Many succeeded in leaving the villages despite legal restrictions on their movement. Some peasants made a lot of money. Readers of Chekhov will not need reminding that the man who bought the cherry orchard was a *nouveau riche* son of a serf. None the less the generalizations do broadly hold good, and the land problem was never far from the preoccupations of the Tsar's ministers. Various proposals were made and postponed. It needed the shock of the 1905 revolution, with its widespread peasant riots and land seizures, to force the government to realize that further delay was out of the question.

The reform was carried out by the Tsar's last efficient and intelligent minister, P. A. Stolypin, and a series of measures effecting these reforms were promulgated in the years 1906–11. Outstanding redemption dues, which had been reduced, were finally abolished. Peasants were now free to leave their communities, to consolidate their holdings as their property, to buy land or to sell it, to move to town or to migrate. Stolypin's object was to encourage the emergence of a class of peasant proprietors who would be prosperous, efficient and politically loyal. This was the so-called 'wager on the strong'. Many go-ahead peasants took advantage of the new opportunities. By 1916 about 2 million households had left the communities and set up private farms, out of 2·7 million who had expressed their desire to do so. This represented some 24 per cent of the households in forty affected provinces of European Russia. Some remained in the villages, others erected farm houses, so-called *khutora*, outside. The process of change was slowed down and then halted by the outbreak of war in 1914. Stolypin himself had been assassinated in 1911. He held the view that his reform, given time, would have provided the empire with a solid social base. We will never know whether he would have been proved right.

Commercial agriculture, conducted both by the progressive landlords and by the more prosperous peasants, was already developing in some areas even before this reform. It naturally speeded up thereafter. It is not surprising that the greatest number of peasants opted to leave the community in the south-west (i.e. principally in the Ukraine) and in the north Caucasus, where grain was produced for the market. It was in these areas that resistance to Stalinist collectivization was most fierce twenty years later.

The Stolypin reform made possible a major rearrangement of

traditional peasant agriculture, and would certainly have led to greater efficiency and to a substantial strengthening of a prosperous peasant class. In fact this development was already in progress. However, its political and social impact was conditioned by two other significant factors. The first was that the reform did not affect the assets of the landlords and the church. It is true that the richer peasants were purchasing some land from the poorer landlords, but it was no part of the object of this reform to redistribute the landlords' land. Consequently the grievances of the peasants as a whole, and their land hunger, were not assuaged. In fact half of the 89 million hectares of land allocated to the landlords in 1861 had passed into peasant hands by 1916; they owned by then 80 per cent of the land and rented part of the remainder,[9] but resentment against big landlords remained strong. Secondly, the poorer peasants received little benefit from the reform, except perhaps that they were now finding it easier to sell their smallholdings if they wished to leave. The effect was to increase the number of landless peasants and migration to the towns. It also stimulated hostility towards the better-off peasants, which was to be an important factor in the revolutionary period.

Agricultural production rose rapidly in the first years of the century, due partly to favourable weather conditions and partly to the effects of the reform and of better methods. While most peasants were still using outdated methods, including large numbers of wooden ploughs, the more progressive sectors were now beginning to use modern equipment. This process was greatly assisted by a sharp rise in agricultural prices, due in part to a rise in world prices. According to Lyashchenko the net income of agriculture increased by 88·6 per cent in the period 1900–13, representing an increased output in constant prices of 33·8 per cent. The spread of commercialism and of capitalist relations was speeding up. Exports of grain rose very sharply. Thus in the years 1911–13 they were 50 per cent higher than in the years 1901–5, on average. Exports of butter, eggs, flax and other agricultural products were also increasing. A protective tariff encouraged the development of cotton-growing in Central Asia.

But over the long period progress was not great. According to the invaluable Bairoch, productivity per head grew as follows, based on a calculation of millions of calories produced per male agricultural worker:

	1860	1910
Germany	10·5	25·0
Belgium	11·0	18·0
Spain	11·0	8·5
U.S.A.	22·5	42·0
France	14·5	17·0
Italy	5·0	6·5
Japan	?	2·6
U.K.	20·0	23·5
Russia	7·5	11·0
Sweden	10·5	16·0
Switzerland	9·0	17·0

Thus Russia moved one up on the ranking list, overtaking Spain, though only because Spain, so to speak, marched backwards. She fell farther behind the U.S.A. and Germany, but grew faster than either Great Britain or France.

A similar picture emerges from the Goldsmith study already cited. His index for 'all crops' (1896–1900 = 100) shows an increase from an index of 51 in 1861 to an average of 140 in 1911–13 (fifty provinces of European Russia, i.e. excluding not only Siberia and Central Asia but also Finland, Poland and the Caucasus). Allowing for margins of error and misreporting of data, he obtains 'an average rate of growth of very close to 2 per cent' per annum for the whole period (though one cannot exclude 1¾ per cent or 2¼ per cent, depending on interpretation of the data). Livestock production, on which statistics for these years are totally inadequate, went up much more slowly; Goldsmith's estimate is only 1 per cent per annum, making an overall growth rate for agriculture of approximately 1·7 per cent per annum for the period 1860–1914. But such a figure is only a fraction above the rate of population increase. Food consumption per head could hardly have increased at all, if one takes into account the relative increase in the area sown with industrial crops, and also exports of food. Total farm output per head of the whole population rose perhaps ¼ per cent per annum. Goldsmith does not attempt to calculate output per head of population engaged in agriculture, but it is clear that his figures are broadly consistent with the picture derived from Bairoch's study.

This slow progress of agriculture, in which so high a proportion of the people was engaged, helps to explain the modest growth of the national income, already referred to. It is true, however, that food production was beginning to increase rather more rapidly towards the end of the period.

Progress in agricultural education was impressive in the last years of the regime. Thus there were only seventy-five students studying agronomy to degree level in 1895, but by 1912 the number had risen to 3,922;[10] not enough, of course, but evidence of a very big growth rate, which could well have paid off in time. Similarly, there was a most impressive rise in the various forms of agricultural cooperation, showing that, with the break-up of the *mir*, new and better ways of expanding peasant commercial activities were being devised. The membership of rural credit cooperatives rose from 181,000 in 1905 to 7 million in 1914. The number of rural consumer cooperatives rose in the same period from 348 to 8,877.[11] Marketing cooperatives grew rapidly. One of these, which organized peasant butter producers in Siberia, maintained an agency in London and made great progress with exports. Thus the last decade of Tsardom may have been the harbinger of an agricultural leap forward. Needless to say, here too improvement was exceedingly uneven, affecting some parts of the country much more than others. Yet there was evidence here to support the pleas of those who, like Chayanov, believed in the productive potential of peasant agriculture.

SOCIAL AND POLITICAL INSTABILITY

The freeing of the peasants caused an ever-growing flow into the towns, and this was naturally speeded up by the Stolypin reform, since now peasants with little land could sell their holdings to their better-off neighbours. Most of the urban labour force was of extremely recent rural origin, and maintained close links with the villages, where many of their relatives still lived. Many went home annually to help bring in the harvest. The industrial labour force was thus not of high quality; skilled labour accustomed to factory work was relatively very scarce. Drunkenness was common, living conditions exceedingly primitive. This rootless and disorientated labour force found itself concentrated in the very large units of which the modern sector of Russian industry was composed, and provided good material for revolutionary propaganda in the cities.

It has often been stated that the Russians were almost all illiterate before the revolution. This statement is greatly exaggerated. In the 1897 census, it was found that, in European Russia, 35·8 per cent of the men and 12·4 per cent of women were literate. A good measure of progress is the record of literacy among army recruits:

(per cent)	
1875 —	21
1890 —	31
1905 —	58
1913 —	73

(SOURCE: *Bol'shaya Sov. Entsiklopediya*, 2nd edn, Vol. 12, p. 434.)

There was a rapid development of schools and of universities and the standards of science were high indeed. Here too the situation was characterized by extreme regional unevenness. Such medical services as existed were of excellent quality, but there were too few doctors. The death-rate in villages was appallingly high.

Thus it may be said that the Russian Empire in 1913 was in the process of rapid change, that industrialization was making good progress, that agriculture was also changing and growing, but progress was uneven and gave rise to social and political stress, which in turn caused unrest in the cities together with land-hunger and rioting in the villages. A Russian middle class was emerging, but it lacked authority and self-confidence. With rare exceptions, the servants of the autocracy were men of mediocre ability overwhelmed by ever-growing problems and unable to cope with a growing and changing empire. The intelligentsia, given to endless arguing and theorizing, was almost wholly opposed not only to the autocracy but also to the spirit of capitalist enterprise. This was an explosive mixture. It is not only with the help of the evidence of hindsight that we can say that the society and the polity were in process of breaking down. The following is an extract from a book by an Austrian observer, Hugo Ganz, published in 1904 and entitled *The Downfall of Russia*. It is said to be a conversation between Ganz himself and a senior official who asked to remain anonymous:

'What will be the end, then?'

'The end will be that the terror from above will awaken the terror from below, that peasant revolts will break out and assassination will increase.'

'And is there no possibility of organizing the revolution so that it shall not rage senselessly?'

'Impossible. . . .'

'There is no one with whom I have spoken who would fail to paint the future of this country in the darkest colours. Can there be no change of the fatal policy which is ruining the country?'

'Not before a great general catastrophe. When we shall be compelled for the first time partly to repudiate our debts – and that may happen sooner than we now believe – on that day, being no longer able to pay our old debts with new ones – for we shall no longer be able to conceal our internal bankruptcy from foreign countries and from the Emperor – steps will be taken, perhaps. . . .'

'Is there no mistake possible here in what you are saying?'

'Whoever, like myself, has known the state kitchen for the last twenty-five years has no longer any doubts. The autocracy is not equal to the problems of a modern great power, and it would be against all historical precedent to assume that it would voluntarily yield without external pressure to a constitutional form of government.'

'We must wish then, for Russia's sake, that the catastrophe comes as quickly as possible.

'I repeat to you that it is perhaps nearer than we all think or are willing to admit. That is the hope; that is our secret consolation. . . . We are near to collapse, like an athlete with great muscles and perhaps incurable heart weakness. We still maintain ourselves upright by stimulants, by loans, which like all stimulants only help to ruin the system more quickly. With that we are a rich country with all conceivable natural resources, simply ill-governed and prevented from unlocking our resources. But is this the first time that quacks have ruined a Hercules that has fallen into their hands?'

I am far from suggesting that one can find ultimate wisdom in the words of a relatively unknown Austrian writer, or that such alarm and pessimism were universal at the time. The above quotation is merely an attempt to dispel the idea that the sense of doom was, so to speak, superimposed on the period by post-revolutionary analysts. Nor is it intended to deny that it was the outbreak of war in 1914 and its consequences that finally made the peaceful evolution of the empire totally impossible. Who knows, Nicholas II might have been succeeded by an ambitious and dominating Emperor, a kind of new Peter the Great, capable of enforcing the necessary adjustments. The 1905 revolution was apparently surmounted by the autocracy, with little loss of executive power. It is true that they found it necessary to concede the existence of an elective legislature, but in 1907 the franchise was so restricted as to ensure a loyal conservative majority. By 1914 a wave of strikes grew threateningly, and in the short run it was the outbreak of war that relieved the authorities of many of their worries by turning the mobs from subversion into patriotic demonstrations.

Thus the question of whether Russia would have become a modern industrial state but for the war and the revolution is in essence a meaningless one. One may say that statistically the answer is in the affirmative. If the growth rates characteristic of the period 1890–1913 for industry and agriculture were simply projected over the succeeding fifty years, no doubt citizens would be leading a reasonably comfortable existence and would have been spared many dreadful convulsions. However, this assumes not only that the tendencies towards military conflict which existed in Europe had been conjured out of existence, but also that the Imperial authorities would have successfully made the adjustment necessary to govern in an orderly manner a rapidly developing and changing society. There is no need to assume that everything that happened was inevitable because it happened. But there must surely be a limit to the game of what-might-have-been. One can end this chapter with another quotation from Gerschenkron: 'Industrialization, the cost of which was largely defrayed by the peasantry, was itself a threat to political stability and hence to the continuation of the policy of industrialization.'[12]

2. War, Revolution and Revolutionaries

WAR AND BREAKDOWN

Russia entered the war with a weak arms industry and relatively poor communications. The mobilization of a vast army adversely affected the labour situation. Russia was quite unprepared for the huge expenditure of arms and ammunition, particularly artillery, which modern war required. In the first six months the reserves of military hardware had been largely dissipated and ammunition was running dangerously short. When in the spring of 1915 the Germans attacked with vastly superior fire power, the Russians were forced into prolonged retreat. Soldiers found that they were forced to pick up the rifles of fallen comrades. The retreat itself led to the abandonment of considerable stores to the enemy. The disasters in Poland and the general atmosphere of incompetence shook the faith of the people in the regime. The Tsar decided in 1915 to take command of the army himself, thereby taking an unnecessarily direct responsibility for failures and losses. Ministers did not take effective action, and there was constant friction between them and various voluntary organizations, of industrialists and others, who claimed that the government was unable to organize the country for war.

It is true of course that other countries also suffered greatly from inability to cope with the munitions problems. One has only to refer to Lloyd George's War Memoirs to see how acute the difficulties were in Britain. However, British industrial capacity and transport facilities were, in relation to the task in hand, greatly superior to the Russians'. Britain was also able to import arms and equipment from the United States. Russia was virtually cut off from her allies, save for difficult and roundabout routes, and also paid the penalty for the underdevelopment of her engineering and machinery industries. (In 1912, 57 per cent of Russia's industrial equipment was met by imports.[1]) By switching her engineering capacity almost wholly to the production of armaments, she deprived herself of the means of maintaining her existing industrial and transport equipment intact, let alone expanding it. There were grave shortages of spare parts. The already inadequate transport network became increasingly strained as more locomotives and rolling stock went out of use. Of course, attempts were made to remedy these defects, and munitions production became considerably more satisfactory by 1916. Early in 1917 a railway line to the ice-free port of Murmansk was completed, making possible year-round importing of Western munitions. The army's equipment was improving. However, the appalling losses of earlier years had undermined morale in the army, and the population in rear areas was not now in the mood to bear the sacrifices imposed upon it by the strain of war and the chronic difficulties of transport. St Petersburg was particularly likely to be affected by shortages, since it was far from the main food-producing areas and also had to import its coal vast distances by rail; it had been importing British coal, but this was cut off by the war. It was in fact shortage of food in St Petersburg which finally broke the back of the Russian Empire. The troops refused to fire on a rioting crowd, and Tsardom collapsed. To say this is far from adopting a purely economic explanation of the Russian revolution. Other powers too suffered from defeats and from food shortages. Russia herself was in a much worse situation in 1942, both from the standpoint of military losses and of food supplies for rear areas, than she had been in 1917. Thus it is a combination of such factors with the political demoralization of regime and people, and social breakdown, that swept away the Empire.

THE PROVISIONAL GOVERNMENT

The Provisional government faced an appallingly difficult task. In the midst of war and threatening economic collapse and social anarchy, it had somehow to establish itself as the legitimate government. For many reasons, the discussion of which is beyond the scope of this book, it conspicuously failed to do this. It took a number of socially progressive measures, the implementation of which was rendered extremely difficult by the circumstances of the time. It adopted protective labour legislation and approved or tolerated the establishment of factory committees, which were to make the operation of industry so difficult in subsequent months. It even attempted to formulate plans. In an effort to ensure enough bread at fixed prices, it decreed the state monopoly of the purchase of grain in March 1917, but proved incapable of enforcing deliveries at prices which it was prepared to pay, and shortly before its fall the government was compelled to double these prices. It is interesting to note that the elimination of the private trader from the grain trade, which many people imagine was the work of Lenin, was in fact attempted, though with little success, by the Provisional government while Lenin was still on his way to Russia.

The industrial situation continued to deteriorate. There were many causes for this. There were the consequences of past failings in maintenance and replacement and the shortage of spares. There was the cumulative effect of a creeping transport crisis; the railways were beginning to break down, and the non-arrival of essential fuel supplies adversely affected all branches of industry as well as the operation of the railways themselves.

As an acute contemporary observer put it: 'The fuel and raw materials supply position continued to deteriorate, and this did not lead to a sharp and shattering crisis only because, for other reasons, their use was being rapidly curtailed.'[2] The same writer bitterly assailed 'the self-assured ignorance and irresponsibility' of ministers facing the problems of general collapse, though he did not fail to note that one cause for this was the fact that 'under the old regime Russia had been deprived of political and social forms which could regulate the relationships of labour and capital on a modern basis. . . . The working class had no organizational or political experience of responsible and open activity.'[3] The Soviets which were set up alongside the Provisional government showed a lack of realism and fear of responsibility. The authority of the government in all matters was increasingly undermined not only by the existence of the Soviets (more or less representative bodies of workers and soldiers who shared authority with organs of government) but also by the dangerous growth of separatist and autonomist tendencies, notably in the Ukraine and in Transcaucasia. Finally, the increasing militancy of the factory committees was making it more and more difficult for industrial management to operate. Strikes were frequent, workers' demands more and more extreme, and the voices of extremist parties ever louder. Matters were not helped by rapid inflatioh. Communist historians sometimes speak of sabotage by the industrialists, as indeed Lenin did at the time. However, Lenin also denounced 'capitalist ministers', and it seems unlikely that more than a handful of businessmen consciously tried to make the work of the economy and of the government impossible. But they certainly did react adversely to what appeared to them to be unreasonable pressure from the factory committees, and there were many lockouts as well as strikes.

There was also increasing confusion in the villages. The peasants resented having to supply the government with produce for which they were paid in rapidly depreciating roubles. But more fundamental was the demand, ever more loudly voiced, for the distribution of the landlords' land. In principle the moderate socialist parties which dominated the government were in favour of this type of land reform. However, the whole question was particularly explosive because the army was largely composed of peasants, and to raise the issue of land reapportionment in the midst of a war was to risk mass desertion; the troops would go home to ensure that they and their families got their due. This was an important reason for the failure of the Provisional government to enact any swift land reform measures, even though they were perfectly well aware of the danger of not doing so. A second reason for delay was the appalling complexity of any such land reform measures. The majority of the government wished them to be properly prepared and carried out in an orderly manner, and there was by no means any agreement about the question of compensation for the landlords or the principles of reapportionment (see below). It was finally decided to put decision off until the summoning of a constituent assembly. However, spontaneous land seizures were already occurring on an increasing scale in the summer of 1917, egged on by the militant 'left' Socialist Revolutionaries as well as the Bolsheviks.

Both in the towns and in the villages the situation was approaching chaos even without the help of Lenin and the

Bolsheviks. Of course, they tried to make matters worse, since they were unconcerned with an orderly land settlement, industrial production or the military situation. They sought to reap the whirlwind. They contributed to the breakdown but did not cause it. The authority of the government had virtually collapsed for some weeks before the Bolsheviks seized the Winter Palace with a relatively insignificant group of ill-armed Red Guards. Galbraith has said that the man who breaks through a rotting door acquires an unjustified reputation for violence; some credit should be given to the door. The fact that the Bolsheviks took charge of a disintegrating society because it was disintegrating is a fact of great importance, which must be borne in mind in analysing their subsequent actions.

THE IDEAS OF THE BOLSHEVIKS AND MENSHEVIKS

On 7 November 1917 Lenin announced that they would now set about building the socialist society. Before endeavouring to analyse the events that followed, which form the essential subject matter of the book, it is necessary to consider the economic ideas of the Bolsheviks as they had developed up to the seizure of power, and to contrast them with those of their principal rivals.

The body of ideas which came later to be called Leninism may be described as Marxism adapted to the political and economic situation of a relatively backward country, with emphasis on the 'voluntarist' aspect of Marx's doctrines. Marx, it will be recalled, wrote his works in Western Europe; his ideas were in the main related to his experiences in England. He was primarily concerned with how the increasing concentration of capital in the leading capitalist states would bring about ever-growing contrasts between the small group of monopolists who owned the instruments of production and proletarian masses into whose ranks would be pressed dispossessed peasants and the petty bourgeoisie. This would facilitate the take-over of power and holding it on behalf of the bulk of the people, while exercising coercion against the tiny group of exploiters and their retainers ('the dictatorship of the proletariat'). The correctness or otherwise of the analysis of monopoly capitalism is not a matter for discussion here. The point is that Lenin and other Russian Marxists at the turn of the century had the task of reconciling this doctrine with the reality of Russia. It is true that Marx wrote also of the 'Asian mode' of production and noted the relationship between oriental despotisms and control over water, a point developed more recently by Wittfogel in his concept of a 'hydraulic society'. It is also true that Marx did react quite specifically to a question put to him by the Russian revolutionary Vera Zasulich. This arose in the course of debates in Russia itself concerning a possible specifically Russian road to socialism. Might there not be a way of avoiding the capitalist road by utilizing the primitive-socialist potentialities of the peasant communes? Such a question must be seen in the context of the Slavophil doctrines, their idea of a peculiarly Russian experience and world view, whose proponents were in the main hostile to capitalism and to its merchant ideology.

Marx had a good deal of trouble in devising an adequate answer to Zasulich, as is witnessed by the existence of rejected variants of his letter. Finally he replied along the following lines. The expropriation of the peasants which had already taken place in England would also take place in other countries of Western Europe. But the reasons that rendered this inevitable were confined to Western Europe. In the West 'one kind of private property is turned into another kind of private property. In the case of the Russian peasants, however, it would be necessary on the contrary to convert their communal property into private property.' He went on to state that his analysis did not give any grounds or arguments either for or against the Russian communal form. However, he came to the conclusion that 'this *obshchina* is the basis of the social rebirth of Russia, but it would only function as such if it would be possible to remove the disintegrating influences to which it is now subject on all sides, and then to ensure for it the normal conditions of free growth'.[4]

This reply proved so embarrassing to Russian Marxists of all hues that it was not published until 1924. In fact they were distinguished from the non-Marxist populists, and from the later Socialist Revolutionary party, precisely by their belief that capitalist development in Russia had become decisively established and that the kind of short cut via peasant socialism implied by Marx's letter was impracticable. Lenin's first major work, 'The development of capitalism in Russia' (1896-9), devoted much energy and an array of statistics to proving that even among the peasants commercial capitalism was making rapid headway. Lenin's more moderate rivals, the Mensheviks (as they were known from 1903), were just as convinced that capitalism would develop in Russia.

Given these assumptions, there were still at least two possible lines of Marxist thought. One, espoused by most of the Mensheviks, based itself on the evolutionary aspects of Marxist doctrine. If socialism was to grow out of advanced capitalism, then evidently Russia was not and would not for a long time be ripe for socialism. The Tsarist regime was a pre-capitalist and semi-feudal system of oppression. The situation was ripe for a bourgeois-democratic revolution designed to overthrow Tsarism. To this extent the Mensheviks were indeed revolutionaries. But what would happen when Tsarism was overthrown? The Mensheviks saw themselves as the social democratic opposition in a bourgeois-democratic republic. Martov, their leader, could claim that they were in line with what Marx had recommended in Germany in 1850, when that country too was underdeveloped.[5] Social change would in due course create the conditions for socialism. This process would involve industrialization under capitalist auspices. They, like the Bolsheviks, had little support and few supporters in the villages, and so had little hope of a parliamentary majority in a country in which, if their objective of universal suffrage were to be achieved, the vast majority of the electorate would be peasants.

The Bolshevik position before the revolution was distinguished from that of the Mensheviks by tougher language and an emphasis on tightly-knit organization of professional revolutionaries. It was also the case that Lenin saw more clearly the revolutionary potential of the peasants. To do this he had to some extent to go outside the established Marxist tradition. Orthodoxy at the turn of the century was exemplified by the argument of the German theoretician, Karl Kautsky. Following Marx, Kautsky had envisaged a gradual expropriation of the German peasants by Capital. He was also a firm believer, as was Marx, in the technical superiority of large-scale agricultural enterprise. Therefore both on technical grounds, and because the proletarianization of the peasants was a step in the direction of the ultimate socialist take-over, Kautsky opposed the adoption of a programme of support for the interests of the small peasants. If they were doomed by history, and if this was a progressive step in terms of the historical evolution of society, what business had the social democrats to delay this inevitable and progressive evolution? (Not very surprisingly, the social democrats never won much support in rural areas in Germany.) The adoption of such a policy in an overwhelmingly peasant country such as Russia was tantamount to political suicide. A sense of political self-preservation therefore inclined all shades of Marxian opinion to take a line favourable to peasant demands, in particular over the expropriation of the landlords. But Lenin also came to see the possibilities of a seizure of power on the basis of a peasant as well as a working-class revolution, in a country unripe for socialism, yet with a weak and 'cowardly' bourgeoisie. He spoke in 1905 of 'the democratic dictatorship of the proletariat and the peasantry'. It was far from clear what this would mean in terms of practical politics, and at this period Lenin was under no compulsion to demonstrate the practicability of his ideas. However, it did show that he was going beyond the Menshevik conception of a bourgeois-democratic republic in which the capitalists would presumably supply the government parties. He was concerned above all with the problem of power, and showed himself willing to adopt and adapt policies according to the needs of the tactical situation, so as to manoeuvre into the position of being able to seize power when opportunity presented itself. In a peasant country with massive discontent, the attitude of a party to the peasant question would naturally predetermine much of its political efficacy. Once power was seized, the problem of the peasants would loom very large (and as we shall see it did loom very large indeed in the economic and political history of the Soviet Union). Lenin was well aware of this. He saw that while the peasants as a whole were likely to be a revolutionary force so long as land was to be obtained from the landlords, at least the better-off peasants would turn conservative once this aim was achieved. He pinned his hopes on the increasing and deepening differentiation among the peasants. The majority of the poorer peasants, he held, would remain in alliance with a working-class government and perhaps collaborate in a future socialist transformation of society.

There is indeed much more that could be said about Lenin's attitude to the peasants, especially as this underwent changes in detail as circumstances altered. But despite tactical manoeuvring he consistently held to the basic strategy of using the revolutionary potential of the peasant land-hunger as an integral part of the socialist takeover. In a speech which he made a few weeks after Lenin's death in 1924, Zinoviev, for years his close associate, put this first among Lenin's contributions to revolutionary theory and practice. 'This was his attitude to the peasants. Probably this was the greatest discovery of Vladimir Ilyich: the union of the workers' revolution with the peasant war.' In the same speech he made the point again. 'The question of the role of the peasantry, as I have already said, is the basic issue of Bolshevism, Leninism.'[6]

Lenin thus developed the concept of a seizure of power by a conscious socialist minority with the help of non-socialist elements in a country where capitalism was unevenly developed. It was wrong in this view to await the ripening of the social and economic situation; this change would therefore have to be achieved after the seizure of power. Lenin understood that such a transformation after a successful revolution would be an exceedingly difficult one. However, he hoped that Russia would prove to be the weak link in the unevenly developing group of imperialist states, that a revolution in Russia would be the first stage of a world revolution, and that the more advanced countries of western Europe would help in the colossal task of the building of socialism. Lenin asserted that it was the duty of revolutionary socialists to set the example to the laggard proletarians of the more advanced countries, and not to wait passively or merely to act as the left wing of a bourgeois-democratic revolution. The hopes for a world revolution naturally received a powerful impetus during the First World War, whose revolutionary potentialities were greatly stressed by Lenin, even while he denounced the 'social patriotism' of the bulk of European social democracy which, including a large portion of the Mensheviks (and some Bolsheviks too), backed their own governments. Lenin died before he was compelled to face the immense and daunting task of changing backward Russia in isolation from the advanced West European countries, on whose support he had doubtless counted. The Mensheviks held that a seizure of power on behalf of a proletarian party with socialist objectives was premature under Russian conditions and that it would lead to deplorable consequences. Many of them were uneasy about Lenin's agricultural views. They saw many dangers and little socialism in the kind of elemental peasant risings to which Lenin appeared to be pinning his hopes.

The Mensheviks none the less also advocated a programme designed to win peasant support. Considering that nationalization of the land would be an unpopular slogan, they advocated so-called municipalization, i.e. control of the land by local elected authorities, which in rural areas would be peasant authorities. They hoped also by this means to encourage the peasants to work together, to avoid individualist fragmentation, even while the peasant land-hunger would be assuaged. There were differences of opinion among the Bolsheviks in the years before 1917 on the difficult subject of just what to put into the party programme concerning the land. In the end they decided to advocate nationalization, claiming with truth that the principal peasant party also accepted this.

THE SRS AND LENIN'S PEASANT POLICY OF 1917

This party was the Socialist Revolutionaries, which everyone referred to as the SRs.

This once great party, which continued the Narodnik (Populist) tradition, has now been almost forgotten. Its peasant policy was deeply influenced by a non-Marxist socialist conception of peasant democracy, with strong traditionalist egalitarian overtones. It declared itself for equal distribution of the land and therefore opposed the Stolypin 'wager on the strong'. The SRs favoured the expropriation of the landlords. They favoured social ownership of the land; it was to be at the disposal of those who cultivated it. Moreover, the SR programme of 1906 was on record as advocating the prohibition of the purchase or sale of land. Such an outlook in a peasant party on the morrow of the

Stolypin reform, with commercial relations growing in the villages, virtually guaranteed political paralysis and splits within the party. In 1917 the right wing of the party leadership participated in the Provisional government and discouraged illegal peasant seizures of land. It seemed more than likely that, given time, this wing of the party would identify itself with the better-off peasants, and scrap or put into cold storage the anti-commercial and egalitarian principles of their own programme. These principles would demand the redistribution of the land of those peasants who had benefited from the Stolypin reform. The more prosperous and the more efficient peasants were bound to oppose such a policy. By contrast the SR left wing supported direct action and for a while participated in the Bolshevik government. Such was the strength of traditional SR ideology that a peasant congress held in August 1917, still dominated by the SRs, adopted policy resolutions ('242 mandates') which included such provisions as nationalization ('land belongs to the people'), the prohibition of buying and selling of land, its equal distribution and the outlawing of the employment of hired labour. They therefore stood for peasant family smallholdings, and for periodic redistribution of the land in order that its allocation to families should conform to some principle. It was not clear whether this principle was to be the number of mouths to feed or the number of able-bodied persons capable of working the land. It was certain that the right wing of the SRs would do their utmost to prevent a policy of this kind from being implemented. Eminent experts – men like Chayanov, Chelintsev, Kondratiev – were evolving doctrines based on peasant proprietors. Some SRs were bound to back these ideas, which were indeed being expressed in other parties which existed at the time: the *Trudoviki*, the Populist Socialists, Cooperators, etc.

Lenin in 1917 consistently pursued one overriding aim: the seizure of power. His programme varied with the tactical situation, and his peasant programme varied too. In his famous April 1917 'theses', published soon after his return to Russia, there was a plain reference to the superiority of large-scale agriculture, and to the need to convert efficient private estates into large and productive (state) model farms. Lenin also wrote scathingly of the 'petty bourgeois illusions' of the SRs, of the 'helpless, unwittingly naive wishful thinking of down-trodden petty proprietors', which showed itself in the proposed ban on wage-labour.[7] But 'it is not enough to expose theoretically the petty bourgeois illusions of socialization of the land, equalized land tenure, a ban on wage-labour, etc.' The line should be: 'the SRs have betrayed the peasants. They represent a minority of well-to-do farmers .... Only the revolutionary proletariat, only the vanguard that unites it, the Bolshevik Party, can actually carry out the programme of the peasant poor which is put forward by the 242 mandates' (the programme adopted by the peasant conference mentioned above). At the same time Lenin also foresaw the split among the peasants themselves and was laying long-term plans to use it for his own purposes. Thus on 7 and 8 July 1917 one sees him advocating an Agricultural Labourers Union, which would be 'the independent class organization of the rural workers'. He took the opportunity to remind his readers of the 1906 resolution of the party at its fourth congress, which spoke of 'the irreconcilable antithesis between its interests and the interests of the peasant bourgeoisie' and which went on 'to warn it against illusions about the small-holding system, which can never as long as commodity production exists do away with the poverty of the masses'.[8]

However, by the autumn of 1917 it seemed tactically sensible to adopt the SR-inspired programme which had been adopted by the peasant congress in that month. This would represent a most potent bid for mass support among the peasants, and have the further advantage of widening the split in the SR party.

Therefore the Bolsheviks' agricultural and peasant programme was a masterly tactical improvisation, on an original Leninist theme, and contained from the first the seeds of the future conflict with the peasantry.

INDUSTRY, FINANCE, TRADE, PLANNING

Soviet writers indignantly deny Western allegations that Lenin had no idea what to do after seizing power, that he had to make it up as he went along. The evidence on the subject is somewhat mixed. Before 1917 Lenin had made some contributions to economic thought: he had in his early work analysed social-economic statistics and expressed strong views in the 'Development of capitalism in Russia'. However, his writings were not distinguished by any attempts to define just how a socialist industry could or should be run. Nor is this surprising. Neither was Marx given to drawing blueprints of a socialist future, and prior to the war Russian socialists could not realistically envisage the sort of situation in which they could have power to act. Lenin, of course, had views about tactics, about the function of workers' immediate demands in relation to the revolutionary aims of a highly disciplined party organization. It was not until 1917, on his return to Russia through Germany in the famous sealed train, that we find Lenin's ideas on industry and planning taking some sort of shape. But the shape was decisively affected by the everyday exigencies of the struggle to gain power. Much of what he said and wrote reads like the purest demagogy.

Make the profits of the capitalists public, arrest fifty or a hundred of the biggest millionaires. Just keep them in custody for a few weeks ... for the simple purpose of making them reveal the hidden springs, the fraudulent practices, the filth and greed which even under the new government are costing our country thousands and millions every day. That is the chief cause of our anarchy and ruin!

Thus he spoke to the first congress of Soviets in June 1917.[9] In a similar spirit he returns again and again to 'workers' control', but the Russian work *kontrol'* means not a takeover but inspection and checking (like the French *contrôle des billets*), and his emphasis was on the prevention of sabotage and fraud by the capitalists. Yet now and again kontrol' shades into control, developing into complete regulation of production and distribution by the workers, into the 'nation-wide organization' of the exchange of grain for manufactured goods, etc.[10] But how this was to happen was left undefined. Lenin denied syndicalism: 'Nothing like the ridiculous transfer of the railways to the railwaymen, or the tanneries to the tanners.'[11] The cure-all was to be 'all power to the Soviets', though how (or whether) they are to operate railways and tanneries is not stated.

In the same month of June 1917 he was writing: 'Everyone agrees that the immediate introduction of socialism in Russia is impossible.'[12] Perhaps the most complex 'programme' of the months before the seizure of power may be found in Lenin's 'The impending catastrophe and how to combat it', published as a pamphlet at the end of October 1917 and written a month earlier. It begins dramatically with the words 'Unavoidable catastrophe is threatening Russia'. The railways were breaking down, famine was threatening, the capitalists were sabotaging; the following measures should be taken, he declared:

(a) Centralization and nationalization of banking.

(b) The nationalization of the 'syndicates', i.e. of the main capitalist associations (for sugar, oil, iron, coal, etc.).

(c) Abolition of commercial secrecy.

(d) Compulsory 'syndicalization' of industry, i.e. that independent firms should form part of syndicates.

(e) Compulsory membership of consumer cooperatives, this measure being related to the strict enforcement of wartime rationing regulations (rationing having been introduced in cities in 1916).

He explained what he meant by nationalization of syndicates: 'Transform reactionary bureaucratic regulation [i.e. by the Provisional government] into revolutionary democratic regulation by simple decrees providing for the summoning of a congress of employees, engineers, directors and shareholders, the introduction of uniform accountancy, control [kontrol'] by the workers' unions, etc.'[13] This sounds as if he had in mind effective control over the syndicates rather than expropriation of capitalists and nationalization of the actual firms. But perhaps this was simply his view of what ought immediately to be done, and not a programme for the party.

This was a time when some of Lenin's thoughts were somewhat Utopian. Thus in 'State and Revolution' we can read:

We, the workers, shall organize industrial production on the basis of what capitalism has already created. . . . We shall reduce the role of state officials to that of simply carrying out instructions as responsible, revocable, modestly paid 'foremen and accountants' (of course with the aid of technicians of all sorts). . . . The function of control and accountancy, becoming more and more simple, will be performed by each in turn, will become a habit and will finally die out as the *special* function of a special section of the population. . . . To organize the *whole* economy on the lines of the postal service so that technicians, foremen and accountants, as well as *all* officials, shall receive salaries no higher than workers' wages, all under the leadership of the armed proletariat, that is our immediate aim.[14]

Lenin was greatly impressed with the German war economy. He thought that the concentration of state-capitalist power gave rise to possibilities of direct socialist takeover of the levels of economic power. One finds this expressed with particular clarity in 'Can the Bolsheviks retain state power?', also written on the eve of the revolution. It is worth quoting from this at some length:

This brings us to another aspect of the question of the state apparatus. In addition to the chiefly 'oppressive' apparatus – the standing army, the police and the bureaucracy – the modern state possesses an apparatus which has extremely close connexions with the banks and syndicates, an apparatus which performs an enormous amount of accounting and registration work, if it may be expressed this way. This apparatus must not, and should not, be smashed. It must be wrested from the control of the capitalists; the capitalists and the wires they pull must be *cut off*, *lopped off*, *chopped away from* this apparatus; it must be *subordinated* to the proletarian Soviets; it must be expanded, made more comprehensive, and nation-wide. And this *can* be done by utilizing the achievements already made by large-scale capitalism (in the same way as the proletarian revolution can, in general, reach its goal only by utilizing these achievements).

Capitalism has created an accounting *apparatus* in the shape of the banks, syndicates, postal service, consumers' societies, and office employees' unions. *Without big banks socialism would be impossible.*

The big banks *are* the state apparatus which we *need* to bring about socialism, and which we *take ready-made* from capitalism; our task here is merely to *lop off* what *capitalistically mutilates* this excellent apparatus, to make it even *bigger*, even more democratic, even more comprehensive. Quantity will be transformed into quality. A single State Bank, the biggest of the big, with branches in every rural district, in every factory, will constitute as much as nine-tenths of the *socialist* apparatus. This will be country-wide *book-keeping*, country-wide *accounting* of the production and distribution of goods, this will be so to speak, something in the nature of the *skeleton* of socialist society.

We can *lay hold of* and *set in motion* this *state apparatus* (which is not fully a state apparatus under capitalism, but which will be so with us, under socialism) at one stroke, by a single decree, because the actual work of book-keeping, control, registering, accounting and counting is performed by *employees*, the majority of whom themselves lead a proletarian or semi-proletarian existence.

By a single decree of the proletarian government these employees can and must be transferred to the status of state employees, in the same way as the watchdogs of capitalism, like Briand and other bourgeois ministers, by a single decree transfer railwaymen on strike to the status of state employees. We shall need many more state employees of this kind, and more *can* be obtained, because capitalism has simplified the work of accounting and control, has reduced it to a comparatively simple system of *book-keeping*, which any literate person can do. [Emphases are Lenin's throughout.]

This is followed by the oddly ambiguous statement: 'The important thing will not be even the confiscation of the capitalists' property, but country-wide, all-embracing workers' control [kontrol' again] over the capitalists and their possible supporters. Confiscation alone leads nowhere, as it does not contain the element of organization, of accounting for proper distribution. Instead of confiscation, we could easily impose a *fair* tax. . . .' He goes on to insist that the rich should work, that it would be right to put poor and homeless families into their houses.[15]

Still in October 1917 Lenin wrote of the revision of the party programme. Here we find him criticizing Bukharin, a young and able colleague who had studied economics in Vienna and who will figure prominently in the chapters that follow. Bukharin in 1917–20 was one of those who suggested an extremely radical line of instant socialism. Lenin was more cautious. True, 'we are not at all afraid of stepping beyond the bounds of the bourgeois system; on the contrary we declare quite clearly, definitely and openly that we shall march towards socialism, that our road will be through a Soviet Republic, through nationalization of banks and syndicates, through workers' control, through universal labour conscription, through nationalization of the land. . . .' But later, 'experience will tell us a lot more. . . . Nationalize banks and syndicates . . . *and then we shall see*'[16] (his emphasis).

Lenin did, however, speak of 'not being able to nationalize

petty enterprises with one or two hired labourers'. The implication would seem to be nationalization of larger ones. However, this was not spelled out.

Dobb, in his very full survey of Russian development published in 1929, rightly emphasized that economic policy at this stage was subordinated to the political objectives: to break the power of the bourgeoisie, to seize the state machine, to take over the levers of economic power. Details were left for subsequent improvisation.[17]

A month after the revolution, Lenin himself wrote that 'there was not and could not be a definite plan for the organization of economic life'.[18] There was a political strategy, there were general socialist objectives, there was ruthless determination. And, last but far from least, was war, disorganization and growing chaos. We must never for a moment forget that Lenin and his followers, and his opponents too, were operating in an abnormal and indeed desperate situation. Who knows what reforms, policies, remedies, they might have proposed in less troubled times? But in less troubled times they would not have been in power.

Most other Bolshevik intellectuals were too busy with political issues. However, it is worth noting that the Bolsheviks at this time were far from being a monolithic body. There were plenty of Utopian illusions, some ideas verging on anarcho-syndicalism, some ignorant tough-guys, some intellectuals with gentlemanly scruples, fanatics, dogmatists. It is hardly surprising that there were splits and factions in the first years of Soviet power.

On 7 November 1917 Lenin declared to the Congress of Soviets: 'We must now set about building a proletarian socialist state in Russia. Long live the world socialist revolution!'

And now our story really begins.

3. War Communism

THE FIRST MONTHS OF POWER

This chapter covers the period from the Bolshevik seizure of power until the promulgation of 'NEP', i.e. the period from November 1917 until the middle of 1921. Politically and militarily these were stirring, dramatic years. In January 1918 the Constituent Assembly, with its SR majority, sat for one day and was dispersed. Tortuous negotiations with the Germans eventually ended with the onerous peace of Brest-Litovsk (April 1918), followed by a revolt by the left SRs, terror and counter-terror, and a vast civil war. Allied intervention contributed to a series of disasters, which for a time left the Bolsheviks in control of only Central Russia. Victory was eventually won, with immense effort and sacrifice, but in 1920 came war with Poland and a last attack by the White armies from the Crimea. By the end of 1920 victory was won, the enemies now being hunger, cold, anarchy and ruin, as well as some bands of rebels of various hues to mop up.

'War communism' is the name commonly given to the period of extreme communization which began in the middle of 1918, i.e. eight months after the revolution had triumphed. It is therefore necessary to trace the events of the intervening period. Did fully-fledged war communism arise out of a series of improvisations, due to the exigencies of war and collapse, or was it consciously introduced as a deliberate leap into socialism, and ascribed to the war emergency when its failure was found to be discreditable to the regime? Both schools of thought exist. Which is right? Or are they perhaps both right?

In interpreting the events of 1917–21, it is important to bear in mind the following. Firstly, there was a good deal of anarchy, of sheer elemental chaos, in the situation of Russia in those years. Orders by the centre might be obeyed, but quite probably the local authorities, even if communist-controlled, pleased themselves. Orders were in any case all too often confused and contradictory, through sheer inexperience or because the civil service machine was all too effectively smashed. Lenin himself wrote: 'Such is the sad state of our decrees; they are signed and then we ourselves forget about them and fail to carry them out.'[1] Therefore much that happened was not due to central orders at all, and many of these orders were due to desperate efforts to cope with confusion and anarchy.

Secondly, all the events of 1917–21 were, naturally, dominated by the war and civil war, by destruction and fighting, by depleted supplies and paralysed transport, by the needs of the front and the priorities of battle, and last but not least by the loss of vital industrial and agricultural areas to various enemies. The policies of the Soviet government in these years cannot, of course, be considered in isolation from these conditions.

Thirdly, while we have already noted some Utopian and unrealistic passages in Lenin's ideas even before the revolution, and while his comrades were even more prone than he to illusions of all kinds, we must allow for the interaction of Bolshevik ideas with the desperate situation in which they found themselves. To take one example among many: rationing and the banning of private trade in foodstuffs were essential features of the period, and came to be regarded as good in themselves. Yet both these measures were common enough among belligerent nations, and in fact the Provisional government had endeavoured somewhat ineffectively to do just these things. It is interesting to note that H. G. Wells, who visited Russia in 1920, laid great stress on just these points, in explaining Bolshevik policy. It is not only the Bolsheviks who made a virtue out of necessity; anyone old enough to recall 1948 will remember how Labour politicians in Britain extolled rationing (so much fairer to ration by coupon than by the purse). This 'ideology' delayed the abolition of rationing in Britain. Yet its cause was, obviously, the war. One could hardly imagine that Labour in power in peacetime would be 'ideologically' committed to introduce rationing. Or, to put it another way: actions taken in abnormal circumstances for practical reasons are often clothed in ideological garb and are justified by reference to high principles. It is all too easy then to conclude, with documentary evidence to prove it, that the action was due to a principle.

This is not to say that principles ('ideology') had nothing to do with it. Indeed, it is quite clear that Lenin and his friends approached practical issues with a whole number of *idées fixes*, and that these influenced their behaviour. The consequences of actions inspired by ideas could influence events by further worsening the objective situation and therefore rendering further action necessary on empirical grounds. And so on. There was a process of *interaction* between circumstances and ideas.

EARLY MEASURES

The legislation of the first months of Soviet rule sought to implement the short-term programme outlined by the Bolsheviks before the seizure of power.

The land decree of 8 November 1917, adopted by the Congress of Soviets and embodied in a law promulgated in February 1918, followed the lines of the programme, in this instance 'borrowed' from the radical wing of the S Rs. Local committees and Soviets were to supervise land distribution. Land was nationalized,[2] the right to use it belonged to the peasants. None should have more than he alone could cultivate, since the hiring of labour was to be forbidden. Some attempt was made to define the size of holdings. But in fact neither the Bolsheviks nor the S Rs, nor any political force, could tell the peasants what to do. Each village made its own arrangements, which varied widely between and within regions. Some of the better-off peasants grabbed more land. Others, including many who had consolidated their holdings under Stolypin, had their land taken away and put back into the common pool. The average size of holdings diminished, and the number of peasant households with land increased, as some very poor or landless peasants gained from the redistribution. We shall see in the next chapter what effect this great convulsion had upon the shape of Russian agriculture. It is sufficient at this stage to emphasize that it was not in fact a reform undertaken by the authorities, it was a more or less elemental act by peasants, with government organs accepting and by implication legitimizing what was happening. Army deserters, often with weapons, joined in the process of land allocation. The forces of peasant traditionalism, egalitarianism, commercialism, the interests of richer and poorer peasants, clashed in varying degrees and in different ways in thousands of villages where authority had broken down. Despite efforts to prevent it, the land seizures were accompanied by many acts of senseless violence: the landlords' cattle were sometimes slaughtered, the landlords' houses, barns or stables destroyed.

The land law of February 1918 did refer to productive efficiency, better technique, land reallocation and even to the development of a collective system of agriculture. But all this remained on paper. The Bolsheviks could not even attempt to impose a settlement. They had no administrative apparatus, they had practically no party members in the villages. They had come to power on the flood-tide of peasant revolt. All they could do in the first years was to try to keep themselves from being swept away. Their actions, as we shall see, were directed in the main to obtaining the food without which the towns and the army would starve, indeed did starve.

On 27 November 1917 came a decree on 'workers' control'. Factory committees, which existed already under the Provisional government, were given stronger powers. They could 'actively interfere ... in all aspects of production and distribution of products. The organs of workers' control were granted the right to supervise production, to lay down minimum output indicators for the enterprise, to obtain data on costs. ... The owners of enterprises had to make available to the organs of workers' control all accounts and documents. Commercial secrecy was abolished. The decisions of workers' control organs were binding on owners of enterprises': in these words a Soviet textbook summarizes the decree.[3] This appeared to put the seal of legality on growing syndicalist, not to say anarchic, tendencies which had been increasingly manifesting themselves for months before the Bolshevik seizure of power. The trade unions were at least nationally organized, and so one could conceive of 'workers' control' exercised by them becoming, or emerging into, some sort of national plan for resource allocation. The trade unions, however, were in these first months not under Bolshevik control, whereas many of the factory committees were. Yet the latter, despite this control, were bound to reflect only the sectional interest of the factory workers. The local leaders had neither the training nor the sense of responsibility to 'supervise' and 'control' production and distribution. They could and did sell off materials, pilfer, disobey instructions. Of course, discipline had to be reimposed. Carr comments that 'as a weapon of destruction workers' control rendered indisputable service to the revolutionary cause'. It could only add to the already fast-spreading chaos. This was the more certain because the decree also insisted that the management's operational instructions were to be binding. This was, therefore, still kontrol', not full control. But the degree of control was sufficient to inhibit the management from effective action, and divided responsibility meant irresponsibility; indiscipline, and even violence towards technical staff, made work virtually impossible. The railways were operated for the first months by the railway trade union, independently of the Soviet government; the union was not Bolshevik-controlled, and decided to run the railways as the railwaymen thought fit. It was only after some delicate negotiations, plus some outright chicanery and finally the threat of direct violence (March 1918), that the railways were finally placed under the Soviet regime's authority, with workers' control ended. Amid all the multifarious causes of breakdown and confusion, 'the onset of industrial chaos, radiating from the capitals throughout Soviet territory, defies any precise record', as Carr says.[4] Of course, workers' control was only one of many causes of this. But no remedy was possible which did not involve the stern subordination of the committees to some authority and discipline.

On 20 November 1917 the State Bank was seized by armed detachments, because its employees had refused to issue money to what was, in their view, an illegal band of interlopers calling themselves the Council of People's Commissars. On 27 December all private banks were nationalized, and, along with the State Bank, amalgamated into the People's Bank of the Russian Republic. In February 1918, all shareholders in banks were expropriated, and all foreign debts repudiated.

VSNKh

On 15 December 1917 the Supreme Council of National Economy was set up. This was known by its initial letters,

VSNKH (or *Vesenkha*) (and by these letters it shall be called in this book). In examining its powers at the time of its creation, we shall find some evidence of the view held at this time of the role of central planning and the intentions with regard to the nationalization of industry and trade.

VSNKH's task was defined as follows:

The organization of the national economy and state finance. With this object VSNKH elaborates general norms and the plan for regulating the economic life of the country, reconciles and unites the activities of central and local regulating agencies [the council on fuel, metal, transport, central food supply committee, and others of the appropriate peoples' commissariats: of trade and industry, food supplies, agriculture, finance, army and navy, etc.], the all-Russian council of workers' control, and also the related activities of factory and trade-union working-class organizations.

VSNKH was to have 'the right of confiscation, requisition, sequestration, compulsory syndication of the various branches of industry, trade and other measures in the area of production, distribution and state finance'.

VSNKH was attached to (*pri*) the Council of People's Commissars, as a species of economic cabinet, and the members were to be representatives of the relevant commissariats plus the workers' councils, plus some others. The full Council seldom met, and a bureau, initially of fifteen members, was responsible for day-to-day work. It had the power to issue orders on economic affairs, which were (in theory) binding on everyone, including the people's commissariats whose functions it partially duplicated. Regional councils (SNKH or *sovnarkhozy*, yet another abbreviation) administered and controlled the economy locally, under the guidance of VSNKH and in close association with local soviets and workers' councils. By May 1918 there were 7 zonal, 38 provincial and 69 district sovnarkhozy.[5] Very soon VSNKH 'sprouted' departments (*glavki*), for controlling particular activities and sectors, bearing such names as *Tsentromylo*, *Tsentrotextil'*, *Glavneft'*, *Glavspichki*, *Glavles*, concerned respectively with soap, textiles, oil, matches, timber, etc. Duplication with the people's commissariats for trade and industry was ended in January 1918 by their liquidation. With the progress of nationalization the various departments of VSNKH took command of the nationalized sectors of the economy. Its structure was repeatedly changed in the years that followed, but there seems little point in boring the reader with reorganizational catalogues. Much more important and interesting is what its functions were and how they changed.

NATIONALIZATION

As the wording of the original decree showed, VSNKH was supposed to guide and coordinate, but it was certainly not clear how closely it would plan and control industry or trade, or how much of these activities would be nationalized. It is true that there were declarations of intent which suggested that all-round nationalization was the policy. Thus in the 'Declaration of rights of the working and exploited people', published on 17 January 1918 and modelled on the Declaration of the Rights of Man of the French Revolution, some of the laws on workers' control and VSNKH were seen as 'guaranteeing the power of the working people over the exploiters and as a first step towards the complete conversion of the factories, mines, railways and other means of production and transport into the property of the workers' and peasants' state'. However, this declaration proposed no time-table, and would have been consistent with a prolonged existence of a mixed economy. Certainly in its first few months of existence the organs of VSNKH included some managers and even owners. Thus the rules for *Tsentrotextil'*, adopted on 1 April 1918, included in the departmental council fifteen representatives of the (private) employers. The Soviet history textbook from which the above data were derived comments: 'Lenin took a positive view of attempts to make agreements with capitalists on definite conditions favourable to the working class. He repeatedly said and wrote this.'[6] Serious negotiations for collaboration were undertaken with a leading 'capitalist' magnate, Meshchersky. And in any case the various glavki (Tsentrotextil', Glavspichki, and the rest) corresponded closely with the analogous syndicates set up by private business before the war, and used for purposes of control by previous governments during the war. The offices and much of their staff were the same.

Nationalization did indeed begin. The railways (already, in the main, in the hands of the state under the Tsars) and the merchant fleet were nationalized by January 1918, but, with these exceptions, individual plants were nationalized, not industries – at first. Such nationalization was due to a number of factors. At this period, it would certainly be wrong to assume that local Soviets, even communist-controlled, acted because the centre told them to. The large majority (over two-thirds) of nationalizations were local, until June 1918, and may have been due to genuinely local decisions. These in turn could have been due to over-enthusiasm, or to real or imagined sabotage, or to the refusal of employers to accept orders from workers' councils. In view of the prevailing chaos, it is only too likely that many employers found conditions intolerable and tried to get out.

The central authorities were alarmed by the extent of unauthorized nationalization and on 19 January 1918 it was decreed that no expropriation should take place without the specific authority of VSNKH. Clearly, no one took very much notice of this, since on 27 April 1918 the same prohibition was repeated, this time with financial teeth: there would be no money issued to any enterprises which were nationalized without the authorization of VSNKH.[7] It is not clear, on the evidence, that all-round nationalization was already seen as an immediate aim when VSNKH was set up. In fact it would seem that Lenin and his colleagues were playing it by ear. The first leaders of VSNKH, men like Obolensky, Kritsman, Larin, Milyutin, were young intellectual enthusiasts, with little grip on the realities of administration. And in any case much of Russia was outside the authority of the government.

Kritsman in his remarkable article on 'The heroic period of the great Russian revolution',[8] refers to the pre-June period as one of 'elemental-chaotic proletarian nationalization from below'. He added: 'Were it not for external factors, the expropriation of capital would not have taken place in June 1918.' It was hoped that 'capital (i.e. capitalists) would be in some sense in the service of the proletarian state'.

It may be added that the whole question of the intentions of the government at the beginning of 1918 is a matter in dispute among Soviet scholars of the period. Thus Venediktov and several others claim that the party did have a basic plan of nationalization for all the major branches of industry. There was a resolution to this effect passed by the sixth party congress, and it is true also that Lenin in December 1917 spoke of 'declaring all limited companies to be state property'.[9] None the less the evidence, though mixed, is still consistent with the intention to maintain a mixed economy for a considerable period. It may not be out of place to recall that, at about the same time, the Labour Party in Britain was also passing resolutions advocating the nationalization of the means of production.

By June 1918 there were still only 487 nationalized enterprises.[10] The great leap into war communism must be dated from the end of June 1918 with the promulgation of the nationalization decree, affecting in principle all factories, as distinct from small workshops.

To find an explanation of this apparent switch in (or very rapid speed-up of) policy, it is necessary to examine three relevant matters: agriculture, trade and the military situation.

THE SLIDE INTO WAR COMMUNISM

It has already been explained that the peasants had seized the land and redivided it according to their own lights. The splitting up of farms had a disorganizing effect on production, as also did a struggle among the peasants themselves about who was to get what. This struggle was given every encouragement by the Soviet government. Already on 15 February 1918 Lenin was speaking of 'ruthless war against the kulaks'[11] (i.e. the better-off peasants). All this was taking place under conditions of growing hunger, and ever-wilder inflation. The peasants, understandably, sought to obtain a better price for their food. Rationing had been introduced in towns in 1916, but the prices paid to peasants fell far behind the general rise in prices of consumers' goods, and a very great shortage of such goods further discouraged sales through official channels. The peasants tended naturally to evade the state monopoly of grain purchase, thus encouraging the development of a flourishing black market. The Provisional government sought to combat this, in vain. The Bolsheviks at first did no better. Lenin's writings show the inability to cope with the (to him) destructive 'petty-bourgeois flood', which threatened to sweep away effective control. For Lenin, trade at free prices was equivalent to 'monstrous speculation', hoarding was considered sabotage. He informed the Petrograd Soviet on 27 January 1918 that there should be mass searches of stores and houses: 'We can't expect to get anywhere unless we resort to terrorism: speculators must be shot on the spot.' Yet the very next paragraph reads: 'The rich sections of the population must be left without bread for several days because they [have stocks and] ... can afford to pay speculators the higher price.'[12] The winter of 1917–18 was a terrible one. In Petrograd the bread ration fell early in 1918 to a mere 50 grams (2 oz.) a day even for workers.[13] Many had to leave the city, and factories closed for lack of labour. Hunger became a matter of the utmost gravity.

The collapse of production and transport and the disruption of existing market relations was accompanied by an effort to ration through the state (and retail cooperative) organs and by a resolute attempt to suppress free trade in essentials. Private trade in a wide range of consumers' goods was forbidden. However, lack of goods to sell and of an effective distribution mechanism made confusion worse than ever. The cooperative movement was called upon to help, but in 1918 it was still controlled by men hostile to the Bolsheviks, and it was not an effective part of the official system.

All this had a logic of its own, the more so as conditions worsened sharply. Shortages grew ever more acute, as the civil war spread over Russia in the summer of 1918, and as the effect of the Brest-Litovsk treaty was felt: the temporary secession of the Ukraine in particular struck a heavy blow at the already disorganized economy. Between July 1918 and the end of 1919, much of Russia was directly affected by civil war. Railways were disrupted, bridges blown up, stores destroyed. The Soviet-held territory was cut off from essential sources of materials and food. There were typhus epidemics. Stern control over the prevailing anarchy was seen to be vital, and control has a logic of its own.

Referring, or purporting to refer, to another revolution, a present-day Soviet commentator writes:

While strictly regulating maximum consumption and at the same time preserving private bourgeois property and the money economy, the Jacobin state could not help but introduce further coercive and plainly terrorist measures. It was not possible, by any means, to compel the factory owner and the individual peasant to produce, while simultaneously ruining him by requisitions and restricting his links with the market. To put into effect laws contrary to all private interests ... it was necessary to strengthen the dictatorship of the central authority, to systematize it, to cover all France with police and military, to abolish all freedoms, to control through a central supply commission all agricultural and industrial production, endlessly to resort to requisitions, to seize hold of transport and trade, to create everywhere a new bureaucracy in order to operate an immense supply apparatus, to limit consumption by ration cards, to resort to house-searches, fill prisons with suspects, cause the guillotine to be constantly at work. Political terror merged with economic terror, and went in step with it.[14]

This picture shows a common logic in operation in France in 1793 and Russia in 1918. Given the conditions under which privately-owned industry was to operate, given also the rationing not only of consumer necessities but also (as supplies ran down) of many vital materials and fuels, there was a fatally logical escalation in the degree of state control, state operation and finally also state ownership. No doubt there was also some pressure from those party zealots who believed that the revolution had to go much further much more quickly. No doubt too that Lenin's repeated and eloquent words about bourgeois exploiters, and his use of 'workers' control' as a deliberate disorganizing device for weakening the bourgeoisie, contributed to the nationalization drive when it came, as it also contributed to making the work of private management utterly impossible, even with the best will in the world. (However, there is no reason why they should have shown good-will to a regime which had usurped power and publicly announced that their ruin was a good and desirable objective.)

DISCIPLINE VERSUS SYNDICALISM

A conflict with the left communists broke out in the very first months, over the question of discipline and control. For Lenin, workers' control was a tactical device, just as in the army a revolt against officers and propaganda in favour of an elective command was an effective means of disrupting the old military structure. But once power was achieved, Lenin quickly became a firm supporter of discipline and order. We find him speaking of

the establishment of strictest responsibility for executive functions and absolutely businesslike disciplined voluntary fulfilment of the assignments and decrees necessary for the economic mechanism to function like clockwork. It was impossible to pass to this at once; some months ago it would have been pedantry or even malicious provocation to demand it [written on 28 March 1918].

Here he found himself in opposition to the left communists, who also opposed him because of his willingness to sign a particularly unfavourable treaty with the Germans.

Undoubtedly the opinion is very widely held that there can be no question of compatibility [of one-man managerial authority with democratic organization]. Nothing can be more mistaken than this opinion. . . . Neither railways nor transport, nor large-scale machinery and enterprises in general can function correctly without a single will linking the entire working personnel into an economic organ operating with the precision of clockwork. Socialism owes its origin to large-scale machine industry. If the masses of the working people in introducing socialism prove incapable of adapting their institutions in the way that large-scale machine industry should work, then there can be no question of introducing socialism. . . . The slogan of practical ability and businesslike methods has enjoyed little popularity among revolutionaries. One can even say that no slogan has been less popular among them. It is quite understandable that as long as the revolutionaries' task consisted of destroying the old capitalist order they were bound to reject and ridicule such a slogan. For at that time the slogan in practice concealed the endeavour in one form or another to come to terms with capitalism or to weaken the proletariat's attack on the foundations of capitalism, to weaken the revolutionary struggle against capitalism. Quite clearly things were bound to undergo a radical change after the proletariat had conquered and consolidated its power and work had begun on a wide scale for laying the foundations of a new, i.e. socialist society.[15]

It is in line with this policy that a decision was taken in March 1918 to take the railways away from 'workers' control' and place them under semi-military command.

Such policies and measures were opposed by Bukharin, Radek, Obolensky and others. They resented Lenin's emphasis not only on discipline but also on the need for material incentives, piece-work and specially favourable conditions for the employment of bourgeois specialists. They accused Lenin of moving towards state capitalism. Lenin replied with eloquence in an article entitled 'Left-wing childishness'. He refused to regard the accusation of state capitalism as an accusation at all. If state capitalism were established this would represent an advance on the existing situation. The real conflict, he asserted, was not between state capitalism and socialism, but between both state capitalism and socialism on the one hand and the menacing alliance of the petty bourgeoisie with private capitalism on the other. The left opposition continued throughout this period to criticize measures designed to strengthen discipline through centralization and one-man management, and we shall find basically the same issues being debated again in 1920–21.

Needless to say, his left opponents quoted many of Lenin's words against him. Had he not written in 'State and Revolution' that specialists should not be paid more than workers? Had he not extolled workers' control? Was his present policy not plainly inconsistent with doctrine? The workers would not under-stand.

Lenin succeeded in curbing some of the excesses of the workers' councils by having them merged with the trade unions, which were gradually being brought under firm party control. But the experience of workers' councils was defended by him, and is still defended by Soviet economic historians, as a necessary, if materially destructive, stage of the revolution. He found it more difficult to curb the excesses of his own colleagues, and, though (as he later admitted) also being over-sanguine himself on occasion, he repeatedly was having to combat what he called the 'infantile disorder' of leftism.

It may well be, as Dobb argued,[16] that Lenin had no intention at first to launch into the extremes of war communism, that he was driven by emergencies of war, hunger and chaos into an attempt to control everything from the centre. Certainly we can-not disprove this proposition. We can say that his own policies contributed to the chaos, of course. He boasted that these had 'destroyed the discipline of capitalist society'. In doing so, he had for a time helped to destroy all order. It is also the case that a proposal, seriously mooted in the spring of 1918, to have mixed state-and-capitalist enterprises, was rejected. And it was presumably with Lenin's approval that the chief of VSNKн in May 1918, Milyutin, spoke of 'completing the nationalization of industry'. But we will return to this point later.

'PRODRAZVERSTKA' AND STATE MONOPOLY OF TRADE

The slide into war communism was stimulated by the food shortages and the failure of efforts to procure food, especially grain, from the peasants at official prices. Attempts were made to organize sales of goods to peasants, but this had little effect. In May 1918 the Supply Commissariat (*Narkomprod*) acquired more powers to obtain and distribute food. In the end it proved necessary to use force. Lenin spoke on 24 May 1918 of a 'crusade for bread', and there developed a so-called 'food dictatorship', with the local organs of Narkomprod, with the help of workers' detachments and of the *Cheka* (secret police), seizing stocks held by alleged hoarders. This was merged into the campaign against the so-called rich peasants, kulaks, which Lenin had been advocating as the means of spreading Soviet power into the villages. The poor peasants, whom Lenin regarded as natural allies against the rural bourgeoisie, were urged to help in the task. On 11 June 1918 the decree on 'committees of the poor' (*kombedy*) in the villages was issued. One of their principal tasks was 'the removal of surplus grain from the kulaks'. The class war was to be bitterly fought in the villages, and many real or alleged kulaks had some land, equipment and livestock, as well as 'surplus' grain, confiscated. This step, said Lenin, 'was a tremendously important turning point in the entire development and structure of our revolution'. By stages, the compulsory deliveries of food were systematized and given the name of *prodrazverstka*. This untranslatable term is derived from the noun *prodovol'stvie*, meaning foodstuffs, and the verb, *razverstat'*, which literally means to distribute or sub-allocate (tasks or obligations, for instance). It came to mean a policy in which each peasant household was ordered to deliver its surplus to the state. In some cases this was outright confiscation, in others it was virtual confiscation, since the nominal prices paid were very low and there was practically nothing that could be bought with the money. The state demanded all that the peasant had, over and above an ill-defined minimum requirement for himself and his family. The peasants naturally resisted, and either hid their grain or sought to dispose of it through a black market or through illegal barter deals which continued throughout the period. To combat this the government sent workers' detachments to find and seize grain and to punish the hoarders, and it also sought to utilize the committees of the poor peasants, to set them upon their richer neighbours and so to try and discover grain hoards. Thus the process of grain confiscation went hand in hand with the effort to fan class warfare in the villages. A bitter struggle was waged between the government and the peasants and among the peasants themselves. Armed detachments sought to prevent the illegal movement of food to urban markets, although in many cases this was the only way in which food could in fact reach the towns, owing to the inefficiency and inadequacy of the official collection and distribution network.

Peasants resented prodrazverstka deeply, and numerous riots broke out. Some parts of the country were in the hands of so-called 'greens', who were against both 'reds' and 'whites' in the civil war and stood for peasant rights. Some of them were of semi-anarchist complexion, notably a powerful peasant anarchist movement in the Ukraine led by Nestor Makhno, who was a major force in 1919. Others were little better than bandits. In his novel *Julio Jurenito*, Ilya Ehrenburg painted a sarcastic picture of the peasant attitudes of the time. Peasants, he wrote, were all for liquidating communists, officers, Jews.

The main thing, however, was to burn all the towns, for that's where trouble and dissension began. But before burning them it would be necessary to salvage any property that might come in useful, roofing for instance . . . men's coats, pianos. That was their programme. As for tactics, the most important thing was to have a small cannon in the village and about a dozen machine-guns. Don't allow strangers to come near, and replace exchange of goods by raids on trains and requisitioning of passengers' baggage, which was far more sensible.[17]

Yet in the end fear of a return of the landlords kept enough peasants loyal to the Bolshevik cause to ensure their ultimate victory in the civil war. For in most 'white' areas the landlords did come back, and peasants who had seized their land were often punished.

However, the peasants could see little sense in producing farm surpluses which would be taken from them by requisition squads. Sowings were reduced. Production fell. It became ever more difficult to find surpluses, though the government's procurement organizations became more efficient and the requisitioning detachments more ruthless as time went on. Actual state procurements of grain, according to official sources, did increase. Thus in the agricultural year 1917–18 total procurements amounted to 30 million poods and in 1918–19 to 110 million poods.[18] Lenin declared that 'this success clearly speaks of a slow but definite improvement in our affairs in the sense of the victory of communism over capitalism'. It did not mean that conditions in the towns in fact improved. Throughout this period it was in fact quite impossible to live on the official rations, and the majority of the supplies even of bread came through the black market. The government was never able to prevent this market from functioning, but did sufficiently disrupt it to make food shortages worse. There arose a class of people known as *meshochniki*, or men with sacks, who moved foodstuffs and dodged the guards who tried to stop illegal trade. Many townsmen abandoned their work and moved to the country where at least there was some food. Many of the workers, being of recent peasant origin, were able to rejoin their relatives in the villages. There was a spectacular decline in the population of big cities, especially those which, like Petrograd, were far from sources of food. The townsmen who remained shivered hungrily in their unheated dwellings. The so-called bourgeoisie were often deprived even of such small rations as workers had, and had to sell off their belongings in order to buy black market food. A famous Soviet humourist described a barter deal in which a peasant acquired a grand piano in exchange for a sack of grain. The piano was too large for the peasant's hut and so it was cut into two and part of it stored in an outhouse.

Kritsman described the existence of two economies, one

legal and the other illegal. Despite all the efforts to requisition bread grain, in 1918–19 60 per cent of its consumption in cities passed through illegal channels. He estimated that in January 1919 in the provincial (*gubernskie*) capitals – i.e. most large towns – only 19 per cent of all food came through official channels; the figure rose to 31 per cent in April 1919, and fluctuated thereafter; it was only 29 per cent in April 1920. This illustrates most clearly the limitations of the government's 'political' grip, the extent to which it was struggling with forces it could not control, for all its ruthlessness. Lenin could cajole and threaten, the detachments of the Cheka could confiscate and shoot. Yet at certain moments even the government itself was compelled to 'legalize' illegal trade. For example, in September 1918 the wicked speculators and meshochniki were authorized to take sacks weighing up to 1½ poods (54 lbs) to Petrograd and Moscow, and in this month, according to Kritsman, they supplied four times more than did the official supply organization.

The government tried to encourage various forms of rural cooperation, varying from the loosest associations for the joint cultivation of the soil to fully-fledged communes and state farms. More will be said about these various types in the next chapter. It is sufficient here to note that even at the height of war communism all these varieties of collective or cooperative farming covered only a tiny minority of households. In other words, they had little immediate significance for the agricultural situation. They were, however, regarded as politically important. Thus a decree published in February 1919 spoke of a transition to collective farming. No such transition occurred in 1919. But all this helped to set a precedent for subsequent events.

Lenin did see in these still ineffective moves towards collectivism the path to the future. As already pointed out, he was aware of the limitations of small-holder agriculture and conscious of the political difficulties which would arise from the dominance of a private peasant economy. It is interesting in this connexion to quote the evidence of H. G. Wells, who saw Lenin in 1920. The following is his report:

'Even now,' said Lenin, 'not all the agricultural production of Russia is peasant production. We have in places large-scale agriculture. The government is already running big estates with workers instead of peasants where conditions are favourable. That can spread. It can be extended first to one province, then another. The peasants of the other provinces, selfish and illiterate, will not know what is happening until their turn comes!' It may be difficult to defeat the Russian peasant en masse; but in detail there is no difficulty at all. At the mention of the peasant, Lenin's head came nearer mine; his manner became confidential. As if after all the peasant *might* overhear.[19]

One sees here a hint of Stalin's later deviousness, not to say plain dishonesty, in respect of his peasant policy. However, this should not lead us to conclude, as Stalin later wished us to conclude, that the collectivization drive of 1930 represented Leninist policies. It is true that Lenin and his more far-sighted colleagues already in 1918 saw not only the acute problem of persuading peasants to part with food in the critical days of the civil war, but also a long-term contradiction between peasant individualism and the socialist transformation of society. However, as we shall see, Lenin drew lessons from the bitter experience of the war communism period, and in his last years counselled care and moderation.

THE MONEY ILLUSION AND ECONOMIC COLLAPSE

In nightmare conditions of civil war, mismanagement, chaos, hunger and breakdown, the rouble collapsed. The bulk of state expenditure was met through the printing press. Free market prices rose month by month. I myself recall as a small child giving a banknote of considerable face value to a beggar, who returned it to me saying that it was valueless. From March 1919 state enterprises were wholly financed from the budget, i.e. they obtained from the budget all the money they needed, and paid their receipts into the budget. Most transactions between state enterprises were of a book-keeping nature only and not for cash. All this was a gradual process. It began with cash advances by VSNKH to meet wages payments and other expenses for those enterprises which happened to have run out of liquid resources. This practice spread. At first, many of the advances were supposed to be credits and not grants. However, in the general conditions of chaos and collapse, the practice of meeting the running expenses of the economy out of the budget became almost universal and cash payments gradually lost their significance. Typical of the views held at this time was a resolution of the second all-Russian congress of SNKH (economic councils) to the effect that

state industrial enterprises should deliver their products to other state enterprises and institutions on the instructions of the appropriate organs of VSNKH without payment, and in the same way should obtain all the supplies they require, and that the railways and the state merchant fleet should transport gratis the goods of all state enterprises. In making this proposal, the congress expressed the desire to see the final elimination of any influence of money upon the relations of economic units.[20]

This policy was gradually brought into full effect during 1919. This led to what was called the 'naturalization' of economic relations. To cite Venediktov again:

Enterprises in fact made no payment for materials and services obtained from other state institutions, since all expenditures took place by book-keeping and took the form simply of the transfer of working capital allocations from one account to another. The next step was the gradual abolition of monetary charges levied on state institutions for communal services, first in Moscow and later throughout the country. At the same time workers and employees and their families and also some other strata of the population were no longer charged for foodstuffs and consumer goods, for postal and transport services, for housing and communal services, etc. This extended not only to the state sector but also to the working elements of the town and some groups of rural residents, families of soldiers and invalids, etc.[21]

This entire process reached its apogee at the end of 1920 and was undoubtedly deeply influenced by the ideology which was so widespread among the party during the period of war communism. Indeed Venediktov himself noted in his book that some of the most extreme measures in this direction were taken after final victory had been achieved in the civil war.

In other words money lost its effective function within the state sector of the economy, and had precious little function at all.

In 1919–20 workers' wages were largely paid in kind, the meagre ration being free. Overcrowded tramcars and trains, insofar as there were any, were free also, as were municipal services. By 1920 there was even an attempt at a moneyless budget. This has been well described by R. W. Davies. As he put it:

When it proved impossible to stabilize the currency and a centralized war communism economy began to be established, the earlier cautions about the dangers of the rapid transition to a moneyless system were heard less often. News spread that the civil war system of complete state ownership and the abolition of the market was the full socialism of Marx and Engels, and that money was therefore an anachronism. And this view was strengthened by the inflation which seemed in any case to make the abolition of money inevitable. By the middle of 1920 the view that the time was ripe for the complete establishment of a moneyless system was almost universally accepted and attention was turned to the problems of operating an economy in kind. In the sphere of the budget the central problem became the replacement of the money by a budget or balance of state income and expenditure in kind (a material budget), a unified plan for utilizing the material resources of the economy.

Since the various goods had to be expressed in some common denominator, there were discussions about finding such a denominator in labour units. War communism ended before some of these ideas could find any practical expression.[22]

Lenin himself, in writing or approving the draft programme of the Communist Party in 1919, included the following phrases:

To continue undeviatingly to replace trade by planned, governmentally-organized distribution of products. The aim is to organize the whole population into producers' and consumers' communes. . . . [The party] will strive for the most rapid carrying out of the most radical measures preparing the abolition of money.

A Russian commentator has noted the contrast between these words and Lenin's own insistence on a very different policy two years later, emphasizing that experience taught him that this was the wrong road.[23]

As money lost all value, private trade was declared illegal and the nationalization of practically all industrial enterprises was undertaken, voices came to be raised among the communists that they were even now in the process of establishing a true socialist economy. The most intelligent ideologist of left communists, Bukharin, devised a theory to the effect that in revolutions there is an inevitable mass destruction of means of production, and that by thereby destroying the structure and social habits of the past it would be possible to build from scratch the true socialist Russia. Markets, money, buying and selling, these characteristics of capitalism would swiftly vanish. So would economics, a science related to commodity exchange and to private property in the means of production.[24] Of course Bukharin and his friends were well aware of the appalling shortage of goods of every kind, and did emphasize the necessity of increasing production. However, they retained at this period a Utopian and optimistic set of ideas concerning a leap into socialism, which would seem to have had little to do with the reality of hunger and cold. Chaos increased. Industrial production fell rapidly. The destructive civil war disrupted communications and made life more difficult still, while calling upon the remnants of industry to supply virtually everything for the needs of the front.

Shortage of food was perhaps the key problem. The government's policy towards the peasants gave no hope of any improvement, since it provided no incentives to produce. Early in 1919 the government wound up the committees of the poor which had wrought such havoc. But the policy of requisitions and armed detachments was maintained unchanged through 1919 and 1920. Measures that made sense, if at all, only in terms of the emergency and disruption came to be regarded as good in themselves. Yet the vast majority of the people obviously yearned for greater freedom of trade, and the authorities knew it very well. They obstinately refused to contemplate such a surrender, as they saw it, to the petty-bourgeois instincts of the masses. However, their position became more vulnerable with the sharp and continuous decline in the numbers of townspeople, the halving within two years of the working-class population itself, the proletariat in whose name the communist party exercised its dictatorship. There were 2·6 million workers in

1917, 1·2 million in 1920.[25] Chaos and misery were unbearable, or rather would become unbearable as soon as the civil war which gave them some conceivable *raison d'être* was over.

It is hard for a prosaic writer without literary gifts to picture the state of Russia at this period. It is true that many remote and isolated villages lived their lives as usual, but over most of the country normal life had become impossible. H. G. Wells, in his book quoted above, spoke of

harsh and terrible realities. ... Our dominant impression of things Russian is an impression of a vast irreparable breakdown. The great monarchy which was here in 1914, the administrative, social, financial and commercial systems connected with it have, under the strains of six years of incessant war, fallen down and been smashed utterly. Never in all history has there been so great a debâcle before. The fact of the revolution is to our minds altogether dwarfed by the fact of this downfall. ... The Russian part of the old civilized world that existed before 1914 fell and is now gone. ... Amid this vast disorganization an emergency government supported by a disciplined party of perhaps 150,000 adherents – the Communist Party – has taken control. It has – at the price of much shooting – suppressed brigandage, established a sort of order and security in the exhausted towns and set up a crude rationing system.

Wells gives a frightening picture of Petrograd, in which all wooden houses were pulled down for fuel and even the wooden pavings had been used for the same purpose. All this accords with the picture of the period which one obtains from reading such a novel as *Dr Zhivago*. Another Russian novel, Gladkov's *Cement*, described the effect of the chaos of the time on the operations of a factory. Amid desperate shortages of materials and fuel, the remaining workers made cigarette lighters out of pilfered metal in order to have something to barter for food. The tremendous shake-up and disruption of these years left a scar on the memories and consciences of millions of people, and it is not surprising that the experiences of the period with their grandeur and miseries played such an important part in subsequent literature. The following is the statistical expression of collapse:

	1913	1921
Gross output of all industry (index)	100	31
Large-scale industry (index)	100	21
Coal (million tons)	29	9
Oil (million tons)	9·2	3·8
Electricity (million Kwhs)	2039	520
Pig iron (million tons)	4·2	0·1
Steel (million tons)	4·3	0·2
Bricks (millions)	2·1	0·01
Sugar (million tons)	1·3	0·05
Railway tonnage carried (millions)	132·4	39·4
Agricultural production (index)	100	60
Imports ('1913' roubles)	1374	208
Exports ('1913' roubles)	1520	20

(SOURCES: *Promyshlennost' SSSR* (Moscow, 1964), p. 32. *Vneshnyaya torgovlya SSSR za 1918–40 gg.* (Moscow, 1960), p. 13. *Narodnoe khozyaistvo SSSR*, 1932, p. XXXIV. *Sotsialisticheskoe stroitel'stvo* (Moscow, 1934), pp. 2, 4. *Etapy ekonomicheskoi politiki SSSR*, P. Vaisberg (Moscow, 1934) p. 55.)

NOTE: Some of the above figures do not refer to strictly comparable territory.

The collapse of foreign trade was due not only to the prevailing chaos, but also to the blockade maintained during the civil war by the Western powers. There was a sizeable British naval force in 1919–20 in the Gulf of Finland, for example, blockading Leningrad.

VSNKH endeavoured to cope with an impossible job. By September 1919, according to Bukharin, there were under its control 3,300 enterprises, employing about 1·3 million persons, or so the statistical records purported to show, while Bukharin himself thought the number of nationalized enterprises was about 4,000, but presumably the figures here given only relate to those within the purview of VSNKH. Of the above-mentioned enterprises, 1,375 were functioning in September 1919. Amid the general breakdown of transport and communications, the unpredictable movements of the war fronts, the demands of the military for all available supplies and its own clumsy and inexperienced mishandling of materials allocations, VSNKH could only struggle to mitigate where possible the general collapse of economic life. Chaos was increased by arbitrary arrests of real or alleged 'bourgeois', including specialists, deprivation of rations, and so on.

Successive reorganizations considerably expanded its administrative apparatus. In Bukharin's conception VSNKH was in a very real sense acting as a single state firm. In 1919–20, through its various glavki, it distributed such materials as were available, issued orders as to what to produce, which of the desperate needs to satisfy, and in what order. This was indeed, as Bukharin noted, an attempt at total and moneyless planning, though in a disintegrating economy under conditions of civil war, with little effective coordination between VSNKH's own glavki, let alone other controlling bodies. He noted that in September 1919 about 80–90 per cent of large-scale industry had been nationalized, and he correctly foresaw that this figure would reach 100 per cent. However, in view of the fact that he was a leader of the left extremists at this period, it is interesting to note that he, and also Preobrazhensky, his co-author, added the following words:

We must remember that we do not expropriate petty property. Its nationalization is absolutely out of the question, firstly because we would be unable ourselves to organize the scattered small-scale production, and secondly because the Communist Party does not wish to, and must not, offend the many millions of petty proprietors. Their conversion to socialism must take place voluntarily, by their own decision, and not by means of compulsory expropriation. It is particularly important to remember this in areas where small-scale production is predominant.[26]

In line with the above conceptions, the decree of 26 April 1919 specified that there should be no nationalization of any enterprise employing five persons or less (ten persons in the absence of a power-driven machine).

Despite the above, many thousands of small workshops were in fact nationalized, even though the state was quite unable to make them function. The statistics of the period were muddled, to say the least. The invaluable Kritsman gave a number of contradictory figures. Thus VSNKH claimed that on 1 November 1920 there were 4,420 nationalized enterprises, while another source made it 4,547. Yet in August 1920 an industrial census counted over 37,000 nationalized enterprises. Of these, however, over 5,000 employed one worker only. Many of these 'enterprises' were, apparently, windmills! This illustrates the fantastic extremes to which nationalization was pushed in 1919–20, despite the clear impracticability of such action.

Kritsman called the resultant confusion 'the most complete form of proletarian natural-anarchistic economy'. Anarchistic because of conflicts between different administrative instances, and because of lack of any coherent plan. Anarchistic too because of the 'shock' (*udarnyi*) of campaigning methods, by which the authorities rushed from bottleneck to bottleneck, creating new shortages while seeking feverishly to deal with others. He claimed that it was 'heroic'. He knew and said that it was chaotic.

The war emergencies and the transport breakdown were ever-present reasons for tighter control. Already in November 1918 the Council of Workers' and Peasants' Defence was set up to collect and utilize resources for war. This council sprouted a number of committees with strangely abbreviated names such as *Chrezkomsnab* and *Chusosnabarm*, with special powers over defence industries, including those administered by VSNKH.

In March 1920 it became the Council of Labour and Defence (*Sovet truda i oborony*, STO). Lenin was its chairman, and its authority conflicted with, and became superior to, that of VSNKH. The STO became the effective economic cabinet and issued binding decrees on all kinds of things, from nationalization to boots. VSNKH, the STO and the government in 1920 were prone to set up committees to make plans for the expansion of production in future years. For this purpose a number of so-called bourgeois specialists were drawn in. They used the work of such men as Grinevetsky, though this able engineer-planner was an anti-Bolshevik. Some of these plans represent interesting pioneering efforts at thinking out means of developing Russian natural resources on a large scale, even though in the short run nothing whatever could have been done to make a reality of them. The best known of these plans was the so-called GOELRO, the plan for electrification of Russia to which Lenin paid so much attention and which Wells in his book described as senseless dreams amid the universal ruin. This was 'the first long-term development plan in human history'.[27] The plan was presented to a Congress of the Party in Moscow in 1920 by the old Bolshevik engineer, Krzhizhanovsky. He illustrated the plan with a vast map of Russia in which electric light bulbs showed the electrification of the future. Such was the state of Moscow's electricity supply at the time that it was necessary to cut off almost all the city in order to ensure that these lights on the map would not cause overstrain at the power station. The organization responsible for GOELRO was eventually merged with the nascent planning organs, as will be mentioned later on.

By the end of 1918, another body was undertaking the coordination of resource allocation: this was the Commission of Utilization (*Kommisiya ispol'zovaniya*), which, as its title suggests, was concerned with distribution, not production. It was interdepartmental, and tried to reconcile conflicting interests of the various glavki and commissariats. In doing so, it began, however haltingly, the practice of drawing up material balances, later to become an essential feature of Soviet planning.[28]

One gets the impression of utter administrative confusion, described by Vaisberg as 'administrative partisan war'. Central organs, in and out of VSNKh, while enforcing stern centralization were often at odds with one another. No unified plan existed. There was priority for war, and numerous improvisations as the economy staggered from critical shortage to outright breakdown. But, to cite Vaisberg again, 'one must not forget – and this is most important – that under these conditions the party coordinated the multitude of plans and operational decisions of the glavki, replacing the non-existent unified national-economic plan and ensuring military victory'.[29] And within the party the effective body, supreme over all, was the polit-bureau.

During the war communism period the party fervently debated the linked issues of industrial administration and the role of the trade unions. We have seen that the original principles of workers' control involved a species of undefined supervision over the function of management, which gave rise to much indiscipline and strengthened syndicalist tendencies. The merging of the workers' councils with the trade unions could only improve the situation on condition that the trade unions were not behaving as sectional interest organizations and that they would do something effective to impose discipline. This concept of the trade unions was at variance with their representative character. It is true that the unions were increasingly under the control of members of the Bolshevik party. However, these members were not yet behaving like obedient cogs in a machine, and themselves embodied tendencies which could only be described as syndicalist in nature. Many also had, as we have seen, illusions concerning the innate virtue of working-class initiative. Opposition to Lenin crystallized around the issues of one-man management and the role of the unions. Gradually, through 1919 and 1920, Lenin succeeded in having the principle of one-man management in industry introduced, but even as late as March 1920 he admitted his failure to persuade the Bolshevik faction in the trade unions to accept his ideas on this subject. In this respect, as in others, there were no clear rules of conduct. Thus management took the following forms in various places:

(1) A worker in charge, with a specialist assistant and adviser.

(2) A specialist in charge, with a worker-commissar attached to him.

(3) A specialist in charge, and a commissar who had the right to query but not to countermand his orders.

(4) A collegium, with a responsible chairman.[30]

Lenin's principle of 'iron discipline', to which he returned again and again, eventually overcame the concept of a management collegium which included representatives of the workers. The so-called Workers' Opposition, led by Shlyapnikov and Kollontai, thought in terms of trade-union control over the economy. This was not at all Lenin's view. He saw in the party the embodiment of the true interests of the entire working class, and would not allow any counterposing of trade unionist or sectionalist interests to the supreme authority of the party. It is interesting to note, in view of recent Chinese policies, that the Workers' Opposition, in its speeches to the tenth party congress in 1921, advocated that every member of the party, whatever his position, should be an ordinary worker for several months in the year.

At the other extreme Trotsky advocated the militarization of labour. His views arose out of the desperate situation of 1920.

A 'military' attack on the chaos in rail transport did have some success. He took the view that the urgency of the need for reconstruction was such that it justified the creation of a kind of labour army which would work under military discipline. Lenin opposed this view. The trade unions in his opinion still had protective functions, given the bureaucratic deformations from which the Soviet state still suffered.

These debates may seem to be of fundamental importance. It might appear that they were concerned with the 'conscience of the revolution' and the very essence of the nature of the Soviet state. Those who take this view may therefore deplore that so little space has been devoted here to the arguments of the protagonists. The arguments were of course deeply felt. Yet one must ask oneself whether the debates really made much difference to reality. It was natural that some communists would advocate direct working-class control over the factories, and that this would conflict with the need for discipline and order. It was natural, too, that the advocates of discipline and order would triumph, especially in the chaotic conditions of the period. The issue of whether or not the apparatus of economic control should be under the trade unions, which seems so fundamental, was surely in a very real sense a pseudo issue. If the trade-union apparatus had taken over the tasks of running industry, it would have become transformed into another version of VSNKH, the economic department of government. It was quite impossible for the unions to retain their characteristics in this new role. As for the militarization of the economy, Trotsky's mistake was surely to ignore the longings of the demobilized soldiers and overwrought citizens for the status of free workers. The unions were of course incorporated into the system and used as a 'transmission belt', between party and masses.

Nor was Lenin opposed to *ad hoc* militarization. Here, for instance, is one of his draft decrees: 'In a belt stretching 30–50 *versts* on both sides of the railway lines, introduce martial law for labour mobilization to clear the tracks.'[31] He repeatedly urged the mobilization of the bourgeoisie for compulsory labour. A resolution approved by him during the height of the argument with Trotsky favoured 'sound (*zdorovye*) forms of the militarization of labour'.[32] Yet the fact remains that acres of scarce paper and tons of scarcer ink were devoted at this time to this particular discussion, by Lenin and many others.

THE ESSENCE AND ENDING OF WAR COMMUNISM

So we can identify the following characteristics of war communism:

(1) An attempt to ban private manufacture, the nationalization of nearly all industry, the allocation of nearly all material stocks, and of what little output there was, by the state, especially for war purposes.

(2) A ban on private trade, never quite effective anywhere, but spasmodically enforced.

(3) Seizure of peasant surpluses (prodrazverstka).

(4) The partial elimination of money from the state's dealings with its own organizations and the citizens. Free rations, when there was anything to ration.

(5) All these factors combined with terror and arbitrariness, expropriations, requisitions. Efforts to establish discipline, with party control over trade unions. A siege economy with a communist ideology. A partly-organized chaos. Sleepless, leather-jacketed commissars working round the clock in a vain effort to replace the free market.

By the beginning of 1920, the White armies were fleeing on all fronts, and the Bolsheviks were in control of an exhausted country. The time had come to consider the basis on which reconstruction could be achieved. No longer was it possible (or necessary) to subordinate all considerations to the struggle for survival. The means of recovery were now to hand. At the end of 1918 and for much of 1919 the Soviet-held territory of Russia was cut off from most of its customary sources of textile materials (Turkestan and the Baltic states), from oil, from the Donets coal basin, from the wheatlands of the North Caucasus and of the Ukraine, from most of its iron and steel plants. All these had returned to Soviet hands. True, they were in a deplorably rundown or decrepit state. But the resources were available, and needed to be activated. The task of reconstruction had to be tackled and the attention of the party leaders was increasingly devoted to this. The Polish invasion (May 1920), the subsequent Russian advance on Warsaw, and the painful retreat, interrupted for a time the process of re-thinking, by providing yet another reason for emergency measures. But fighting with Poland ended in October 1920.

However, the key problem was the relationship with the peasants, and also the related problems of freedom of trade and of private small-scale industry. It was becoming increasingly clear that the state organs were quite unable to cope with running all sections of industry and with the processes of material allocation, rationing and trade. Requisitioning (prodrazverstka) was bitterly resented by the bulk of the peasants, and agricultural recovery was impossible unless they could be given some incentives and a sense of security. State farms were not an acceptable solution and the peasants were strongly opposed to the transfer to them of any usable land which they themselves wanted. Trotsky may have appeared to be an extreme supporter of discipline, in that he favoured militarization of labour in 1920. Yet the same Trotsky was the first prominent Bolshevik to accept publicly the need to abandon prodrazverstka, to substitute a tax in kind and to allow greater freedom for trade, or at least to barter. He said as much in February 1920. The same view was urged with eloquent indignation by the still-active Mensheviks, especially F. I. Dan. Such proposals were strongly opposed by Lenin, who at this time seemed to have come to regard free trade in grain as a state crime. Yet he was on record as saying that requisitions were a necessary and temporary phase, arising from the emergency situation and the general destitution of the country. He could surely not have imagined that peasants would or could be persuaded (for long) to hand over their surpluses for what he himself described as 'coloured pieces of worthless paper'. He seems to have hoped for some sort of organized product exchange, which would still cut out private traders and keep the peasants wholly dependent on state sources of supply. Certainly the speeches of the time, and also such novels as *Cement* by Gladkov, show that a great many party members were devoted to the proposition that free trade and private manufacturing were sinful, that to allow such things was to surrender to the enemy, and that the wide-spread black market was an evil to be firmly suppressed. Later on, Lenin admitted that he too was affected by the prevailing atmosphere. On 29 April 1920 he said: 'We say that the peasants must give their surplus grain to the workers because under present-day conditions the sale of these surpluses would be a crime. ... As soon as we restore our industry, we will make every effort to satisfy the peasants' needs of urban manufactures.'[33] Yet two days earlier he had said: 'We will not feed those who do not work in Soviet enterprises and institutions',[34] which meant it was a crime for them to eat.

In fact, he seems to have gone right off the rails. Far from modifying the extremes of war communism, the decrees adopted towards the end of 1920 were more extreme than ever. Prodrazverstka was strongly reasserted. Aware of the 'accursed vicious circle' – no industrial production meant no food in towns, no food in towns meant no industrial production – he tried to break out by more ruthless requisitioning. The peasants, he knew, 'needed [industrial] products, not paper money'. He was willing to say that 'we admit ourselves to be despots to the peasants'. But he insisted on their duty to deliver up their surpluses. Even as late as 27 December 1920, speaking to a conference in Moscow, he urged still more attacks on alleged kulaks, a category he refused to define: men on the spot would know; a man who bought a horse for five poods of grain was a kulak, for instance. When a delegate implied that delivery obligations of his area (Stavropol, North Caucasus) might be reduced, so as to avoid 'confiscations ... and so as not to destroy the economy', Lenin told him: 'Act as you acted before. With strict conformity to the decree of the Soviet regime and your communist conscience,' which clearly meant – confiscate.[35] To be fair, on the very same day he found himself in a minority when he proposed the issuing of bonuses in kind to peasant households who produced more. His colleagues thought that bonuses should go only to agricultural associations of various kinds. Lenin replied: 'We have twenty million separate households, which are individually run and cannot be run otherwise. Not to reward them for increasing their productivity would be basically wrong.'[36] On 8 February 1921 Obolensky proposed to the central committee that prodrazverstka should be abolished and Lenin apparently approved.[37] He started making drafts of resolutions on the subject.

Yet as late as 24 February 1921, faced with what he described as kulak risings and a catastrophic situation, he blamed the peasant risings on an S-R conspiracy fomented from abroad. Why? 'The connexion may be seen because the rebellions occur in those regions from which we take bread grains.'[38] Such an absurd statement suggests he was overwrought, or just not thinking. But by then he was about to make up his mind that change was necessary.

In December 1920, too, an attempt was made to control by decree the sowing and harvesting on the twenty million peasant holdings. Of course, this could not be made effective. But the declaration of intent was made, and organizations to carry it out were set up (*posevkomy*, sowing committees), which is a significant index to the party's mentality at the time. The same extremism showed itself in industrial policy. Here too, as on the food front, Lenin in the same year thought in terms of organized action, priorities, suppression of the market. Industry must be started again, food must be provided. 'We must concentrate all our efforts on this task. . . . It has to be solved by military methods, with absolute ruthlessness, and by the absolute suppression of all other interests'; so he said to the executive of the Soviets on 2 February 1920.[39] Perhaps it is in the last few words of this quotation that one finds the clue to the policy pursued in that year. Collapse was total. Priority of reconstruction must also be total. But the great illegal underground market economy, defying all efforts to control it, was sucking resources away, corrupting the apparatus and the proletariat alike. Very well, suppress it totally. So they must have argued. Only a year previously Bukharin and Preobrazhensky, both at this time known for their 'leftism', had quite explicitly stated (in their 'ABC of Communism') that nationalization of small-scale industry would be 'absolutely out of the question' (see page 69). So Lenin would hardly have been acting under pressure from them. This whole leap into extremism in the last months of 1920 plainly perplexed Kritsman, who made no attempt to explain it. He quoted the decree nationalizing all small-scale industry, which was dated as late as 29 November 1920, though by then most of it had already been either nationalized or paralysed. Of course, the administrative organs were quite unable to cope with thousands of tiny productive units. Chaos increased. Thus in that year efforts were made to retain, for the reconstruction period at least, the characteristic methods of war communism. But this proved impossible.

Events forced their hand. Peasant riots grew in intensity as the menace of a White victory in the civil war receded. Bandits roamed wide areas. In some provinces it proved necessary to despatch large armies to suppress rebellion, particularly the Antonov rising in Tambov. Food supplies were gravely endangered. The final straw was the Kronstadt rising, when the sailors rebelled against the miserable conditions of life, and in their slogans reflected the peasants' hostility to the party's policy. This rising began on 28 February 1921. But probably on 8 February, at latest on 24 February 1921, there were clear indications that Lenin had seen that a drastic change of line had become necessary. He expressed willingness to consider the end of prodrazverstka.[40] The rising may, however, have helped to convince even the more extreme of the Bolshevik leadership that a sharp about-turn was a matter of life or death for the regime – and therefore also for them. The New Economic Policy was born.

WHY WAR COMMUNISM?

We must now return to the theme with which we started this chapter. Was war communism a response to the war emergency and to collapse, or did it represent an all-out attempt to leap into socialism? I have already suggested that it could be both these things at once. Perhaps it should also be said that it meant different things to different Bolsheviks, and this is an important element in our understanding their view of the about-turn of 1921. Some felt that the days of 1918–20 were not only heroic and glorious days of struggle, leading to victory against heavy odds, but were also stages towards socialism or even the gateway to full communism. Some of these men were deeply shocked by the retreat, which seemed to them a betrayal of the revolution. Others saw the necessity of the retreat, but were above all concerned with limiting its consequences and resuming the advance at the earliest date. Still others – some of the future right wing among them – looked forward to a prolonged pause, and saw in war communism at best an unavoidable series of excesses. For them a large private sector in small-scale industry and trade, linked with an overwhelmingly private agriculture, was the condition of political security and economic reconstruction, and this would go on for a long time. These attitudes were by no means clear-cut. Lenin himself admitted that he had been over-sanguine about the war communism period. More strikingly still, Bukharin swung from the extreme left to become in the end the ideologist of caution and compromise, as we shall see in the next chapter.

Evidence as to how the Bolsheviks saw the events of 1918–21 may be found in some fascinating debates held in the years 1922–4 at the Socialist (later Communist) Academy. The men who spoke were still able to express frank views, to disagree with one another, speaking of events which they all remembered vividly. This was no official post mortem, or the smooth cleaned-up version of later party histories. It is therefore worth quoting what the various views were, even though all too briefly.

E. A. PREOBRAZHENSKY: You know that in 1918 we introduced nationalization only on a very modest scale, and only the civil war compelled us to go over to nationalization all along the line.[41]

V. P. MILYUTIN (one-time head of VSNKh): In (early) 1918 we had no war communism, no all-round nationalization, and it is wrong to say that there were slogans advocating all-round nationalization. On the contrary, we moved cautiously towards taking over a few trustified sectors of industry.[42]

P. A. BOGDANOV (a minor communist, not the philosopher): Two elements determined war communism. The element of *catastrophe,* this is what made war communism necessary in Russia. . . . The second was a formal element: communism had to be achieved. But who could do it? He who knows how. I will remind you of the Paris Commune. There, in a besieged city, it was necessary to carry out at least some communist measures. . . . I remember in 1918 how Lenin tried to prepare the ground for the thought that for the present we needed only state capitalism. Yes, we were cautious. I recall how the Bolsheviks felt when power fell into their hands. The first feeling was one of disarray (*rasteryannost'*). These were men given by history a gigantic burden to carry. They tried to act cautiously, but military-revolutionary necessity developed and compelled them, life compelled them, to act as they did.[43]

B. GOREV (ex-Menshevik): The most terrible enemy of the proletariat is the petty-bourgeois peasantry, and this enemy must be neutralized. In this sense the experience of the Russian revolution shows that the nationalization of petty trade should be the last phase of the revolution, and not the first. . . . The difficulty was not in the logic of civil war, as comrade Preobrazhensky thinks, but in the fact that the rebellious proletariat demand equality, that is consumer communism [i.e. he gives weight to pressure from below].[44]

V. FIRSOV (minor communist): NEP was arising already in 1918. . . . Then came the period of civil war; 'war communism' appeared. Our movement towards communism thereby slowed down, since socialist construction is impossible when all production potential is utilized unproductively. The war ended. . . . The inevitable NEP[45] appeared, the first stage of our move towards socialism. NEP may be detestable and disagreeable for us, but it is inevitable. Is it a concession, a step back from our ideals? Yes. A move back as against our past? No, since we have nothing to retreat from. We are just beginning to advance. NEP is the advance line of socialist construction.[46]

L. N. KRITSMAN: War communism was a natural-anarchistic economy. Not a socialist form, but 'transitional to socialism'.[47]

YU. LARIN: We had to run the economy in the almost complete absence of normal economic conditions, and so inevitably the planned economy turned simply into the allocation of whatever was available. . . . That is the principal reason why the planned economy under war communism took the form of administrative measures, not of economic regulation but of administrative allocations.[48]

Trotsky, writing in 1920, had this to say:

Once having taken power, it is impossible to accept one set of consequences at will and refuse to accept others. If the capitalist bourgeoisie consciously and malignantly transforms the disorganization of production into a method of political struggle, with the object of restoring power to itself, the proletariat is *obliged* to resort to socialization, independently of whether this is beneficial or otherwise *at the given moment*. And, once having taken over production, the proletariat is obliged, under the pressure of iron necessity, to learn by its own experience a most difficult art – that of organizing a socialist economy. Having mounted the saddle, the rider is obliged to guide the horse – in peril of breaking his neck.[49]

So it would be somewhat over-simple to conclude, with Wiles,[50] that the war-communism period represented a model of fully-fledged quantitative planning, or that the communists only later on came to blame war conditions for actions which they had all along intended.

What did Lenin think? In 1920 he could still talk of reconstruction by enthusiasms – such as voluntary extra work, the so-called *subbotniki* – and by continuing strict centralization of economic life. We have already seen how he seems to have become 'over-enthusiastic' in 1920. Evidence can be multiplied. Thus on 16 October 1920 he telegraphed the Soviet authorities in the Ukraine and urged them to develop collective cultivation, confiscate the money reserves of so-called kulaks (over and above the 'workers' norm', whatever that was), to collect 'fully' all bread-grain surpluses, to confiscate the farm implements of kulaks.[51] True, every now and again he told his comrades that they were going too far, as when, commenting on a draft decree on confiscation of property, he objected to the confiscation of all money held by anyone which exceeded the annual income of a worker, and also the confiscation of all books owned by anyone in excess of 3,000 volumes.[52] But it is clear that by 1920 Lenin was himself going too far, too fast.

He became finally convinced of the necessity of retreat and, true to his nature, he made of the necessity a set of basic principles, of which much more will be said in the next chapter. The new policy was to be carried through 'seriously and for a long time'. In his notes he has left us some interesting insights into his thought-processes. One such note reads: '1794 versus 1921.'[53] The Jacobins, in the French revolution, had found that the terror and economic centralization had lost their *raison d'être* with the victory of 1794. The beneficiaries of the revolution, especially the more prosperous peasants, had pressed for relaxation and freedom to make money. This had swept away Robespierre, and the whole revolution moved to the right after 'Thermidor' (the month of Robespierre's downfall). All Russian revolutionaries had the example of France vividly before them. Lenin's notes show that he intended to carry out the *economic* retreat to avoid a head-on clash with the forces that broke Robespierre. Robespierre, in his view, failed to take into account the class nature of his enemies, had struck out against individuals and had been swept away in the end. He, Lenin, would avoid such political consequences by keeping the levers of political power firmly in the hands of a disciplined party. So it was not a coincidence that the beginnings of NEP were accompanied not only by the final ban on all political parties other than the Bolsheviks, but also – at the tenth party congress in March 1921 – by a ban on factional organization within the Bolshevik party itself.

4. NEP

In the years 1921–2, and for a few years thereafter, the entire social-economic balance shifted. The private sector, the 'petty bourgeoisie', came to act in a way that seemed to be in total conflict with the ideology and practice of the war-communism period. The role of the market, in relations with the peasants and even within the state's own economic sector, was dramatically enhanced. Yet when at last the bitter logic of circumstances convinced Lenin of the need for change, the full extent and consequences of the change were not, at first, discerned.

We have seen how, all the way up to February 1921, Lenin kept stubbornly on the course of all-round nationalization, centralization, the elimination of money, and, above all, the maintenance of prodrazverstka. There was no pressure on him from his colleagues to change this policy. Events, rather than the central committee, provided a potent means of persuasion. The first public sign of second thoughts came in a speech at the plenary session of the Moscow Soviet, on 28 February 1921: he saw the point of a delegate's argument to the effect that the peasants needed to know what they had to deliver to the state, i.e. that the seizure of 'surpluses' be replaced by a tax in kind (*prodnalog*); they would consider this proposal.[1] Once this idea was accepted, however reluctantly, it was bound to lead to a reconsideration of the entire basis of the war communism economy. As already mentioned in the last chapter, any hesitations he may have had left on the subject were overcome when the peasant risings in many parts of the country were followed by the Kronstadt sailors' revolt. This occurred during the sessions of the tenth party congress in March 1921. As emergency military measures were improvised to crush the revolt, so Lenin proposed the substitution of food tax (prodnalog) for confiscation of surpluses (prodrazverstka).

The decisive decree went through the party and Soviet organs during March 1921. The tax in kind was fixed substantially below the 'requisitioning' targets of the previous year, and therefore well below the actual needs for produce. Thus the delivery quota for 1920–21 had been 423 million poods, whereas the grain tax in kind for 1921–2 was fixed at 240 million poods. For potatoes the figures were respectively 110 and 60, for meat 25·4 and 6·5, and so on.[2]

In 1924 tax in kind was replaced by a money tax, but this was after the stabilization of the currency was well under way. After payment of tax, the peasants were to be free to use the rest of their produce as they thought fit, and could sell it 'in the local market', in the words of the original decree. But this would make little sense if the object was to encourage sales of food to the food-deficit areas, and so this limitation was quickly forgotten.

And since it was absurd to contemplate peasants travelling hundreds of miles to sell their own goods in remote industrial cities, the legalization of private trade was inevitable too, despite the strong feelings of revulsion which private trade inevitably caused among many party members. Again, at first it was hoped to control such trade strictly, to limit it. 'Freedom of trade', said Lenin to the tenth congress, 'even if at first it is not linked with the white guards, like Kronstadt was, none the less inevitably leads to white-guardism, to the victory of capital, to its full restoration.'[3] Yet in practice the desperate need for free exchange of goods was such that, once trade of any kind was legalized (March 1921), it grew like a snowball and swept away restrictions. Cooperative trading was encouraged, and was particularly successful in selling consumers' goods in the country-side, as well as goods of all kinds in towns alongside the state retail network, which was gradually being built up from the ruins of war communism. However, private traders were allowed gradually to enter into trade deals of almost every kind: selling to peasants, buying from peasants, buying from and selling to state enterprises, selling goods made by state factories as well as those made by a resurgent private manufacturing sector (of which more later).

At first, when the party was persuaded of the need for change, it was thought that the retreat would be limited to the substitution of 'commodity exchange' (*produktoobmen*) for confiscations. Speaking in October 1921, Lenin frankly admitted this had been an error, an illusion. The only way was trade, and the state and party would have to learn to trade. 'What is the use of talking to us about state trade,' argued a delegate. 'They didn't teach us to trade in prison.' Lenin replied: ' ... Were we taught to fight in prison? Were we taught how to administer a state in prison?'[4] The logic of events, or 'the elemental forces of the petty-bourgeois environment', swept aside efforts to restrain them.

None the less, the Party firmly held to the decision to retain in the hands of the state 'the commanding heights' of the economy: banking, foreign trade, large-scale industry. But it was recognized that the attempt totally to nationalize manufacturing was an error. Under the conditions of 1921, with shortages of every kind, especially of fuel, many state-operated enterprises had to be closed, and some of these, as well as small workshops nationalized in the previous two years, were leased to private entrepreneurs and cooperative groups of various kinds, payment being in goods, or taxes in money. Such small enterprises as escaped the nationalization decrees were allowed to re-open. On 17 May 1921 the decree nationalizing all small-scale industry was formally revoked. On 7 July 1921 every citizen was authorized 'freely to undertake handicraft production and also to organize small-scale industrial enterprises (not exceeding ten to twenty workers)'. Leasing of enterprises in the possession of VSNKh was regulated by a decree of 5 July 1921, and leasing continued through 1922. Already early in 1922 over 10,000 enterprises had been leased, on terms of two to five years on payment of ten to fifteen per cent of output, but of these the large majority were windmills. 3,800 were appreciable enterprises, employing fifteen to twenty persons, and fifty per cent of the lessees were private individuals, some of them former owners.[5] By October 1923 the number of leased enterprises had risen to 5,698, employing an average of sixteen workers; of these, 1,770 were in food processing, 1,515 in hides and skins. Outright denationalization was rare: seventy-six enterprises were 'returned to their former proprietors' by the presidium of VSNKh, an unstated number by provincial organs.[6]

The New Economic Policy was universally referred to as NEP, and the 'privateers' who flourished under it were known as 'Nepmen'. It was a form of mixed economy, with an over-whelmingly private agriculture, plus legalized private trade and small-scale private manufacturing. We shall show later on that the 'Nepmen' did make considerable headway. However, the authorities did not allow the creation of big private enterprises, though a total of eighteen private enterprises did employ 'between 200 and 1000 workers' each in 1924–5.[7] The vast majority of those engaged in manufacturing and mining worked for the state.

FAMINE

But the still feeble economy was struck at once by a disaster of the first magnitude. The cumulative effect of years of prodraz-verstka had caused a marked reduction of sowings, and on top of this there was a severe drought in the east and south-east. The result was an appalling famine. The 1920 grain harvest, only fifty-four per cent of the 1909–13 average, was bad enough. In 1921 the harvest was only 37·6 million tons, only 43 per cent of the pre-war average overall, but far worse in the affected areas.[8] Uncounted millions died. Relief measures were taken. The tax in kind had to be waived in the affected provinces. But supplies in the hands of the authorities were far too small to render effective help to the starving. An emergency relief com-mittee was formed, with prominent non-communists, even with anti-communists such as Prokopovich and Kuskova, as partici-pants. (They were soon afterwards arrested.) American aid was accepted, under the auspices of the American Relief Administration. Scarce resources of foreign currency were used to purchase grain. Yet shortage of food, the breakdown of transport and general disorganization limited the effectiveness of relief measures. Diseases such as typhus carried off many. Millions of hungry survivors wandered in search of some sort of food into the more fortunate provinces.

INDUSTRIAL DIFFICULTIES

With the fuel crisis causing the closing of many state-operated factories, 1921 was a nightmare year for people and government alike. Towards the end of the year and in 1922 there emerged a nightmare of a different kind. The leadership decided that the time had come to abandon the system under which state industry had been operating. Hitherto, as has already been mentioned, they had produced regardless of cost, receiving all their money expenses from the state. The various factories were under divisions (glavki) of VSNKH; they produced to their orders, and received materials and fuel (when they received them at all) in order to carry their orders out. Wages had almost lost their significance, and rations and services were mainly free. The result was a monstrous growth of bureaucratic tangle, an unworkable degree of centralization ('glavkism'), waste and inefficiency. It was found necessary to close many enterprises because there was no fuel and no materials. While desperate efforts were made to restore rail transport and the fuel industries – and by the end of 1922 substantial progress had been made in that direction – the opportunity was taken to rebuild state industry on a new commercial basis, to shed surplus staff and to compel more efficient operation by making management pay its way. Wages were once again paid in cash, and in July–August 1921 services were again charged for. Rationing was abolished on 10 November 1921.[9] No more was heard of the abolition of money as a means of leaping into socialism. Much was heard instead of the urgent need for a stable currency, in which calculations and payments could be made. State industry and state trade was henceforth to operate on economic or commercial accounting (*khozraschyot*). Materials and fuel had to be bought. Workers had to be paid. The necessary resources would have to be obtained from sales. No more spoon-feeding, and no easy sources of credit. This stern medicine was contained in a government decree of 9 August 1921. To enable industry to operate in this way, it was necessary to divide it into autonomous units, instead of treating it as if it were part of one great firm of which VSNKH was the board of directors. These units were, in most cases, known as 'trusts', which controlled varying numbers of 'enterprises' (*predpriyatiya*), i.e. factories, workshops, mines. A few large plants were themselves considered each to be the equivalent of a trust. They were all now to operate commercially. At first there were various limitations on their freedom to sell or purchase, though the major part of industry had already during 1921 been told to sell what they could where they could. But by early 1922 the trusts (or large autonomous enterprises) were having to fend for themselves. Profit-making and the avoidance of losses were to be the operational criteria. There was, as a rule, no definite obligation to give priority to supplying the state; if 'privateers' offered better prices, they handled the goods. If private contractors or intermediaries gave better service than the trading organizations which were slowly replacing the materials-allocation bureaucracy of the war communism period, then here too the Nepmen got the business. As Lenin said, communists had to learn to trade. But by 1922–3 (the economic year ended on 30 September) 75 per cent of all retail trade was in private hands.[10] Conditions in 1921–2 were anything but conducive to commercial accounting and orderly trading. There was famine and desperate general shortage. Factories had few reserves, and trusts had little cash. To obtain liquid assets they had to sell, and sell quickly. Yet the general poverty was such that few would buy. Paradoxically, the feeble industrial effort managed to cause a glut on the highly disorganized market. Trusts competed with one another in trying to sell even raw materials and equipment to raise cash. They opened their own stalls in city streets to do so – this was the so-called *razbazarivanie*, or disposing of assets by bazaar methods.

All this led to a relative fall in prices of industrial goods as compared with (very scarce) foodstuffs. (Note the word 'relative'; both were rising fast with the inflation.) Prices often failed to cover costs. Unsaleable goods piled up amid universal shortages. Trusts were unable to pay their inflated staffs, and unemployment grew rapidly, even though over half of the pre-war proletariat had fled from the towns. The apparently vital need for more output co-existed with its unsaleability. Such were the immediate consequences of the sudden immersion of industry in the cold bath of commercial principles. But we shall see that by 1923 the tables were turned.

Lenin was prepared to go to almost any lengths to restore the economy, feeling, with justice, that this was essential for survival. He fought hard to convince doubting comrades that foreign concessions were an entirely proper way out of the problem of reconstruction, and his works and speeches at this period abound in references to this subject. Some said: 'We chased out our own capitalists, and now we call in foreign capitalists.' Lenin insisted that, by letting foreign capitalists operate oilfields, exploit timber resources and so on, the Soviet state would obtain materials of which it stood in desperate need, and some modern equipment would be provided by the concessionaries. A major move in this direction might have led to a big enclave of foreign-owned industry inside Bolshevik Russia. Lenin had exaggerated hopes of this, and gave publicity to a few ambitious capitalists, such as the American businessman Vanderlip, who came to Moscow with various proposals. However, in the end it came to very little. Only forty-two concession agreements were made, and only thirty-one of those functioned, mainly in timber. In 1924–5 only 4,260 workers were engaged in thirteen significant 'concession-enterprises'.[11] All sixty-eight concessions which existed in 1928 accounted for 0·6 per cent of industrial output.[12] Probably the main reason was the acute distrust of the Bolsheviks on the part of the capitalists abroad, which is hardly surprising in view of the chaotic state of Russia at the time, the declared hostility of the Bolsheviks to the capitalist order, and the fact that they had denounced and defaulted on all past Russian debts, confiscated foreign property, etc.

Foreign trade, however, began to grow again. In 1921 this was still deeply affected by immediate emergencies: grain and coal were imported to deal with critical situations. But a more normal trade pattern began slowly to emerge. A trade agreement was signed with Britain in 1922, and other countries followed suit. In fact, in 1922 there were some complaints from trusts about competition from imports, since some consumers' goods were purchased to provide incentives for the labour force. Imported locomotives, farm machinery, electrical and other equipment, contributed greatly to recovery. Exports in 1924–5, though only a little over a third of 1913 levels, were nine times above those of 1921–2.[13]

TRANSPORT

As already indicated, the transport situation in 1921 was appalling. Over half of the available locomotives were described as 'sick' and the repair shops were incapable of coping with their tasks, for lack of manpower, equipment and fuel. Indeed in 1921 the principal bottleneck in the railways was poor supplies of fuel for locomotives, and even the few that were in good health could not run. Great efforts were made to build up stocks of fuel, and scarce foreign currency was used to import locomotives and components. In 1922–3 45 per cent more passengers and 59 per cent more goods were moved than in the previous year. Recovery continued. In 1923–4 rail transport carried 54 per cent of its 1913 traffic. Already in 1926–7 it surpassed 1913 levels. It is interesting to note that the estimate made in 1922 concerning the recovery of rail transport proved to err greatly on the side of caution: by 1926–7 it had been expected that only 62·7 per cent of the pre-war level would be reached. In this and in some other respects, the ability of the economy to do better than bourgeois specialists expected may have influenced the minds and attitudes of the political leadership, and affected their subsequent behaviour when faced with warnings from these specialists about over-optimism.

Road transport consisted at this period almost wholly of horse-drawn vehicles. Even in 1925 the whole vast territory of the Soviet Union contained only 7,448 cars, 5,500 lorries and 263 buses.[14]

CURRENCY REFORM

The logic of NEP required, as we have seen, a stable currency. Meanwhile the rouble continued to depreciate with startling rapidity. The virtual abandonment of price control, under conditions of the most acute scarcity, gave a new twist to the inflationary spiral. During the war communism period, as we have seen, many a Bolshevik leader accepted the proposition that it was possible, or soon would be, to do without money. Now the word 'money' could be used again, instead of such evasive abbreviations as *sovznak* ('Soviet token'). It was one thing to desire currency stabilization, however, and another to achieve it.

At first there were various experiments designed to find a stable unit of account. Thus the budget drawn up in 1922 was in terms of pre-war roubles, the then existing rouble being 60,000 times greater. But rapid depreciation led to a sharp rise in this figure. There were various devices such as the 'gold rouble', once again linked with pre-war purchasing power, a circumstance which led Preobrazhensky to assert that the value of this kind of Soviet money was based on the memory of what prices had been in 1913. Loans were raised and payments demanded in this unit of account, though the actual cash in use was still the rapidly depreciating rouble which poured from the printing presses. (The first loan of the NEP period was levied in terms of rye.) The decision was taken in July 1922 to create a new unit, the *chervonets*, backed by gold, and to pass as quickly as possible to a stable currency, a properly balanced budget and sound finance, based on a gold standard (though without any freedom to buy or sell gold, and with a strict government monopoly of foreign trade and foreign dealings generally). Recently published memoranda by Lenin include a request to the Commissar of Finance, Sokol'nikov, to send him a note setting out his proposals for the free circulation of gold.[15] Even though this did not happen, it is an interesting fact that such ideas were mooted and were contemplated by Lenin himself. Indeed, for a short while the chervonets was quoted and bought and sold in foreign exchange markets. But

through the rest of 1922 and all of 1923, the chervonets and the paper rouble co-existed unhappily, the latter becoming so utterly valueless that Bazarov quipped that 'the time is not far distant when the sum of those nominal roubles will exceed the number of all atoms or electrons of which our planet is composed'.[16] The chervontsy were few, in heavy demand, and only available in large denominations. The rouble or *sovznak* was still legal tender for most purposes. The rouble currency in circulation increased as follows:

	(*milliards*)
January 1921	1,169
1 October 1921	4,529
1 September 1922	696,141
1 January 1923	1,994,464

Already in October 1922 one pre-war kopeck equalled something like 100,000 of these so-called roubles or sovznaki. No wonder 'in villages transactions were accounted in poods of bread grains, which became a general unit of account'.[17] Soviet scholars and politicians earnestly debated the nature and purpose of money.[18]

For nearly two years there was 'bipaperism', a unique phenomenon. The chervonets (= 10 new stabilized roubles) became sole currency in February 1924. One of these new roubles was equal to 50,000 *sovznaki* of 1923 issue, and one *sovznak* rouble of 1923 represented one million *sovznaki* of 1921 issue. So 50,000 million was the approximate devaluation ratio. When finally withdrawn, the *sovznak* circulation reached 809,625,216,667,200,000 '1923' roubles.[18a] It is only right to mention that Germany in 1923 'achieved' even more, if for somewhat different reasons.

This entire operation was carried out under the aegis of the State Bank, created in October 1921, and of the People's Commissariat of Finance (*Narkomfin*), under the energetic and competent commissar, Sokol'nikov. Not very surprisingly, these institutions during the twenties were the cautious conservative guardians of financial orthodoxy, aiming at the preservation of a balance which had been achieved amid so much difficulty.

In 1922 several other banks were created, with the aim particularly of facilitating the necessary credits to industry (Prombank and Electrobank, the latter to 'finance electrification'), to municipal enterprises (Tsekombank) and agriculture (Cooperative bank, with State Bank participation, as well as private shareholders).

The problem of balancing the budget, without which currency stabilization would have remained a pipe-dream, was solved by levying a variety of excise taxes, commuting agricultural tax in kind and the *corvée* into money payments, taxes on private and state enterprises, income and property taxes, and a whole range of others (e.g. a 'military tax' levied on those who had 'no right to serve in the Red Army'), plus voluntary and forced savings, such as a six per cent bond issue which was 'placed by coercion amid capitalist elements'. In the financial year 1923–4 the budget was balanced, in 1924–5 there was a surplus.[19] (The financial and economic year, until 1930, covered the period 1 October–30 September.)

SCISSORS CRISIS

However, 1923 brought a fresh crisis. From a situation in which the terms of trade between village and town were too favourable to the former – though under conditions of famine the peasants were unable to take much advantage of this – the changed circumstances led to an opposite distortion: a rapid move in relative prices in a direction unfavourable to the village, so unfavourable indeed as to discourage agricultural marketings and to constitute a political menace, since the precarious political stability of the regime depended on peasant acquiescence, or at least a decision on their part not to rebel.

The reasons for this remarkably rapid change were the following. Firstly, agricultural production recovered more rapidly than industry. The 1921 famine led to a decline in the area sown in the following year, since there was a shortage of seed and able-bodied peasants in the affected provinces. This is shown in the table on page 94. However, the 1922 harvest was fairly good; by 1923 the sown area reached almost 90 per cent of the pre-war level, and while the harvest was still far below 1913 levels, the shortage of food was no longer desperate. By contrast, the ruined industrial structure took much longer to repair. The same table shows that industry in 1923 was relatively much further below 1913 levels than was agriculture. Industry was handicapped by the destruction of its basic capital, years of neglect in maintenance, shortages of spares and of skilled labour, of knowledgeable management, of fuel, materials, means of transport. The general chaos of previous years had led to a shift away from industrial and towards food crops, so that, for example, the acreage under cotton had fallen from 688,000 hectares in 1913 to 70,000 in 1922. It was impossible to find currency with which to import sufficient materials to restart the textile industry quickly. In 1922 its output was only 26 per cent of pre-war, while agriculture reached 75 per cent.[20]

Secondly, the chaotic conditions in which newly-formed state trusts unloaded their goods and materials in competition with one another did not last. Credits began to be available from the reorganized banking system. VSNKH reacted to the 1921–2 experience by creating 'syndicates' during 1922–3. These were

	1913	*1920*	*1921*	*1922*	*1923*	*1924*	*1925*	*1926*
Industrial (factory) production (million 1926–7 roubles)	10251	1410	2004	2619	4005	4660	7739	11083
Coal (million tons)*	29·0	8·7	8·9	9·5	13·7	16·1	18·1	27·6
Electricity (million Kwhs)	1945	–	520	775	1146	1562	2925	3508
Pig iron (thousand tons)	4216	–	116	188	309	755	1535	2441
Steel (thousand tons)	4231	–	183	392	709	1140	2135	3141
Cotton fabrics (million metres)	2582	–	105	349	691	963	1688	2286
Sown area (million ha.)	1·500	–	90·3	77·7	91·7	98·1	104·3	110·3
Grain harvest (million tons)	80·1†	46·1‡	37·6‡	50·3	56·6	51·4	72·5	76·8
Rail freight carried (million tons)	132·4	–	39·4§	39·9§	58·0§	67·5§	83·4§	–

* Excluding lignite.

† This was an extremely favourable year.

‡ These are Gladkov's figures; some other sources are higher (e.g. 42·3 for 1921).

§ For post-war the 'economic year' (i.e. 1920–21, 1921–2, etc.).

– not available.

(SOURCES: *Sotsialisticheskoe stroitel'stvo SSSR* (1934), pp. 2–3; Gladkov, *Sovetskoe narodnoe khozyaistvo (1921–5)* (Moscow, 1964), pp. 151, 316, 357, 383; E. Lokshin: *Promyshlennost' SSSR, 1940–63* (Moscow, 1964), p. 32; *Nar. khoz.*, 1932, p. 8.)

NOTE: There are minor disparities between various sources for most years.

government disposal agencies which limited or eliminated competition between trusts, and they joined together for the purpose of joint selling. This placed them in a strong position to demand higher prices when, as was often the case, the state sector of industry was the dominant producer of the items in question.

Thirdly, state industry was inefficient, operating far below capacity, with heavy overheads and much overstaffing. Labour productivity was far below pre-war. Costs were therefore high.

Fourthly, the wholesale and retail distribution system was exceedingly inefficient and costly. According to Preobrazhensky, the average trade margin in 1913 was 17·3 per cent. It had now grown to something like 60 per cent.[21]

Fifthly, the government was in fact the principal purchaser of bread grain, despite the substantial role of the Nepmen, and it sought to buy at low prices.

Finally, the peasants were losers in the inflationary race, when money depreciated daily, because even a week's delay in a journey to town to spend the money meant heavy loss. (The peasants seem to have been the last to get the new chervonets currency.)

The 'price scissors' parted, in the sense that industrial prices were above, agricultural prices below, their 1913 prices. On 1 October 1923, in terms of the newly-stabilized currency, industrial prices were 276 per cent of 1913, agricultural prices 89 per cent. The same source shows the stages by which this remarkable shift in relative prices was achieved:

Industrial prices as a ratio of agricultural prices
(1913 relationship = 100)

	Wholesale	Retail
October 1922	131	161
December 1922	141	167
February 1923	169	180
May 1923	215	223
July 1923	202	211
September 1923	294	280
October 1923	310	297

(SOURCE: Gladkov, *Sovetskoe narodnoe khozyaistvo (1921–5)*, (Moscow, 1960), p. 413.)

Thus, by October 1923, when the 'scissors crisis' reached its peak, industrial prices were three times higher, relative to agricultural prices, than they had been before the war. It was hardly surprising that this caused trouble.

To some extent this wide price divergence was self-correcting. The peasants were the principal purchasers of manufactured goods, and state industry experienced severe difficulties in selling. Seasonal factors also led to a rise in agricultural prices in the months that followed. However, the government reacted strongly, and helped to restore a less unhealthy price relationship. Vigorous attempts were made to force prices of state manufactures down; there were decrees controlling prices (or preventing increases without authorization); there were drives to reduce surplus staffs in industry and in the trade networks, to improve and extend the system of consumer cooperatives, to tighten credit so as to compel trusts to unload stocks. Industrial output meanwhile continued to recover rapidly, though still far below pre-war levels (see above). VSNKH, thoroughly alarmed, was exercising its still considerable formal authority over the state sector of the economy, and a combination of all these factors led to a partial closing of the blades of the now-notorious price 'scissors'. During the financial year 1923–4 industrial selling prices fell by 23·3 per cent. A People's Commissariat for Trade was set up, which tried with marked success to enlarge the area of state trade and to sell manufactured goods in rural areas at prices below those charged by Nepmen. By April 1924 the agricultural price index had risen to 92 (1913 = 100) and the industrial index had fallen to 131,[22] in terms of the new stable currency. By then voices were being raised declaring that the blades had come too close together. This formed part of the debate about the future of NEP and the basic strategy of the Soviet regime, which will be discussed in the next chapter.

PLANNING AND CONTROL

So the NEP system of mixed economy weathered the storm and, with the establishment of a stable currency and balanced budgets, entered into calmer waters by 1924. The years 1924 and 1925 may be described as 'High NEP'. Before discussing the issues and arguments which arose and which, in due course, ended the system, we must take a look at the system as it was during this period. How did it work? Was there any planning and, if so, by whom? What was the relative weight of the private sector? How autonomous were the trusts, and what were the powers of VSNKH? What was happening in agriculture? What was the situation of the workers and of the trade unions? To these and similar questions the rest of this chapter will be devoted.

Let us begin with *planning*. VSNKH was decentralized by the 'trustification' of 1921–2, but was still the headquarters of Soviet state industry. Of the 430 trusts functioning in 1922, 172 were subordinated to VSNKH directly or via its local organs (*Promburo*), and 258 to local *sovnarkhozy* (councils of national economy). To take an example, thirty-three trusts in the metal goods industry were subordinate to VSNKH, controlling 316 factories employing 218,344 workers, while the twenty-four 'provincial' trusts ran ninety-five factories employing 12,701 workers, i.e. these were mainly small-scale. Trusts were in total command of 'their' factories, the latter having no financial autonomy and, as a rule, no separate profit-and-loss accounts or separate legal personality. They had, roughly speaking, the same status as a sub-division of a Soviet enterprise in the sixties, or of a sub-unit within a centralized Western corporation. Indeed, the term 'enterprise' (*predpriyatie*) was not used to describe them. Their entire output was planned and sold by the trust, who supplied them with the funds they needed, e.g. to pay the labour force. Gradually this situation changed, but it was not until 1927 that the directors of what came to be called 'enterprises' within the trusts acquired defined rights and duties – though not yet a legal personality. Actual powers of directors varied greatly within different trusts, and there was little attempt at this period at standardization of practices. According to a resolution of 29 July 1922, VSNKH operated its control over industry by:

(a) Methods of an economic character: the financing of industry, the organization of industrial credit, price policy, etc.

(b) Methods of an administrative character: appointment and dismissal of responsible officials of trusts and other trading-and-industrial units, the transfer of material resources from one branch of industry to another, from enterprise to enterprise, and so forth, in conformity with the industrial plan.

(c) Methods of a production-planning character: the drafting of production and disposal plans, inspection and checking on their execution, ensuring the conformity of the industrial plan with the general plan, etc.

That is to say, becoming in substance the Commissariat for Industry and Trade, VSNKH is the real boss (*khozyain*) over the enterprises within its jurisdiction. All talk of the narrowing of VSNKH's functions and the transfer of part of these to improvised (*fakul'tati-*

vnye – perhaps 'optional') industrial units represents vulgar free-tradeism (*fritrederstvo*) [sic!]. . . .[23]

One senses in this a concern to assert the authority of VSNKH, but it would be a great error to take such formulations literally. It is not only that the resolution in question had no legislative effect. Even legal decrees in these days of uncertain 'revolutionary legality' seldom described the real situation, and were in any event badly or vaguely drafted.

Decrees defining and re-defining the functions and internal organization of VSNKH were numerous, and no attempt will be made here to trace the many changes. A decree of 12 November 1923 repeated many of the points cited above, but in a less assertive way, with more emphasis on trusts and on relations with other bodies. Thus point (d) of Section II of the decree gives VSNKH the function of 'formulation of the production plan and budget of industry of all-union significance, the examination of production plans and industrial budgets of union republics, the formulation of a general production plan and budget for the industry of the entire Soviet Union and its submission through Gosplan for confirmation to the Council of Labour and Defence (STO) . . .'. The new emphasis on union republics was due to the formation of the U.S.S.R. as a federal state in 1922. The function of control over its subordinate units is several times qualified by reference to 'conformity to existing legislation', presumably designed to prevent too much interference from above with the trusts.

VSNKH in 1923–6 included the following sections, departments or units:

(a) The chairman and presidium of VSNKH.

(b) Internal administration.

(c) Chief economic administration (G.E.U.) with the Industrial planning commission (*Promplan*) attached to it.

(d) Central administration of state industry (*Tsugprom*) (with numerous industrial sub-divisions).

(e) Chief administration of the armaments industry.

(f) Scientific-technical department.

The metal industry and electricity generation came under the successors of the former glavki (Glavmetall, Glavelektro) which, for some odd reason, were directly subordinated to the presidium of VSNKH and not to Tsugprom.

In addition, other specialist committees were attached to VSNKH.[24]

Republican councils of national economy (sovnarkhozy) were also set up, with powers over less important industries and the right to be consulted by VSNKH. Provincial and regional councils continued to exist, with powers that varied greatly. In some cases they administered all-union industry in their area, in others only purely local industry.

So at this period state industry of all-union significance was under the authority of VSNKH, which (except for metal and electricity) operated its control through the above-named Tsugprom and its industrial sector divisions, sometimes directly and sometimes through local organs. Its general-policy functions were largely duplicated by the chief economic administration (GEU) of VSNKH, its planning functions by Promplan within GEU. Promplan, in turn, duplicated some of the planning functions of Gosplan, which was under the STO, not VSNKH. This clumsy administrative machinery was re-arranged rather more logically in 1926–7, when both the GEU and Tsugprom were abolished, and the industrial-sector divisions were once more given the name and status of glavki and placed directly under the presidium of VSNKH, as also was the planning unit Promplan.

These changes may, however, amount to no more than different labels on the doors of mostly the same officials and offices. They do not begin to show us the actual functions of VSNKH, the extent of central control over trusts by its administrative subdivisions, or the extent to which in fact there was any real planning in the twenties. This is not easy to define or describe. The best and most detailed account is undoubtedly by Carr and Davies.[25] This shows that there was marked inconsistency between industries and dates. Thus some key sectors of heavy industry, which supplied 'strategic' items for the economy, were given orders as to what to produce and for whom, and therefore the appropriate division of VSNKH exerted, *vis-à-vis* the trusts, power of detailed supervision and control, in many respects similar to the powers of ministries *vis-à-vis* enterprises in the later

'Stalin' model. On the other hand, many trusts, especially in the consumers' goods industries, made up their own production plans by reference to the market, and VSNKH did little to interfere. Indeed, in the case of textiles the VSNKH division (Glavtextil') was almost powerless, the trusts dealing above all with the 'textile syndicate', the state wholesaling organization which came to act as agent for supplies to the textile industry and to exercise a dominant role, so much so that Glavtextil' was abolished in 1927 as superfluous. Various organs of VSNKH gave orders on all sorts of issues; many trusts were in effect wholly autonomous most of the time. No clear picture emerges, except that, firstly, control over new investment was much tighter than over current business, and, secondly, control of all kinds became stricter towards the end of the decade, as resources became tighter. There will be more about this in Chapter 6. We must now look briefly at the planning mechanism as it existed at this time.

Gosplan, as we have seen, was set up to assist the Council of Labour and Defence (STO). The latter body was nominally a commission of the Council of Peoples' Commissars, but in fact it was the effective economic-military cabinet, was presided over by Lenin while he was still capable of work, and included among its members the chairman of VSNKH, as well as the commissars of War, Labour, Transport, Agriculture, Supplies and a representative of the trade unions.[26] Its decrees carried legal force, as if they were issued by the government. Gosplan (at first officially known as the state general-planning commission) was set up on 22 February 1921 to 'work out a single general state economic plan and methods and means of implementing it'. The members were appointed by the STO. Its duties were redefined on 21 August 1923, its title being now the more familiar 'state planning commission'. The first paragraph concerning its functions was reprinted almost verbatim as above. It was also to help to prepare the budget, examine 'basic questions concerning currency, credit and banking', industrial location, standardization. It had an essential coordinating function, in that it had the right to examine and express its views on all plans and production programmes put forward by people's commissariats (including VSNKH).

It can be seen that duplication and some rivalry with the planning division of VSNKн inevitably arose. Thus the work of preparing the 'control figures for 1925–6' was undertaken only by Gosplan, and Krzhizhanovsky, Gosplan's head, complained to the fifteenth party congress about the lack of cooperation with other agencies.

However, as already pointed out, the word 'planning' had a very different meaning in 1923–6 to that which it later acquired. There was no fully worked-out production and allocation programme, no 'command economy'. The experts in Gosplan, many of them non-party or former Mensheviks, worked with remarkable originality, struggling with inadequate statistics to create the first 'balance of the national economy' in history, so as to provide some sort of basis for the planning of growth. The pioneering contribution of Russian economics at this period will be the subject of comment in the next chapter. The point is that what emerged from these calculations were not plans in the sense of orders to act, but 'control figures', which were partly a forecast and partly a guide for strategic investment decisions, a basis for discussing and determining priorities. An expert committee of VSNKн, known as OSVOK (Council for the restoration of basic capital) studied the capital assets and needs of various branches of the economy, and produced a series of reports and recommendations (1923–5), which affected government thinking. But neither was this yet operational planning.

So neither Gosplan nor VSNKн could provide, or tried to provide, output plans for all trusts and enterprises, except some in some key sectors. Most trusts made their own arrangements, with only partial supervision from above. They varied greatly in size and *modus operandi*. Some were very large indeed, such as the Baku oil organization, which, being responsible for the bulk of Soviet oil production, was closely linked with the relevant sub-division of VSNKн. Others were more remote from the control of VSNKн, and made contracts freely with other trusts, with private traders, cooperatives and the like, negotiated credit arrangements with the bank, or with each other, guided very largely by the profit motive, with spasmodic efforts at price control by the Centre, though there was a marked tendency towards tighter control later in the decade, as we shall see.

Dzerzhinsky, chairman of VSNKH and head of the Cheka (police), wrote in 1924: 'We have almost every trust doing just what it pleases, it is its own boss, its own Gosplan, its own Glavmetall [the metals division of VSNKH], it is its own VSNKH, and if anything does not work out right it hides behind the backs, and receives the support, of local organs.'[27] He sought, with only modest success, a tighter degree of control over the trusts.

Already some of the trusts themselves were felt to be too unwieldy and too large for economic operation. Thus one finds the twelfth party congress (1923) urging trusts to give more initiative to the enterprises comprising them, also to calculate profitability and to issue bonuses at enterprise level. But little was changed until much later.

PRICES, MARKETS AND PRIVATE ENTERPRISE

With the coming of NEP, the problem of price control became particularly acute, especially under conditions of scarcity of goods, when wages were again being paid in money, and free rations and services were ending. A decree of 5 August 1921 set up a Prices Committee attached to the Commissariat of Finance. It had power to fix wholesale and retail prices for goods made or sold by state enterprises, as well as prices at which government agencies were to buy from others, for instance private peasants. However, these controls were largely ignored, and in 1922 were replaced by so-called 'approximate' (*orientirovochnye*) prices, which soon came to be regarded as minima, so the trusts pleased themselves. Exceptions were the selling prices of some monopolized commodities, such as salt, tobacco, kerosene and matches, and where prices were genuinely fixed by the government, though private traders sold them, when they could get them, at any price which the local market could bear. A Commission on Trade attached to the STO, set up in 1922, endeavoured with some success to establish direct links between state industry and consumer cooperatives, and to cut out private commission agents, but had no effective way of controlling prices.[28]

The co-existence of private and state (plus a largely autonomous cooperative) sectors, under conditions of inflation, transport breakdowns and administrative inefficiency, led to some very substantial price fluctuations. We have already referred to the price 'scissors' crisis of the autumn of 1923 and the great difficulties experienced at this time by state industry in selling their high-cost products, which became even dearer when they finally reached the consumer, especially in rural areas, through many private intermediaries. The Nepman was almost the sole seller in many rural areas in 1923. Where a rural cooperative existed, it was exceedingly inefficient. The following table, showing the figures for October 1923, illustrates this:

	Cloth	Nails	Kerosene	Sugar	Salt
Trust (manufacturers') price	100	100	100	100	100
Provincial cooperative price	174	136	100	162	107
Village cooperative price	243	177	128	222	121

(SOURCE: Malafeyev, *Istoriya tsenoobrazovaniya v SSSR* (Moscow, 1964), p. 53.)

With such colossal margins, the cooperatives could hardly compete effectively with Nepmen.

Even in Moscow, where in 1922 a 'commodity market centre' (*tovarnaya birzha*) was set up under official auspices, the Nepmen controlled 14 per cent of purely wholesale trade, 50 per cent of mixed wholesale-retail trade and 83 per cent of retail, cooperatives taking 10 per cent and the state only 7 per cent. 'Wholesale trade in textiles in the country as a whole until March 1923 was at least 50 per cent in the hands of private capital.' Private trade in 1922–3 constituted 78 per cent of all retail trade, the proportion falling to 57·7 per cent in 1923–4, 42·5 per cent in 1924–5, 42·3 per cent in 1925–6 and 36·9 per cent in 1926–7.[29]

However, this falling percentage at first represented an increase in absolute volume, within a rapidly-growing total trade turnover. Both wholesale and retail private trade rose by about 50 per cent in 1925–6. It was in the next year that the absolute volume of private trade began to fall.

Private trade filled the gap left by the inadequacy of the state and cooperative network. As already noted, it dealt with goods produced by state enterprises, as well as the bulk of the products of private industry and handicrafts – of which more in a moment. From 1923, at an increasing rate, the government expanded the state and cooperative network, and their share in the trade turn-over rose constantly. At first this was by competition, and diverting more state-produced goods through 'official' chan-nels. Later on, the Nepmen were squeezed out by methods of less agreeable kinds, as we shall see.

As for industrial production, in 1924–5 the situation was as follows: 'Large-scale' industry – roughly coterminous with factory as distinct from workshop production – was over-whelmingly in state hands. A mere 1·82 per cent was private. However, the total output of small-scale and handicrafts industry was divided as follows:

	1923–4	1924–5	1925–6	1926–7
	(percentage of total output)			
State	2·2	2·6	2·5	2·3
Cooperative	8·1	20·4	19·8	20·2
Private	89·7	77·0	77·7	77·5

(SOURCE: Gladkov, *Sovetskoe narodnoe khozyaistvo (1921–5)*, (Moscow 1960), p. 201.)

In 1925, the following were employed in the above:

State industry	30,644
Cooperatives	127,162
Private craftsmen*	2,285,161
Private employment†	270,823
Total, small-scale	2,713,790

* Not employing labour outside family.
† Employing labour, or employed.
(SOURCE: *ibid.*, p. 204.)

This shows the domination of the private sector in small-scale industry, and also just how small-scale this industry was, with so large a proportion of it conducted on an individual or family basis. These private activities were greatly dependent on supplies of materials from state industry, and most of the workshops were leased from the state. A Soviet historian commented: 'This placed in the hands of the state a powerful weapon for controlling small-scale production.'[30]

Employment in private industry rose as follows:

	(*per cent*)
1924–5	+13
1925–6	+20
1926–7	+2 or 5

This was followed by a catastrophic fall, the reason for which will be discussed later. 'Basic funds' (capital) in the private sector as a whole increased in each of the years in question.[31] However, the government was rapidly achieving an effective dominance, and, once the state and especially the cooperative trading network was effectively extended into rural areas, it could squeeze out private trade by starving it of manufactured goods, and private industry by starving it of fuel and raw materials, as and when it chose. In fact state-encouraged and state-supported cooperative retail trade multiplied exceedingly rapidly. In 1922–3 its share in retail turnover was only 10 per cent (as against 75 per cent for private trade); in 1926–7 its turnover had risen nineteen times in constant prices, the picture evolving as follows:[32]

	1922–3	1926–7
	(millions of roubles)	
State	512	1817
Cooperative	368	6838
Private	2680	5063
Total	3560	13718

The figures (in roubles of constant value) show the extremely rapid rate of recovery which characterized the first years of NEP.

AGRICULTURE AND THE PEASANTS

In agriculture, the private sector was predominant throughout the NEP period. There were few collectives and communes in 1918–20, in subsequent years there were even fewer. Even as late as 1927 the situation was as follows:

	Sown area (per cent)
State farms	1·1
Collective farms (all types)	0·6
Individual peasants	98·3

(SOURCE: *Etapy ekonomicheskoi politiki SSSR*, edited by P. Vaisberg (Moscow, 1934), p. 35.)

The effect of the land reapportionment of 1917–18, together with the further measures of 'class war in the villages' in 1918–20, led to the elimination not only of the landlords' estates but also of many of the larger peasant holdings too. Millions of landless labourers, or ex-peasants who had returned from town in the days of war communism, had acquired land. Therefore the number of peasant farms rose sharply. From roughly 17 or 18 million on comparable territory before 1917 (exact comparisons do not seem to be available) the number of family holdings rose to 23 million in 1924 and 25 million in 1927. Poorer peasants had gained land at the expense of their neighbours. The average size of landholding fell, the extremes of riches and poverty diminished.

The years of revolution had undone much of the positive effects of the Stolypin reform. The traditional peasant community had presided over most of the land redistribution in 1918. It now had more power than at any time since 1906 to redistribute peasant holdings, to insist on traditional strip cultivation. In 1925 over 90 per cent of the peasants belonged to village communities. It could be asserted that 350,000 such communities controlled the economic life of the village.[33] Thus the effect of the revolution was, in a technical sense, reactionary. Land was sometimes scattered over dozens – sometimes a hundred – strips located in widely separated fields, and subject to redistribution. The three-field system was predominant, and modern crop rotations little used, as they did not fit the *mir* arrangements. According to evidence cited by Lewin, even as late as 1928, 5·5 million households still used the *sokha* (wooden plough), and half the grain harvest was reaped by sickle or scythe. Forty per cent was threshed with flails. Such modernization as had begun earlier in the century was largely lost. All this set big problems before the regime.

The situation was exacerbated by class attitudes. The simple categories of 'kulak', middle and poor peasant, to which should be added the landless labourer (*batrak*), were in reality anything but simple. Much was written and more was spoken about 'peasant stratification', and statisticians and analysts laboured to fit Marxist-Leninist definitions to complex phenomena which refused to conform to the prepared labels. Since the whole question was political dynamite, disinterested research was at a discount, or risky. As Carr put it: 'It was no longer true that class analysis determined policy. Policy determined what form of class analysis was appropriate to the given situation.'[34]

An admirable account of the situation may be found in the work of Moshe Lewin, and what follows is largely based on his researches. The difficulty was that each category shaded into the other. The 'landless' may have had some land, but so little that one or more members of the household spent much of the year working for another peasant, or for the *mir* (e.g. as shepherds), or in seasonal employment away from the village. The 'poor peasant' (*bednyak*) by definition had not enough land to feed his family, and he too hired himself out for part of the year, or members of his family did. It must be remembered that the peasant 'household' (*dvor*) usually included several individuals of working age, and this too made classification somewhat complex. The 'middle peasant' was often very poor indeed by any reasonable standard, and at the lower levels was indistinguishable from the poor peasant, in that members of the household frequently had to hire themselves out, but some also made use of hired labour themselves. Many so-called middle peasants were among those who had no horse (40 per cent of all peasants in the Ukraine, 48 per cent of those in Tambov province, etc.). Sometimes the analysts used the term *malomoshchnyi* (weak), a term which covered the less well-favoured middle peasant as well as the poor. Those above this ill-defined level would be classed as *zazhitochnye* (prosperous), or *krepkie* (strong), and this would include some arbitrary proportion of middle peasants as well as the so-called kulaks.

What, then, was a kulak? This too was a category re-defined by statisticians to suit the political circumstances, or re-defined by politicians who ordered statisticians to produce appropriate figures to prove their point. Kulaks were generally deemed to number somewhere between 5 and 7 per cent of the peasantry. Yet by far the larger number of these were, by Western standards, poor. Two horses and two cows, and enough land to ensure a square meal the year round and something to sell, might qualify a peasant for the designation kulak. Yet, according to data collected by Lewin, only 1 per cent of the total number of peasant households employed more than one labourer.

Some kulaks were able to act as usurers (the word *kulak*, 'fist', was originally a term of abuse related to this particular function). Some were able to lend their poorer neighbours grain when, as often happened, they ran out of food in the spring. They were also able to hold stocks of food and so benefit from higher prices, whereas the less well-off peasants had to sell quickly after the harvest, when prices were low. They often had such machinery as was available, and benefited from hiring it out, as well as hiring out a horse or bull or other scarce beast. Kulaks had initiative, they had commercial sense. They excited jealousy at times, but they were also what every ambitious peasant wanted to become.

The numbers of alleged kulaks seemed to be growing:

	1922	1923	1924	1925
Percentage of total households:				
leasing land	2·8	3·3	4·2	6·1
employing labour	1·0	1·0	1·7	1·9

(SOURCE: Gladkov (1921–5), p. 271.)

But some of the 'employers' only used hired labour for a very few months in the year.

The government's dilemma from the first was this. There was the attempted 'class' analysis. All peasants had a kind of dual allegiance, being semi-proletarian, semi-petty-bourgeois. They vacillated, they could provide fertile ground for the resurrection of capitalism. On the other hand, the U.S.S.R. was the republic of *workers and peasants*. NEP was based on the link or alliance (*smychka*) between them. The political conclusion was drawn, remembering Lenin's old dicta, that the poorer peasants would be allies, the middle peasants perhaps tolerant associates; the kulaks represented, by definition, the enemy, the menace. If any middle peasants, by their success, became kulaks, then they too joined the camp of those who would see the Soviet regime as a hostile force, and who would be busily secreting or engendering capitalism and political opposition.

However, to define any peasant who made good as an enemy would make nonsense of the principle of the smychka. Even more important, perhaps, was the fact that it made economic nonsense to penalize success when, above all, more production and more marketings were needed. Here, underlying all the factional disputes in the party over the peasant, which will be treated in the next chapters, were some very real perplexities and dilemmas.

Meanwhile recovery had to take place within an institutional setting of a semi-medieval character, under which the vast majority of small-holder peasants had little opportunity, incentive or resources to improve their methods of production or to use machinery. Their way forward as individual peasants would only be through consolidating holdings, increasing commercial production, and in general following the path which two million peasants had begun to tread on the morrow of the Stolypin reforms (see Chapter 1). But for Bolsheviks this was a path which led – or might lead – to capitalism.

Lenin, in one of his last works, saw a way out: cooperation and also mechanization. Gradually, cooperation would wean the peasant from individualism. Shown the advantages of mechanized cultivation, shown the power of the tractor, he would in time become converted to socialism. But for a long time to come he had to be handled carefully, and certainly he must not be forced, he must be patiently persuaded. This was accepted by all factions in the first years of NEP. Rural consumer cooperatives were favoured because, *inter alia,* they accustomed the peasants to cooperation. They were therefore deemed to be an instalment of 'Lenin's cooperative plan'. As for producers' cooperatives, little was said about them, and even less done, as may be seen from the figures cited on page 106. Typical of the period, and unquestioned at the time, was the assertion of N. Meshcheryakov, in a speech on a most formal occasion, an academic meeting in honour of Lenin shortly after his death:

'Lenin was a determined enemy of any sort of coercion in the area of agriculture. Lenin said that only a fool could think of coercion in relation to the peasantry. . . . There is only one road, that of persuasion.'[35] It matters little in the present context that this was hardly a correct characterization of what Lenin actually did in 1918–21. This is how his views were interpreted in 1924.

The peasants were handicapped not merely by the obsolete system of landholding. The First World War and the civil war had carried away many of the horses, which was one reason why many even of the 'middle peasants' did not have one. There were heavy losses in other categories of livestock. Sowings of industrial crops had fallen sharply, partly as a result of industrial collapse, partly because of the priority given by the peasants to subsistence crops.

It was a characteristic of the Russian peasant to produce mainly for subsistence. In pre-revolutionary times, according to Soviet calculations, landlords and kulaks provided between them 71 per cent of marketed grain. The middle and poor peasants, numerically far superior, managed less than 29 per cent and consumed most of their production themselves (only 14·7 per cent of their crops were marketed). The revolution had greatly added to the middle and poor peasants, and greatly reduced the number and scope of operations of the kulaks. So inevitably there arose a contrast between total output, which recovered first, and marketings which lagged behind.

The size of sown area and harvests and numbers of livestock showed a rapid rise from the famine conditions of 1921. The following are the relevant figures:

	(*1913*)	*1922*	*1925*
Sown area (million hectares)	(105·0)	77·7	104·3
Grain harvest (million tons)	(80·0)	50·3	72·5
Horses (million head)	(35·5*)	24·1	27·1
Cattle (million head)	(58·9*)	45·8	62·1
Pigs (million head)	(20·3*)	12·0	21·8

* 1916

(SOURCE: *Sotsialisticheskoe stroitel'stvo* (Moscow, 1934), p. 4.)

By contrast, grain marketings remained below pre-war. The extent of the shortfall was deliberately exaggerated by Stalin.

He claimed that in 1927 the proportion of grain marketed was only half of what it had been in 1913. This figure has been repeated by numerous authors, Soviet as well as Western. The Soviet writer Moshkov as well as the American analyst J. Karcz have pointed out that Stalin substantially understated the level of marketings in the late twenties, and he also distorted the picture by comparing them with the single year 1913, which was abnormally favourable. Both Moshkov and Karcz would agree that marketed grain amounted to 16·7 million tons (about 25 per cent of production) in the years 1909–13, and that it was something of the order of 16 million tons as an average of the years 1926–8, or roughly 21 per cent of total output, and not 13 per cent as claimed by Stalin and others. However, R. W. Davies has shown that, on a narrow definition of marketed grain, Stalin may have been right.[36] Even if one allows fully for Stalin's statistical devices, the fact remains that marketings were below pre-war, while the need for grain was increasing and would quite obviously increase rapidly as industrialization got under way.

The principal reason for low marketings has already been mentioned: the shift to a small-peasant subsistence-type economy. In 1925 the peasants were eating better, selling less. Contributory reasons, however, were unfavourable terms of trade with the towns (especially in the 'scissors' crisis period, and the blades of the scissors, as we shall see, parted again in due course), the efforts made to expand livestock numbers (livestock have to be fed), the greater attractiveness of industrial crops (after 1923) and, last but not least, the results of deliberate government efforts to control grain purchase prices. Thus as early as 1924–5 'decisive measures were taken to bring order into the system of [grain] procurements . . . to strengthen planning and regulation, to push out the private traders. In 1925–6 measures were taken which reduced private trade in grain. . . .'[37] Already the government was the dominant purchaser, buying 75 per cent of marketed grain in 1925–6. With the abandonment of tax in kind in 1924, a money tax being substituted, the government was more than ever interested in paying as low a price as possible for *the* staple crop, grain. We shall see the effects of this policy on later developments.

Exports were a principal sufferer from the low marketings, as the following figures show (in 1913, a good harvest year, grain constituted 40 per cent of total exports):

	(millions of tons)
1913	12·0
1925–6	2·0
1926–7	2·1
1927–8	0·3

The peasants on their part reacted variously to the new situation and the opportunities it afforded. They naturally welcomed free trade and the private trader. Some better-off peasants tried to expand their activities. Opportunities were not lacking, when so many of the poorer peasants had insufficient land on which to live, and leased it officially or unofficially to a more prosperous neighbour and went to work for him. There are contradictory statistics on all this, no doubt due to the influence of factionalism on statisticians. This confused Soviet researchers in later years. Thus whereas, as we have seen, Gladkov's book gave peasant households hiring labour as 1·9 per cent of the total, another historian gave the figure as 7·6 per cent for the same year, 1925. He also recorded 1·7 million landless labourers,[38] whereas Gladkov pointed out that 450,000 of these were employed as herdsmen by village communities and not by kulaks. But obviously peasant differentiation deepened; a richer peasant class was emerging.

This was, of course, a quite natural and inevitable development within the peasantry, and the government did not like it, the more so as the traditional peasant meeting of elders (*skhod*) was a much more effective authority than the feeble and ineffective rural Soviet. The party was exceedingly weak in the villages. However, up to and including 1925, priority was given to encouraging output and sales, and therefore the less poor peasants received some encouragement, to the dismay of party activists in villages, and of those poor peasants who expected special favours. This, as will be shown later, was the subject of much bitter argument.

As for the peasant attitude to the regime, this must have been very mixed. True, they had gained land. However, they were well aware that they had seized the land themselves. Furthermore they had been much shocked by their experience of requisitions in the war communism period, and had little understanding of or sympathy for the typical, overwhelmingly urban, communist officials. The party leaders' many pronouncements about the 'petty-bourgeois sea' that surrounded them must have given ground for uneasiness among those who read such pronouncements. Events would show that there were grounds for this distrust. The peasant, in any case, had the natural ambition to acquire more land, another horse, a cow, perhaps a labourer. Then, he knew, he would be classed as a kulak, publicly reviled as a menace and a bloodsucker, perhaps taxed heavily, and always conscious of appearing to be a menace in the eyes of at least some of the men in the Kremlin. Peasants, for the regime, were 'the accursed problem'. One imagines that most of the peasants reciprocated such sentiments. The fact that urban employment opportunities remained modest until the end of the twenties had one other consequence: the already considerable degree of rural over-population, in relation to land resources, was growing worse as the natural increase of the population replaced the heavy losses from war, famine and pestilence.

THE URBAN WORKERS

What of the workers, the 'proletariat' in whose name the party exercised its dictatorship? The fall in their numbers during the period of war communism has already been noted. According to Soviet sources,[39] the total number of wage- and salary-earners had fallen from roughly 11 million in 1913 to only 6½ million in 1921–2. The number of industrial workers more than halved in the same period.

With the coming of NEP and of recovery, conditions altered drastically, but at first it seemed to be a matter of jumping out of the frying pan into the fire. With the abolition of free rations and services the workers had to buy everything with a wage of minuscule proportions, which was being eaten away by raging inflation. Of course, the shortages of all kinds made a standard of life well below pre-revolutionary levels inevitable. However, during war communism, at least, the worker was favoured, usually receiving higher rations and sometimes a pair of requisitioned bourgeois boots or trousers into the bargain. Now came NEP. In 'real' terms (i.e. in constant roubles) wages in 1922 came to 9·47 roubles per month (hundreds of thousands, or even millions, of the then existing sovznak roubles). The 1913 wage level in those same prices was 25 roubles.[40] Private traders charged prices which dismayed the underpaid proletarian, and made profits which made him wince. The conversion of workers' pay into purely money wages did not happen at once. At its lowest, in the first quarter of 1921, only 6·8 per cent of 'wages' were paid in money, the rest being issued free in the form of goods and services. Only in the middle of 1922 did over half of the wage take the form of money, and even in the first quarter of 1923, 20 per cent of the wage was in kind.[41] There were strikes, discontent, complaints. The trade unions, still responsive to pressure, made representations.

Conditions improved as more goods became available. The average monthly wage of workers in constant roubles was as follows:

	Monthly (roubles)	Hourly (kopecks)
1913	30·49	14·2
1920–21	10·15	5·4
1921–2	12·15	7·3
1922–3	15·88	8·9
1923–4	20·75	11·7
1924–5	25·18	14·3
1925–6	28·57	16·6

(SOURCE: Gladkov, *Sovetskoe narodnoe khozyaistvo* (*1921–5*) (Moscow, 1960), p. 536.

The statistics are of doubtful accuracy, given the great variety of local prices and the inflationary twists and turns, but they do reflect the consequences of rapid recovery of production. Farm prices, as we have seen, were relatively low after 1922. 1925 was a particularly favourable year, as all statistics show.

The comparison with 1913 is more favourable if one takes social services into account. These, for the 'proletariat', were relatively generous. Already the provisional government had adopted advanced labour legislation, and the labour code of the war communism period would have been very advanced indeed if it had been possible to make a reality of it in the chaos of the time. The labour legislation of 1922 reasserted some of the principles of past decrees, and laid down some new ones. Workers were entitled to an eight-hour day (less in heavy work), two weeks' holiday with pay, social insurance benefits (including sick pay, unemployment pay, medical aid). Collective agreements between management and unions would regulate wages and working conditions. A disputes commission, with the union strongly represented, would consider grievances. The regime could point with pride to such legal enactments; they were well ahead of their time.[42]

However, a grave problem emerged: that of unemployment. It rose particularly rapidly in 1923, when trusts were finding it difficult to sell their goods, and when the government's policy was to encourage profit-making, and the elimination of surplus jobs and featherbedding. With so much pre-war industrial capacity still out of use there was not enough work to go round, even for the reduced urban labour force. Unemployment reached 1·24 million in January 1924, and fell to 950,000 in the next year, but began rising again to reach a figure of 1·6 million in 1929.[43] This might not at first sight seem serious, since unemployment in Britain was of about this magnitude at the time, and Britain was a much smaller country. However, this would be a misleading impression. The vast majority of the population of the Soviet Union were peasants, and peasants – whether or not they are in fact lying on the stove and counting fleas – are not considered to be unemployed. Taken as a percentage of the *employed* labour force, 1·46 million was a very high figure indeed; there were only 8½ million 'workers and employees' in 1924, still well below 1913 levels. Thus the problem of unemployment was serious, and remained so until the end of NEP and of the decade, when a very different policy rapidly caused an acute shortage of labour.

The social consequences of high unemployment affected the young people in particular, partly because the progressive labour legislation of 1922 gave them special privileges (including a six-hour day), which naturally made the employers (be they state managers or private Nepmen) think twice about employing them. Youth was already much shaken by its experience in the civil war. The number of orphan vagabonds (*besprizornye*) was a menace to public order. The fact that there were so few honest jobs for them to do was of no help in rehabilitating them. Crime rates were high.

Wage determination was supposed to be centralized: a unified seventeen-grade wage schedule was adopted in 1922. But this was unenforceable and was systematically ignored in favour of local bargaining. The trade unions had, on a formal level, increased authority to protect the workers and advance their interests. However, the union leaders, among whom Tomsky was particularly prominent, were in a dilemma. They were senior party officials, with the closest links with the party leadership. Times were very hard, it was everyone's duty to pull together. The boat should not be rocked. Strikes would delay recovery and were to be discouraged. It is true that in 1919–20 the trade-union communists functioned as a quasi-autonomous group, as witness their rejection of Lenin's views on managerial authority (see Chapter 3). However, party discipline had perceptibly tightened since then, particularly since Stalin had assumed the general secretaryship of the Party in 1922. With Lenin out of action from the end of 1922, his successors were increasingly intolerant of dissent. In any case, the exercise of power in a hostile environment ('around us – the petty-bourgeois sea') predisposed them to stand together *vis-à-vis* the unpredictable and fickle masses, even though they could and did fight each other in the corridors of power.

The unions did have a clear and legitimate role as far as private employers were concerned. But they did not use their power to fight them, as this would have been contrary to the spirit of NEP; the dearth of all commodities required everyone to produce more.

In state enterprises, the trade-union committee secretary played a significant role alongside the manager and the party secretary. This kind of 'triumvirate' management persisted through most of the twenties. No doubt it was a step forward from 'workers' control', but it was still not an efficient method. However, since all the *spetsy* (specialists) and most managers were bound to be bourgeois and so of doubtful loyalty, this was a way of ensuring that each was provided with two watchdogs.

Given that tradition required some more direct workers' participation, there were also 'production councils' (*proizvodstvennye soveshchaniya*) and, in 1924, 'production commissions', the latter being permanent consultative bodies within factories, representative of the employees.[44] They seem not to have had much effect on management practices, but they did have an educational purpose: out of the workers that came forward in these bodies were chosen many future managers of a new type; they were sent off on training courses which provided new cadres, no longer of 'bourgeois' origin. In Leningrad alone, by April 1925, 900 such individuals were nominated for training for 'various administrative, technical and economic posts'.[45]

1925 may be said to have been the high point of NEP. The party's peasant policy was at its most tolerant; real or alleged kulaks were allowed to increase output. The humourist Zoshchenko wrote a story at this time about a kulak whose house was burning down; the (poor-peasant) fire brigade did not rush to put the fire out, but neither did the kulak. He was insured! So 'the fire brigade was deported for left-wing deviation' (the leftists, as will be seen, were accusing the leadership of being soft on kulaks). At the same time, while expanding state trade, the government was allowing Nepmen to function, as traders and as petty manufacturers and artisans, without much let or hindrance. A nationalized industry was operating on a largely decentralized basis. There was much talk of plans and planning, but no command economy. Growth was rapid, and so the system seemed to be succeeding beyond reasonable expectation. But this growth was based to a great extent on the reactivating of existing capacity, the re-absorption of available factory labour. Thus, according to Kviring,[46] gross investments in 1924–5 (385 million roubles) were not greatly in excess of depreciation allowances (277 million roubles), and yet the rate of growth in production in that and the next year was very rapid. Further progress would require a much greater investment effort, devoted more to building new plant than to repairing and renovating old ones. Just as to repair old 'sick' locomotives and to re-lay damaged track cost much less than building a totally new railway, so, as the period of reconstruction came gradually to an end, the rate of return on investment was found to decline substantially. Far-sighted party leaders could discern new problems and pitfalls ahead, calling perhaps for new policies.

Some of these leaders were already looking with distaste at the NEP compromise. Their arguments will form an integral part of the next chapter.

5. The Great Debate

WHAT WAS NEP?

In the previous chapter we have been giving an outline of the evolution of the economic system within the general pattern and assumptions of NEP. It is now necessary to pass from a description of an evolving situation to an analysis of the discussions of the twenties, which were of the very greatest political and economic importance. They have a significance far beyond the period and location in which they took place. Many developing countries today face similar problems: the financing of capital accumulation, the strategy of economic growth in industrialization, the role of the peasantry after land reform in the context of development, these problems have arisen in many places outside Russia. It may be said that they arose in Russia first, or rather that politicians and economists first became conscious of such problems in the Soviet Union. Far be it from me to suggest that the Stalin solution of such problems is a model for any developing country. Elsewhere the answers to the questions posed may be very different. It remains both interesting and important to study the thought patterns of politicians and economists in the Soviet Union in these years, when relatively frank discussion was still possible, and hard-hitting debates took place in academic and political circles. Although Stalin's political machine already controlled much of what was going on, it was still possible to conduct a genuine public debate on burning issues.

First of all, what exactly was the nature of NEP?

Lenin had left behind, in his articles and speeches, a number of interpretations which were by no means consistent. His successors, busily engaged in a political struggle and anxious to preserve the Bolshevik regime, all tried to present themselves as orthodox Leninists. So it would be proper to return in thought to 1921, to the end of Chapter 3. It was already shown there that Lenin had been thinking of 1794, of the need to avoid

Robespierre's fate by timely retreat. He used the word 'retreat' repeatedly. In referring to war communism he used the parallel of Port Arthur, which had been unsuccessfully attacked by the Japanese at the beginning of the siege. They then withdrew, re-grouped their forces and resumed their assault more methodically, succeeding in the end. The moral seemed to be that some unsuccessful attacks were a necessary pre-condition to a victorious advance, since otherwise the right road could not be found. On either of these two interpretations, NEP represented a forced and highly undesirable retreat, and logically the next step should be to re-group and to resume the advance in due course.

Yet at other times Lenin hotly denied that NEP was undesirable. He argued that, but for the necessities of war, it would have been possible to continue the much milder policies which were begun early in 1918. In still another mood, Lenin would point to the errors and stupidities of the war communism period, to excessively sweeping nationalization, over-centralization, etc. However, on this view NEP could hardly be regarded as retreat at all. If the economic system of 1918–21 was either a forced reaction to an emergency situation or an error, then a return to the *status quo ante* June 1918 was a return to the correct road, and not a withdrawal in the face of the superior forces of the enemy.

What then was in Lenin's mind when NEP was fully established and fate removed from him the power of movement and coherent speech? Did he, as Bukharin believed, draw from the horrors and excesses of war communism a cautious, gradualist conclusion? NEP, he asserted, was intended 'seriously and for a long time'. How long is a long time? Lenin himself answered this question by hinting that twenty-five years would be a rather pessimistic view.[1] Lenin and all his comrades must have believed that the advance would be resumed, otherwise they had no *raison d'être* as Bolsheviks at all. They were bound to regard the ultimate achievement of socialism as the one possible justification for their being in power. But when was the advance to be resumed? At what speed? In what direction? Above all, what was to be done to convert or transform the peasant majority of the population, and how was the industrialization of Russia to be pursued after the period of reconstruction had come to an end? Questions such as these interacted with political issues, concerned with the power position of individuals and factions and the succession to Lenin.

In disentangling the various strands of the argument it is important to distinguish between a number of aspects of a highly complex situation. There was first and foremost the basic dilemma of a Bolshevik revolution, triumphant in the name of Marxism and the dictatorship of the proletariat in an over-whelmingly peasant country. The party had power, the pre-revolutionary productive capacity had been, or was being, restored. By 1925–6 the party had to face a vast question of political economy: how to transform the entire social-economic situation by deliberate action from above. If this was to be done by planning, then by what kind, enforced by what mechanism? A large increase in savings, in accumulation of capital, would be necessary. Who was to bear the sacrifices, and how severe would these sacrifices be?

Another aspect of the problem concerned national security. Of course Lenin and Stalin were not the first to see the intimate relationship between industrial development and military potential. Witte, thirty years earlier, had known all about this and had been greatly influenced by such considerations. How-ever, the isolation of the Bolshevik revolution, alone in a world dominated by 'imperialist' powers, lent a special urgency to the situation, as this was seen by the leadership. The extent and the vigour of intervention during the civil war were exaggerated, but the fact that intervention had occurred strengthened the predisposition to believe that another series of 'imperialist' conspiracies would soon challenge the security of the Soviet state. In the middle twenties the Western powers were quiescent, but the recurrent alarms about the plots of 'Chamberlain and Poincaré' were only partly a matter of deliberate invention for political reasons. The fears seem to have been genuine. The breaking-off of diplomatic relations by Britain in 1927 lent some substance to these fears. There was also nervousness about the Japanese in the Far East. The importance of these attitudes in the present context lay in their impact on the speed of industrialization and on its direction and pattern. The higher the projected rate of growth, the greater the savings and sacrifices. Equally clearly, the more weight that was given to national security considerations, the more the priority afforded to military might and economic independence. This meant more attention to heavy industry, to steel, coal, machinery, at the expense of consumers' goods. But this meant even more sacrifices, which had one further consequence which proved to be of the very greatest significance. If the emphasis in investment was to be on heavy industry, then the peasants could not be offered material incentives sufficient to persuade them to sell more produce. This in turn severely limited the power of manoeuvre of an industrializing Soviet government, within the context of the agricultural settlement bequeathed by the revolution.

THE PEASANTS AND ACCUMULATION

There has already been some discussion in the previous chapter of the consequences of the land settlement of 1917–18. Agricultural production recovered fairly rapidly, but there was a persistent shortage in marketed produce, and the towns could only be fed at the cost of a drastic reduction in exports of grain. Yet urbanization called for a substantial increase in off-farm consumption of food and also for a large export surplus to pay for essential imports of capital goods. Could this problem be resolved within the traditional peasant methods of production, the three-field system, strips, tiny holdings? Was this not a bottleneck which would hold back the entire economic development of Russia? NEP was based on the so-called smychka with the peasants, the word implying a link, cooperation, harmony. Yet Lenin knew and said that a market-orientated private peasantry generated capitalism. It is true that Lenin, in his last year of political life, also said that the peasants had to be shown the advantages of socialism and cooperation, that they should not be coerced. This presented his successors with a very complex question. It was from the better-off peasants that marketable surpluses would come. Any peasant who specialized in providing marketable surpluses would increase his income. Would this be a dangerous growth of potentially or actually capitalistic elements? What alternative was there, without being in breach of both the assumptions of NEP and of the principles of the smychka? Bukharin, who had been a leader of the left wing during the war communism period, became the best known of the leaders of the moderates (the future 'right-wing deviation') in the twenties. He reasoned as follows: NEP is to be persisted with for a long time, for a generation at least. It is out of the question to use force against the peasants. While support for the poor peasants may be politically preferable from the Bolshevik point of view, it is from the middle and better-off peasants that the needed farm surpluses will come, and in no circumstances must they be antagonised. On the contrary, they must be encouraged. The alternative policy would bring back the black days of confiscations, and gravely endanger the Bolshevik hold on political power, since it would lead to peasant rebellion. Bukharin was in favour of building socialism, but only at a pace which the individual peasant producers could be persuaded to accept. In his view, greater prosperity among the peasants, more commercial production, was not only essential but was also not dangerous. In the process of time these peasants too would 'grow into socialism'. Following his own logic, he launched in April 1925 the slogan 'Get rich'. The Russian word for this, *obogaschaites*, was the exact translation of a slogan coined in the 1840s by Guizot, the minister of Louis Phillipe of France: *enrichissez-vous*.

Parallels with French revolutionary history were never far from the minds of Bolshevik intellectuals. Guizot was a bourgeois statesman *par excellence*. The slogan was too much. While at this period Stalin was in political alliance with Bukharin, and favoured tax concessions to the more prosperous peasants, which were in fact accorded in 1925, he never committed himself as far as did Bukharin to the logic of his peasant policy. He declared to the fourteenth party conference in the same month: 'The slogan "get rich" is not our slogan.' Bukharin was forced to withdraw the offending words and to admit that kulaks were an evil to be limited and squeezed.

However, this retreat from the logic of his policy made Bukharin's entire position untenable. A kulak is a prosperous peasant. A middle peasant is a less prosperous peasant. The official line in 1925-7 accepted the middle peasant as the indispensable provider of farm surpluses, while making political gestures towards the poor peasant (there were bitter complaints that the poor peasants' interests were in fact neglected). But this meant that any middle peasant who was successful in developing commercial sales, who sought to expand his holding by leasing or his production by employing a couple of labourers, would speedily convert himself into a kulak. Agriculture had to succeed, and yet it could not be allowed to succeed on the basis of a prosperous private peasant. This was not a sensible or logical policy. Yet what was the alternative?

This kind of dilemma has been faced in other developing countries. There is a tendency for the same people to demand both land reform and industrialization. Yet land reform often has the effect, at least in the short term, of reducing the volume of marketable production, and sometimes of total production, because an egalitarian land redistribution strengthens the traditional subsistence sector. The problem can in principle be resolved by the emergence of a commercially-minded peasant minority, though no one who knows anything about agricultural problems in developing countries would fall into the error of supposing that there is an easy solution. In the special case of the Soviet Union under Bolshevik rule, an advance in this direction came up against an ideological/political barrier.

The logic of the Bukharin approach necessarily involved an emphasis on the production (or importation) of goods the peasants wanted. The same logic called for relatively slow growth, since progress would be limited by the peasants' willingness to save and to supply the state with food surpluses. Bukharin himself spoke of 'riding into socialism on a peasant nag'. But could the peasant nag be persuaded to go in the right direction? Would the Party be able to control it? It must not be forgotten that Soviet power in the villages was weak, and that traditional peasant communal institutions were in effective command; within them the more prosperous peasants tended to become dominant as natural village leaders and because so many of their poorer neighbours depended on them.

The so-called left opposition challenged the validity of Bukharin's policies. As might be expected, their arguments were deeply influenced not only by their views on the particular issues but also by the logic of factional struggle. It is this same logic which in the middle twenties led to a temporary alliance between Stalin and the Bukharin faction. Zinoviev and Kamenev in 1923 found it politic to support both Stalin and Bukharin in a struggle against the left opposition. In 1925, Zinoviev and Kamenev joined the Trotsky group, and thereupon saw virtues in the left opposition's case which had quite escaped their notice two years earlier. However, these and other policy zigzags should not cause us to suppose that the perplexing issues faced by all these men were unreal. Issues are often adopted by political men for their own purposes.

The most cogent theoretical statement of the opposition's case, and one which lights up most vividly the nature of the difficulties which faced the regime, came from the pen of Preobrazhensky. He had been a collaborator of Bukharin's in 1918, but unlike Bukharin he accepted NEP with many reservations and clearly wished to resume the offensive against the hated private sector at the earliest date. He therefore emphasized the dangers which the regime ran if it were to persist for long in the course set in 1921-4. Already in 1923 he was quoting with approval the concept of 'primitive socialist accumulation', and his lectures on the subject in 1924 at the Communist Academy were later expanded and published as a book.[2]

Primitive (or initial) capitalist accumulation was described by Marx using British models. Capital was accumulated through the expropriation of the peasantry, by the enclosures of agricultural land, colonial exploitation, the Highland clearances in Scotland. The resultant concentrations of capital came to be invested in industrial development. Applying this analysis to the situation of the U.S.S.R., Preobrazhensky pointed out that there were no colonies to exploit and the peasants could not be expropriated, and yet the necessary socialist accumulation had to come from somewhere. It would be necessary not only in order to finance industrialization, but also to expand the socialist sector of the economy at the expense of the private sector. Clearly the necessary resources could not arise wholly or even mainly within the socialist sector of the economy. Apart from the fact that it was too small to bear the burden by itself, it was wrong and politically dangerous that the sacrifices should be borne by the working class employed by nationalized industries.

Resources would therefore have to be obtained from the private sector. The bulk of the private sector were the peasants. Preobrazhensky saw that the necessary capital would not be provided by voluntary savings. The better-off peasants were very unlikely to lend sufficient money to the government, and the Nepmen in the cities naturally used whatever capital they possessed to make hay while the sun shone, realistically fearing that it might not shine for long. Resources would doubtless have to be obtained by taxation, but most of all through unequal exchange, by 'exploitation' of the private sector. The state should use its position as the supplier of the bulk of industrial goods, and as the foreign trade monopolist, to pump resources out of the private sector and so finance the state's investments into the expanding socialist industrial sector. Preobrazhensky never failed to emphasize the importance of this conflict between socialist and capitalist elements, and he wrote of the struggle between 'the law of value' and the principle of primitive socialist accumulation, i.e. between the forces of the market and those of the socialist state expanding the socialist sectors.

Bukharin and other leaders of the party majority strongly counter-attacked. This doctrine, in their view, was threatening the alliance between workers and peasants. The word 'exploitation' and the principle of unequal exchange were severely criticized. After all, this was 1924, when the country was only just correcting the excessively unfavourable terms of trade for peasants which characterized the 'scissors' crisis of 1923 (see previous chapter). Every effort was still being made to compel a further relative reduction in the prices charged by state industry. Was this the time to speak of unequal exchange? It may well be that some reasoned privately thus: 'Of course we will have to exploit the peasants in due time, but for goodness' sake let us keep quiet about it now.'

Trotsky, Preobrazhensky and their followers developed two further criticisms. Firstly, they held that the official line had been too favourable to the better-off peasants. They spoke loudly of the kulak danger and envisaged the degeneration of the party into some sort of adjunct of the NEP bourgeoisie. (We shall see that this degeneration, such as it was, was of a very different kind, but this was not apparent to the opposition at the time.)

The party majority was under continuous attack from the left for being soft on the kulaks. Secondly, the 'left' opposition contended that the party's industrialization programme was too modest, that a major campaign to build up industry far beyond the levels of 1913 should be launched forthwith. Both Stalin and Bukharin argued at this time that higher growth rates and additional investments advocated by the left opposition represented an adventurist and unpractical policy, which would endanger hard-won financial stability and impose intolerable sacrifices. This would be inconsistent with the principles of NEP. It is true that the same Stalin was a few years later advocating tempos which were far more ambitious and ruthless than any which the opposition had proposed.

The entire controversy was linked with a famous debate on 'Socialism in one country'. One must introduce a warning here. It might appear logical that the industrializers and 'primitive accumulationists' should favour this slogan, while the cautious Stalin–Bukharin gradualists would be against it. In fact the reverse was the case. Stalin and Bukharin said that they would build socialism in one country, even though Bukharin did say that the process would be long and slow. Trotsky and his friends cited Lenin to show that this was impossible and unorthodox. The factional considerations which led to the arguments around this political slogan are outside the scope of this book. However, there is an economic point to be made which is relevant to an understanding of the logic of the left position. Preobrazhensky himself recognized, in one of the last articles in which he was allowed to express his own point of view, that under conditions of isolation Russia's problem was virtually insoluble as he envisaged it. He listed the many contradictions. He drew the conclusion that all this showed 'how closely our development towards socialism is connected with the necessity of making a breach in our socialist isolation; not only for political but also for economic reasons we must be aided in the future by the material resources of other socialist countries'.[3] Within the political-economic assumptions of NEP, it would indeed be impossible simultaneously to fight the kulaks, raise prices charged to peasants, increase off-farm surpluses and greatly to raise the levels of capital accumulation. Yet Preobrazhensky never faced up to the possibility of resolving the dilemmas through coercion, through expropriating the peasantry. A way would be provided by a revolution in more advanced Western countries, which would come to the aid of developing Russia and so mitigate the harshness of Russian industrialization. Therefore the argument of the left opposition over the slogan 'Socialism in one country' to some extent reflected their disbelief in the practicability of any solution to the dilemmas of the twenties so long as the U.S.S.R. remained isolated. This was politically a weak point in their arguments, and it is hardly surprising that Stalin and his colleagues attacked their alleged lack of faith in socialism and in Russia.

Of course the above represents only the roughest summary of arguments advanced. There were plenty of other points of disagreement and also plenty of other protagonists. Thus there was the People's Commissariat for Finance (Narkomfin) which was devoted above all to financial soundness. In their economic policies the Commissar, Sokol'nikov, and the head of the State Bank, Shanin, could be identified as on the extreme right, because of their insistence that any industrial project be sound and profitable. Yet Sokol'nikov was a political supporter of Zinoviev, who had joined the 'left' opposition. There is of course no inherent reason why one should not support a conservative investment policy, owing to consciousness of the acute shortage of capital, and at the same time inveigh against the dangers of kulaks in villages. The lines of controversy were by no means clear-cut. It must be emphasized also that the protagonists shared many common assumptions. All took for granted the necessity of the retention of sole political power by their party. All took for granted the necessity of industrialization and were under no illusions concerning the limitations of individual peasant agriculture. Peasant cooperation and collectivization were regarded by all as desirable aims. The difference lay in tempos, methods, the assessment of dangers, the strategy to be followed in pursuit of aims very largely held in common. Soviet historians are fond of contrasting the policies of the majority ('the party') with the negative, defeatist, anti-industrializing, pro-peasant policies of various oppositions. Such a picture is a most distorted one. The fourteenth congress of the party meeting in 1925, passed resolutions favouring industrialization, while the fifteenth congress (1927) declared in favour of collectivization and of the five-year plan. However, these resolutions were adopted with the support of the future right-wing opposition. Indeed the 'industrial plan' resolution in 1927 was introduced by Rykov, who was Bukharin's most influential supporter.

Trotsky's views have also become distorted, not least by those who purport to be Trotskyists. Until 1925, when he was expelled from the Politbureau, he remained wholly within the assumptions of N E P. In his speech to the 12th party congress in 1923, and on other occasions, he accepted the need to encourage the private peasant producer, and advocated more effective and comprehensive planning for the peasant market. He was better able than his ally Preobrazhensky to see the necessity for the co-existence of plan and market, and perhaps this helps to explain why Preobrazhensky broke with Trotsky when, in 1928, Stalin turned 'left'.[4]

Does this mean that the argument was concerned only or mainly with who should wield political power? Such a conclusion would be totally misleading. The policy differences were deeply felt. It is true that Stalin later on stole many of the clothes of the left opposition, but Bukharin's entire vision of Soviet development differed radically from that which came to be adopted by Stalin, despite the fact that they shared some common aims. Bukharin wished to preserve N E P for a long time yet. Stalin destroyed it. The right opposition were horrified by Stalin's peasant policies and by his industrialization strategy, as well as by his political methods. There were deep and sincerely held policy differences.

SOME ORIGINAL ECONOMIC IDEAS

We have so far been emphasizing differences of opinion among professional politicians. But the debates and controversies of the twenties contain much more that is of interest for the economic historian, and perhaps most particularly for the historian of economic thought. Development economics could be said to have been born here.

It is not that the Soviet economists, planners and statesmen

were more intelligent or imaginative than their Western con-
temporaries. It is just that the institutional and political
circumstances posed problems in Russia which demanded
consideration. Even at the height of NEP, the bulk of investment
capital was in the hands of the state. Even the most moderate
protagonist of the NEP compromise was bound to reflect on
the next step, and thus on the development strategy to be followed.
In the West, economic theory did not even discuss investment
criteria. Such matters were encompassed within the theory of
market equilibrium, and, since the very notions of development
and growth were absent from the discussions, the idea of any
deliberate policy with regard to investment was absent too, the
more so as the bulk of capital assets and investment resources
were in the private sector, and so were not subject to public
policy, even if one could be imagined. Therefore the Soviet
theorists and practitioners found themselves in the role of
pioneers. Whatever weaknesses there may have been in their
thinking and their practice, it must be emphasized that they could
have learnt nothing useful from the West, which did not begin
to discuss these issues until 1945, or even 1955.

As this is not a history of economic thought, these matters
will be dealt with briefly. The interested reader is referred to the
works listed in the Bibliography, and, so far as theory is
concerned, particularly to J. M. Collette,[5] on whose research
much of what follows is based.

There was, first, the issue of agriculture *versus* industry, and
the linked questions of foreign trade and comparative advantage.
The Soviet Union was a high-cost and inefficient producer of
industrial goods in general, of equipment in particular. As the
'scissors' crisis underlined, it was a relatively much lower-cost
agricultural producer. It might then appear to follow that the
correct policy was to invest in agriculture, to increase marketable
surpluses, to expand exports, and thereby to obtain industrial
goods, especially capital goods, abroad. Shanin was a particularly
strong supporter of this view. Of course, in the very short run
this was the only possible tactic. In the absence of an adequate
capital goods industry there was no alternative to buying
machinery abroad in exchange for food and timber. But whereas
the more radical 'industrializers' wished to orientate these

purchases towards the rapid creation of the U.S.S.R.'s own heavy industry, Shanin and his friends envisaged a longer-term dependence on the developed West. This argument is familiar enough among development economists today; the role of agriculture is often the subject of debate.

Then there was the question of the conflict between short-term investment criteria and strategy for development. This was related to the important issue of unemployment and the surplus labour which, by universal consent, was available in the villages. Capital was acutely scarce. There was a shortage of most goods, in the aftermath of the civil war. Therefore, argued some, the most important thing was to use scarce capital to the best advantage, this being defined as maximizing employment and producing as much as possible for the least possible expenditure of capital. This policy, advocated by P. P. Maslov, led to the following practical recommendations:

(a) Capital-saving, labour-using investments were to be preferred.

(b) While one should also seek to choose investments which promoted a high or quick rate of return, the fact was recognized (twenty years before W. Arthur Lewis) that the existence of surplus labour was not adequately reflected in the wage rates and social benefits which entered into costs of production.

(c) Heavy industry required massive investments which matured over a long period. Therefore considerations of economy of scarce capital, the expansion of production in time of universal shortage and the problem of unemployment all called for priority to be given to light industry and agriculture.

However, as we know, this approach seldom satisfies the development planner. Already in 1926, Bernstein-Kogan was conscious of a dilemma or contradiction. This is how Collette summarized his argument: 'Either save investment resources by maintaining the lowest possible capital intensity, and so condemn the Soviet economy to long-term stagnation; or else renew the existing equipment intensively, develop the basic industries, install a powerful infrastructure and lower considerably the rate of return on investment.'[6] There was much discussion of the use of interest as a means of time-discount, as an aid to choice; but in itself it could not resolve the problem. The talented economist Bazarov, also in 1926, asserted the need to divide the economy into two – a priority sector (e.g. electrification, transport) to which the criteria of rates of return were not to be strictly applied, and the rest of the economy.

DEVELOPMENT STRATEGIES

As with NEP so with theories of development, 1926 was a year in which the atmosphere changed. Perhaps this was due to the virtual completion in that year of the restoration of the pre-war economy, and a consciousness that a new investment policy was necessary. This too found its reflection in the politics of the time, with resolutions favouring industrialization high on the agenda. Needless to say, the arguments of the economists were also related to the political factions, either directly or indirectly. Thus the argument in favour of investment in agriculture and the consumers' goods industry would naturally fit into the Bukharin approach to NEP. Equally clearly, it would find little sympathy among the supporters of the left opposition, or, when he moved left, from Stalin. It is because of this (sometimes unwanted) association of theoretical arguments with factional struggles that so many of the able economists who expressed original ideas in the twenties died in prison in the thirties.

All planning, in the sense of deliberate decision-making affecting the use of resources, must represent some sort of compromise between two principles, which in Soviet discussions came to be known as 'genetic' and 'teleological'. The first lays stress on the existing situation: market forces, relative scarcities of factors, rates of return, profitability. The second reflects a desire to change the proportions and size of the economy, to maximize growth, to emphasize strategy of development rather than adaptation to circumstances. The conflict between the two attitudes, on both the theoretical and the practical-political planes, increased in the second half of the decade. Naturally, neither side to the argument was unaware of the need for some sort of reconciliation of the opposing principles, though the original and intelligent ex-Menshevik economist V. Groman did assert that there was some 'natural' relationship between agriculture and industry which remains (or should remain) constant over time. Most of the protagonists were concerned with relative emphasis. Surely, as Bazarov pointed out, any plan which ignored the existing situation was doomed to failure; any plan which saw no further than the demands of the immediate present was patently inadequate. The changing emphasis after 1926 led more and more to the stress on drastic change, and as party policy veered towards rapid industrialization one heard more and more voices advocate the priority of heavy industry, asserting the criterion of maximizing growth. The economist of this period whose name is now most familiar was Fel'dman, whose growth model has been introduced to Western readers by E. Domar.[7] Evidently, if the objective is the most rapid industrialization, the investment choices in any developing country are bound to be based on principles quite different to those which would minimize unemployment or economize scarce capital in the short term. As Collette has pointed out, this type of thinking was found in the West only after 1955.

Much was made, by the supporters of a sharp rise in investments, of the phrase 'extinguishing curve' (*zatukhayushchaya krivaya*). This was the forecast of a reduced rate of growth, made by a committee of V S N K H, already referred to under its abbreviation O S V O K. Its conclusions followed logically from the inevitable slowdown which would be the consequence of the end of reconstruction. In part they assumed both a rise in the capital-output ratio and a fall in the volume of investment. But the idea of a slowdown was unacceptable to the political leadership; and indeed contradicted the dynamism and optimism without which the party rank-and-file would lose much of their drive and morale. It was sharply rejected. The fact that these and other bourgeois specialists were so very cautious in their prognostications later encouraged Stalin and his colleagues to ignore 'moderate' advice.

A 'strategic' decision, much discussed at the time, concerned the so-called Ural-Kuznetsk combine. This was an immense project linking the iron ore of the Urals with the excellent coking coal of the Kuzbas, a thousand miles away in Central Siberia. It was a long-term project *par excellence*. It would lock up a great deal of capital. It could not be justified by rate-of-return calculations. It might have vast external effects in the long run. It would – it did – save the situation militarily in the event of invasion by 'imperialist' powers. The issues are fully discussed in several works in the West.[8] (The Ural-Kuznetsk combine project was in fact begun in 1930.)

The issue of balanced *versus* unbalanced growth, known to most economists from the (recent) work of Hirschman and Nurkse, was also discussed in Russia at this period. In part this was included in the 'genetic-teleological' debate, and in part it came up as a problem of how to tackle existing or anticipated bottlenecks. Bukharin in particular advocated a careful attention to balance, and warned of the consequences of neglect of this factor which, after 1928, tended to discredit this approach, since it was associated with right-wing heresy. This was regrettable, since it affected the fate of another of the Soviet innovations: the 'balance of the national economy'. Using data from 1923–4, a group of gifted men led by Popov and Groman created the 'grandfather' of the input-output tables of later years. They invented a new idea, without which planning could hardly begin. It was necessary to trace the interconnections of the sectors composing the economy; to discover how much fuel was (or would be) needed to produce a given quantity of metal, to take just one example. The attempt was in many ways inadequate, many of the necessary data were missing. But it was the first such attempt, if one excepts Quesnay's *Tableau économique*, to which Soviet economic literature makes frequent reference.

The twenties were an intellectually exciting period. Not only were there debates among Bolshevik leaders and intellectuals, among whom were men of great eloquence and wit, but quite independent ideas were put forward by men who were not Bolsheviks at all. Gosplan and VSNKH experts included many former Mensheviks, later to be accused of being plotters and saboteurs. Men like Groman, Bazarov and Ginzburg contributed significantly to policy debates. Ex-populists, ex-SRs, were active too, for example the famous economist Kondratiev, the agricultural experts Chayanov and Chelintsev. Even non-socialists, like Litoshenko and Kutler, could raise their voices. There was a one-party state, there were no legal means of organizing an opposition, but conditions were far from resembling the monolithic thirties. The communists were very weakly represented at this time among the planners. Thus in 1924, out of 527 employees of Gosplan, only forty-nine were party members, and twenty-three of these were drivers, watchmen, typists, etc.[9]

The great debate, or more properly debates, must be seen as taking place at many different levels. There was the political struggle for power. There was the conflict at the political level between advocates of different policies towards the peasants, or on industrialization rates (tempos), or 'socialism in one country'. There were discussions and proposals put up by experts on investment criteria and growth strategies. Theory and practice, expert judgement and politics, interacted in various ways. Thus, not surprisingly, political men who were not allowed to express open dissent in a political way did so in their capacity as experts, just as others did so as novelists and poets. Politicians used experts, and selected statistics to suit their arguments, which eventually proved very dangerous for the experts. Thus the apparently abstract argument about peasant stratification became political dynamite, inevitably linked with the question of the kulak danger and the steps which could or should be taken to combat it. Even so statistical an issue as the volume of marketed grain became highly 'political', as we have seen.

6. The End of NEP

POLICY CHANGES AND THEIR CAUSES

NEP reached its apogee in 1925. During 1926 the absolute growth of the private sector outside agriculture came to a halt. As already shown in Chapter 4, its relative position began to decline earlier; however, until this date it could be said that it was the general understanding that private enterprise had its legitimate part to play in Soviet life. The decline of NEP cannot be dated precisely, the more so as official statements on the subject were ambiguous or deliberately misleading. Thus Stalin even in 1929 was still indignantly denying rumours to the effect that NEP was to be ended, and in fact as late as 1931 the tenth anniversary of NEP was the occasion for statements that it was still in operation. The five-year plan, in its optimal variant adopted as late as the spring of 1929, envisaged an increase over the five years in the national income generated in the private sector by as much as 23·9 per cent.[1] Yet, as we shall see, the offensive against the Nepmen outside agriculture had already been raging at this date for some time, and the offensive against the private peasants was about to begin. The following tables show the decline in legal private activities:

Trade

	Total private turnover (million roubles)	Per cent of total trade
1924–5	3300	42·5
1925–6	4963	42·3
1926–7	5063	36·9
1928	3406	22·5
1929	2273	13·5
1930	1043	5·6
1931	–	–

(SOURCE: Malafeyev: *Istoriya tsenoobrazovaniya v SSSR* (Moscow, 1964), p. 134. The author draws attention to the large volume of illegal and unrecorded trade in and after 1929.)

The share of the private sector in the national income is said to have declined as follows:

	1925–6	*1926–7*	*1928*	*1929*	*1930*	*1931*	*1932*
				(per cent)			
Socialized	45·9	48·7	52·7	61·0	72·2	81·5	90·7
Private	54·1	51·1	47·3	39·0	27·8	18·5	9·3

(SOURCES: 1925–6 and 1926–7, *Narodnoe Khozyaistvo SSSR, 1932*, pp. xlvi–xlvii, 1928–32. E. Kviring, *Problemy ekonomiki*, Nos. 10–12 (1931) p. 5.)

As already indicated earlier, the state had a potential stranglehold on the private trading and manufacturing sectors, in that supplies of raw materials and of goods to sell depended greatly on state industry. Therefore a simple administrative decision could change the situation, even without legal or tax measures specifically directed at the Nepmen. It was not until 1930 that private trade became (*de facto*) the crime of speculation and the employment of labour for private gain became in fact illegal.[2] But well before that date the squeeze was on. For example, there was a steady increase in surcharges on transport of private goods by rail: 'By 1926 the surcharges for transporting private goods reached fifty to 100 per cent, and in subsequent years were raised up to 400 per cent for some goods.'[3] The year 1926 also saw the first of a series of fiscal measures designed to make private trade less profitable: thus by decree of 18 June 'a temporary state tax on super-profits' was imposed on Nepmen. It was the first of many. On 9 April of the same year the central committee plenum decided that flour mills in private hands should be 'drastically curtailed'. An ever sharper tone crept into party pronouncements about Nepmen and kulaks.

Taxes on better-off peasants were increased, as the following table shows:

	1925–6	*1926–7*
Poor peasants (roubles per annum)	1·83	0·90
Middle peasants	13·25	17·77
Kulaks	63·60	100·77

(SOURCE: G. Maryakhin, *Voprosy istorii*, No. 4 (1967) p. 27.)

An amendment to the criminal code adopted in 1926 'envisaged imprisonment for up to three years with total or partial confiscation of property for those guilty of evil-intentioned (*zlostny*) increases in prices of commodities through purchase, hoarding, or non-placing on the market'.[4] This is the famous Article 107 of the Code, which was to be used by Stalin two years later. At the time at which it was promulgated it remained very largely a dead letter. However, the fact that it was adopted in 1926 is an indication of the evolution of official opinion with regard to private trade in general.

Why this shift of policy? Official party histories tend to underplay the change. The resumption of the offensive against the private sector, according to such a view, was inherent in the very concept of NEP from the first. The state was now stronger, better able to run trade and industry, and in a position to begin to provide the capital equipment which would ultimately revolutionize social and productive relations in agriculture. Such an interpretation is not wholly wrong. For large numbers of party members NEP was a forced compromise with the hated enemy, who was to be attacked as soon as conditions were ripe, using any weapons at the party's command. However, to see things in this way is gravely to underestimate the extent of the change of policy which was taking shape gradually from 1926. This culminated at the end of the decade in what can best be described as the Soviet great leap forward, involving the destruction of the last bastion of private enterprise in the great collectivization campaign.

The causes of the change of policy were numerous and complex, and interacted with one another.

Firstly, we must note the close relationship between ambitious investment programmes and the end of NEP. As has already been noted, 1925–6 saw the end of what might broadly be called the reconstruction or restoration period. It is true that the metallurgical industry was still below its 1913 level, and that some other industries, notably electricity generation, coal mining and some branches of engineering, were already above it. The essential point was that henceforth further efforts to increase industrial production would cause increasing strain. In December 1925 a resolution of the fourteenth party congress called for industrialization, and also for the victory of the socialist sector. The concentration of resources in the hands of the state seemed inconsistent with the activities of Nepmen, who competed for resources and diverted them from the priority tasks of the moment, and indeed made profits out of the shortages which an investment programme would cause.

A second and very important feature of the situation, to which sufficient weight is seldom attached in histories of the period, was the price policy pursued by the government. We have seen in Chapter 4 that its response to the 'scissors' crisis of 1923 was to press the state trusts to reduce their costs and their prices. This policy was continued in subsequent years, although both urban and rural incomes rose faster than the volume of output (though the latter did rise rapidly). Faced with strong market pressure for price increases, the government obstinately persisted with its policy of price cuts, and in order to make them effective extended price control over an ever wider portion of state industry and state and cooperative trade. As might have been predicted, this speedily gave rise to the phenomenon of 'goods famine'. These words occur repeatedly in official and unofficial pronouncements from 1926 onwards. A talented young economist, who was to contribute greatly to the rebirth of Soviet economics forty years later, described the situation as follows: 'Commodities no longer seek buyers, the buyers seek commodities. ... There are long queues in front of some shops. In private trade prices are significantly higher than the selling prices of the state trusts, 100 per cent or 200 per cent higher for some commodities. Limitations on purchase are introduced: the goods most in demand are sold by state and cooperative shops not to all those willing to buy them, but to selected categories of buyers, for instance members [of the cooperatives] or members of trade unions. If in big cities the goods shortage has taken such acute forms, the situation in the villages is worse still.' This is due to price policy. 'The shortage of goods, therefore, occurs only when prices cease to carry out their function in balancing supply and demand, when they become inert and unresponsive to market forces.' Surplus purchasing power moved into the area in which price controls did not operate, and so private trade took on the characteristics of speculation, since it was profitable to buy state-produced goods for resale, thereby in effect transferring resources to the private sector. More seriously, it had a grave adverse effect on peasant purchasing power. Even though the reduction in prices charged by state trusts and state trade may have been originally motivated by the desire to close the price 'scissors', i.e. to improve the peasants' terms of trade, the effect was quite different in practice. Since the official prices were below equilibrium prices, those closest to the factories got at the goods first. 'Towns are closer to the sources of industrial goods, the villages are further away. Therefore the towns appear to obtain a larger share of industrial goods than they would have obtained at prices which balance supply and demand. The policy of low prices not only failed to lower prices for the village, but on the contrary it lowered prices for the towns at the cost of raising them for the village, and by a large percentage too.' The young critic patiently pointed out that this policy was absurd, and would become the more absurd as state investments increased, since they would generate incomes which would constitute still heavier demand on the available consumer goods and services.[5]

This advice was ignored. A whole series of decrees and declarations demanded further cuts in prices. On 2 July 1926 the STO (Council of Labour and Defence) issued a decree the title of which must surely seem unsound even to the dimmest first-year student of economics: 'The reduction in retail prices of goods in short supply made by state industry'. The reduction was to be 10 per cent, and there was another and similar reduction ordered on 16 February 1927.

The pursuit of such a policy was plainly inconsistent with the logic of NEP, and was bound to lead to an attack on private traders who were selling manufactured goods in the villages at prices more than double those charged by state and cooperative shops, in which, however, the goods could not be obtained. Likewise the additional profit made by the still-legal private manufacturers and craftsmen would hardly appeal to the political and fiscal authorities.

Thirdly, a similar kind of blunder distorted agricultural production and procurements. The peasants' willingness to sell was already adversely affected by the goods famine. It was further affected by the government's efforts to lower procurement prices. Taking advantage of a run of reasonably good harvests, the state endeavoured to economize in its expenditure on purchases of farm produce. In the agricultural year 1926–7 the general level of state procurement prices fell by about 6 per cent compared with the previous year. No such reduction was justified by the market situation. But worse still, for ever mesmerized by the key role of grain, the government cut prices for this particular crop much more severely, by as much as 20–25 per cent. The result, as might have been foreseen, was a reluctance to sell grain to the state, a tendency to concentrate on other crops and on livestock, for which prices were more favourable, and the emergence of a large gap between official state prices and those paid to the peasants for their produce by the still legal Nepmen traders. We shall have much more to say about the consequences of this later in the decade.

Why were these price policies adopted? A little reflection clearly shows them to have been totally out of line with the principles of NEP, and indeed the continuance of such price policies was bound to lead to grave conflict and confusion, even in the absence of other factors. No doubt much can be explained by a combination of obstinate blindness with a built-in dislike of market forces which characterized many Bolsheviks. However, part of the obstinacy may have had a political explanation. Trotsky and his friends had favoured increasing the level of savings and investments, and the logic of their position demanded somewhat higher, not lower, prices of manufactured goods, since in this way the state would obtain the necessary revenue. This in itself would have been an argument for Stalin to oppose such a move, and he doubtless saw political advantages in contrasting his policy of lower prices with that of the opposition. The pressure to increase grain prices, in turn, came especially from Bukharin and his friends of the future 'right-wing deviation'. Here again, Stalin had political reasons for obstinacy. This is not mere surmise. Stalin repeatedly accused various oppositionists of wanting higher prices. The party plenum held on 7–12 February 1927, in reaffirming the need for lower prices all round, emphasized that 'in the problem of prices are interlinked all the basic economic and political problems of the Soviet state'. The policy chosen was basically hostile to market forces in industry, trade and agriculture. Either the policy would have to be amended, or the market and its manifestations would have to be destroyed.

The survival of NEP was conceivable only if this price policy was altered. It was not altered, and NEP was doomed. Only after the decisive defeat of the 'privateers' was a very different price policy followed, this time in order to finance the spectacular expansion of state industry. But much more of this later on.

The approach of many, if not most, party members to the whole question of trade and prices is well expressed in a pamphlet by a leading official of the *Tsentrosoyuz* (central consumer cooperative) organization. For him, the whole point of the spread of cooperative shops was that this would combat and gradually squeeze out the private trader. He was well aware of the fact that the latter could charge much higher prices, owing to the 'goods famine'. However, he drew the conclusion that in 'tearing the mass of commodities out of the hands of elemental market forces, of private trading speculation', consumer cooperatives were helping to 'dig defensive trenches around socialist industry'. It was a matter of 'who shall beat whom' (*kto kovo*), part of the battle for socialism. The fact that cooperatives supplied goods cheaper than the private traders, i.e. that they sold at below supply-and-demand prices, was to be used to persuade the peasants to sign contracts undertaking to sell farm produce (at state-fixed prices) in exchange for an undertaking to supply them with manufactured goods (at state-fixed prices), all these prices being below the market level. 'What would it matter if the privateer-speculator paid this or that peasant a few roubles extra per hundredweight of wheat? After all, this would be offset by his bandit-speculationist overcharging in supplying the peasants with the goods they need.' The last years of the twenties did see an attempt to make contracts of this kind (*kontraktatsiya*) and the author of the pamphlet evidently hoped that these contracts would form the basis of relations between town and village, and grow into 'the higher form of products exchange'.[6] Obviously, such thought-patterns as these were unlikely to be affected by economists' arguments about market equilibrium.

INDUSTRIALIZATION AND THE DRAFT
FIVE-YEAR PLANS

These were the years of the political triumph of Stalin, and this was accompanied by a change in the whole atmosphere of Soviet life. Whether in literature or in philosophy, in the party's own internal arrangements or in the sphere of economics, the line became one of stern imposition of conformity, centralized authority, suppression of uncontrolled initiatives. The Nepmen were thus to some extent victims of a more general tendency. This same tendency led, as did the logic of the industrialization drive, to growing centralization of the state's own planning mechanism. As already noted in Chapter 4, the intention to plan effectively from above was voiced even during the period of high NEP. There was a procedure by which the industrial and financial plans of trusts (*promfinplany*) were approved by VSNKH. At first this had little practical effect, but it did provide a mechanism by which control could be tightened. As early as 1925 one finds in planning documents such declarations as: 'The state is becoming the real master of its industry. . . . The industrial plan must be constructed not from below but from above.' (I owe the reference to R. W. Davies.) Procedures for price control existed from the early days of NEP and they began to be used to keep prices of some basic industrial materials and fuels, and also of freight transport, at low levels. Subsidies were often called for. A goods famine therefore developed in the field of producers' goods too. This led by 1926–7 to a more systematic control of production and distribution of some key commodities, particularly metals, by VSNKH. It was only logical that this should result in closer integration between Gosplan's perspective plans and the VSNKH's current operations. Gradually the extent of administrative controls increased, the role of market forces declined. While many branches of industrial activity still operated with very considerable autonomy until the end of the decade, the contours of the future command economy were becoming increasingly visible, and the acute strains and shortages of the next years would lead to much tighter and systematic centralized planning. The plenum of the central committee held in April 1926, emphasizing the need for more capital accumulation, also spoke of 'the strengthening of the planning and the introduction of a regime of planned discipline into the activities of all state organs'. The fifteenth party conference (26 October to 3 November 1926) declared for 'the strengthening of the economic hegemony of large-scale socialist industry over the entire economy of the country', and spoke of the necessity of striving to achieve and surpass the most advanced capitalist countries 'in a relatively minimal historical period'.

To do this it was necessary to formulate a long-term plan, and so we must now consider the five-year plan. Early drafts of such a plan were widely discussed among economists during the middle twenties.

Preparatory work for a long-term plan began in earnest in 1927. On 8 June 1927 a decree by the Council of People's Commissars called for the creation of 'a united all-union plan, which, being the expression of economic unity of the Soviet Union, would facilitate the maximum development of economic regions on the basis of their specialization, ... and the maximum utilization of their resources for the purpose of industrialization of the country'. The role of Gosplan was strengthened, and the republican Gosplans placed under its authority. But the expansion of industrial investment did not await the formulation of a long-term plan. In the economic year 1926–7 the total volume of investments increased by 31·7 per cent, but investments in new construction more than doubled.[7] Persistent shortages of metal caused vigorous efforts to expand production of iron and steel, and also ore mining. The great Dnieper Dam was begun, and so was the Turksib railway. While the volume of investments in this year was soon greatly to be surpassed, they imposed a strain on the economy which reacted on the availability of resources for other purposes and so contributed to the 'goods famine'.

Meanwhile the country's leading experts, belonging to various currents of Bolshevik opinion and those belonging to no party, were hard at work formulating a five-year plan. The teleological school increasingly obtained the upper hand, and so the specialists in Gosplan and VSNKH were under continuous pressure to adopt ambitious growth targets. These pressures and revisions, as well as the truly colossal nature of the task, explains why a five-year plan which was to operate with effect from October 1928 was submitted to the approval of the sixteenth party conference in April 1929, i.e. when implementation was already in full swing. It proved necessary to formulate the plan in two versions: the initial variant and the optimal variant. As may be seen from the table below, the initial variant was optimistic enough. It was rejected by the sixteenth party conference in favour of the more ambitious version. We shall see that this in its turn was replaced by yet more fantastic targets.

The formation of the plan required an immense amount of detailed work, for which there was no precedent. True there had been 'control figures' for earlier years, and a balance of the national economy had been drawn up, as was mentioned in the previous chapter. However, the huge task of drafting a five-year development plan to transform the economic structure of Russia required much more information about inter-industry links than could be available in the then existing state of infor-

First five-year plan

	1927–8 actual	1932–3 first version	(per cent inc.)	1932–3 'optimal version'	(per cent inc.)
Aggregates					
Employed labour force (million)	11·3	14·8	(30·2)	15·8	(38·9)
Investments (all) (1926–7 prices milliard roubles)	8·2	20·8	(151)	27·7	(228)
National income (milliard roubles)	8·2	44·4	(82)	49·7	(103)
Industrial production (milliard roubles)	18·3	38·1	(130)	43·2	(180)
of which:					
Producers' goods (milliard roubles)	6·0	15·5	(161)	18·1	(204)
Consumers' goods (milliard roubles)	12·3	22·6	(83)	25·1	(103)
Agricultural production (milliard roubles)	16·6	23·9	(44)	25·8	(55)
Consumption:					
Non-agricultural (index)	100	152·0		171·4	
Agricultural population (index)	100	151·6		167·4	

First five-year plan (cont.)

	1927–8 actual	1932–3 first version	(per cent inc.)	1932–3 'optimal version'	(per cent inc.)
Industrial output targets					
Electricity (milliard Kwhs)	5·05	17·0	(236)	22·0	(335)
Hard coal (million tons)	35·4	68·0	(92)	75·0	(111)
Oil (million tons)	11·7	19·0	(62)	22·0	(88)
Iron ore (million tons)	5·7	15·0	(163)	19·0	(233)
Pig iron (million tons)	3·3	8·0	(142)	10·0	(203)
Steel (million tons)	4·0	8·3	(107)	10·4	(160)
Machinery (million roubles)	1822	?	—	4688	(157)
Superphosphates (million tons)	0·15	2·6	(16·3)	3·4	(21·7)
Wool cloth (million metres)	97	192	(98)	270	(178)

(SOURCE: *Pyatiletnii plan* (3rd edition, 1930), pp. 129 ff. Machinery figures from 'Fulfilment of first five-year plan', p. 273.)

mation and statistics. The detailed targets therefore included much that was insecurely based, and contemporary comment made no secret of this. One of the authors of the plan, G. Grin'ko, writing in February 1929, showed that many of the detailed calculations available at this date were still based on the first variant of the plan, and Grin'ko himself treated the lower variant as 'so to speak a guaranteed minimum within the optimal variant', which would be achieved if the favourable assumptions underlying the more optimistic version proved to be ill-founded.[8]

The plan as adopted was, to say the least, over-optimistic. Miracles seldom occur in economic life, and in the absence of divine intervention it is hard to imagine how one would expect simultaneous increases of investment and consumption, not to speak of the output of industry, agriculture and labour productivity, by such tremendous percentages. Efficiency in labour and management was to be such that costs and prices were supposed to be substantially reduced during the five years. It is hard to see how anybody could have regarded this as realistic at the time, let alone in retrospect.[9] Yet the optimal variant was shortly afterwards replaced by a still more fantastic series of targets. However, we will defer consideration of this phase until the next chapter.

The five-year plan as adopted far exceeded in the scale of its investments the demand of the defeated left opposition. In 1926 they had been denouncing as far too modest the plans adopted by the Stalin–Bukharin majority. The latter's line had been that Trotsky and his friends were demanding a tempo of growth which would be inconsistent with economic and political balance, with the smychka. They may well have been right. Certain it was that the higher tempos now adopted were inconsistent with the maintenance of the alliance between Stalin and Bukharin. The latter published a veiled attack on excessive and unbalanced growth rates in his 'Notes of an economist' (*Zametki ekonomista*, in *Pravda*, 30 September 1928). Nor were such tempos reconcilable with NEP, and especially with the existing situation in the villages and in agriculture.

The heavy financial expenditures which began in 1927 were partly financed by the placing of industrialization loans, at relatively high rates of interest (thus bonds issued on 1 June 1927 to the value of 200 million roubles, repayable in ten years, carried a 12 per cent interest rate). The need for revenue also encouraged the regime to impose heavier taxes on the Nepmen and kulaks. The policy of charging relatively low prices for the products of state industry was persisted with, however. Inflationary pressure grew, and with it the gap between official and free prices for the products of both industry and agriculture.

Did Stalin adopt a plan which he knew to be impossible, as a political manoeuvre? Did Kuibyshev, the chairman of VSNKH, or Strumilin, a leading party planner, adopt propaganda plans? It is difficult to say. Planning as a technique was hardly born. Over-optimism, which contributed to the excesses of the 1929–33 period, had already infected the leadership. Productive capacity, human energy, the consequences of a great drive, the effects of enthusiasm, were all over-estimated. It was believed that 'there was no fortress that the Bolsheviks could not take'. The voices which called for caution, which drew attention to difficulties and bottlenecks, were thought to be those of former Mensheviks or SRs, or bourgeois specialists. The latter were discredited and suspect following a show trial, the so-called *Shakhty* affair, held in 1928, which purported to show that a group of such specialists were wreckers and deviationists in the pay of foreign powers. Caution, and emphasis on balance, were also attributes of the right-wing opposition; they were denounced by Stalin with increasing vehemence in and after 1928. Experts who gave unwelcome advice were thrust aside; men whose advice was more congenial were put in their places. Perhaps many of the party leaders were genuinely carried away. Some may also have seen some advantage in the mobilizing force of a plan which would cause maximum effort, and so in the end achieve more than a sound, solid, balanced plan would do.

THE PEASANT PROBLEM AGAIN: THE PROCUREMENT CRISIS

However, it is now time to return to the peasants. It had not escaped the notice of Stalin that the ambitious industrial investment plans and the existing structure of peasant agriculture were inconsistent with one another. In his speech to the fifteenth party congress (December 1927), he spoke about the relatively slow rate of development of agriculture, advanced familiar reasons to explain this backwardness, and then said:

> What is the way out? The way out is to turn the small and scattered peasant farms into large united farms based on cultivation of the land in common, to go over to collective cultivation of the land on the basis of a new higher technique. The way out is to unite the small and dwarf peasant farms gradually but surely, not by pressure but by example and persuasion, into large farms based on common, cooperative, collective cultivation of the land. . . . There is no other way out.[10]

In the resolution adopted by the fifteenth congress one finds the following words: 'At the present time the task of uniting and transforming the small individual peasant holdings into large collectives must become the principal task of the party in the villages.' Yet, for reasons which have already been explained, this was not understood to mean the imminence of a revolution from above. The general desirability of collective agriculture was not in dispute. If it was to be done voluntarily and by example, there was no danger of anything particularly drastic happening quickly.

Undoubtedly what brought matters to a head was the problem of marketings, in particular of state procurements. Every year the leaders watched anxiously as deliveries mounted in the autumn and winter, wondering if there would be enough to feed the towns and the army and, who knows, something for export too. Attention was particularly concentrated on grain, the key crop since bread was the staff of life in Russia, and because over 80 per cent of all sown land was sown to grain.

Difficulties accumulated after 1926. Some of the reasons have already been mentioned. There was the 'goods famine'. There was the reduction of procurement prices in 1926, affecting particularly grain. This naturally discouraged marketings. The peasants tried to sell grain to the still-surviving private traders rather than to the state procurement agencies, to hold grain in expectation of higher prices, or to feed it to livestock. As Stalin subsequently pointed out, the better-off peasants were able to manoeuvre more effectively than their poorer neighbours to take advantage of any possibility of obtaining better terms. The government reacted by measures against Nepmen, by streamlining the state procurement apparatus to avoid a situation in which different procurement agencies bid against one another, and also by measures against kulaks, who were held responsible for the shortages. Thus the resolution of the fifteenth congress instructed the Central Committee to devise higher and more progressive taxes on the more prosperous peasants. The tone of party pronouncements on the kulak question became sharper. The idea of liquidating them as a class was not yet born, or at least not yet mentioned. However, they were to be limited, penalized and generally discouraged. This was a policy hardly designed to encourage the more ambitious peasants to expand production or investment.

As already mentioned, the fifteenth congress advocated the spread of collectivization. Of the various kinds of agricultural producers' associations, the most promising seemed to be the loosest the TOZ (the letters stand for 'Association for the

Joint Cultivation of the Land '). In these associations the members retained ownership of their tools and implements, most of their livestock and control over their land. They simply carried out some of the farm work jointly. The more advanced forms of producers' cooperation which then existed were thought to be unattractive to the peasants; for instance in the decree of 16 March 1927 it was laid down that the TOZ was to be favoured. However, in 1927 all the various types of collectives and co-operatives accounted for only a tiny proportion of peasants and of agricultural production, and a few inefficient state farms made little difference to the general picture.

Collective and state agriculture in 1928

	Percentage of sown area
Individual peasants	97·3
Collective farms	1·2 (of which about 0·7 TOZ)
State farms	1·5

(SOURCE: *Sotsialisticheskoe stroitel'stvo SSSR* (Moscow, 1935), p. xxxix.)

It was considered axiomatic that the peasant problem was to be handled cautiously. Coercion was excluded. Had not Lenin, and before him Engels, warned about the need for patience and the preservation of the voluntary principle? The poorer peasants could be allies, the middle peasants should be befriended, or at least neutralized, the kulaks could and should be restricted or taxed. There were still no grounds to suppose that a storm would break over the heads of the entire peasantry.

But the storm clouds were gathering, and the procurement difficulties of 1927 led to the first flash of lightning. Procurements had apparently been going more or less according to plan until December of that year. Then there was trouble. It became abundantly clear that they would be well below the previous year's level, insufficient to meet the needs of the towns and the army.

The shortfall in grain procurements may be seen from the fact that by January 1928 the state had succeeded in purchasing only 300 million poods, as against 428 million on the same date in the previous year. The shortfall was particularly great in Siberia, the

Volga and the Urals, where the harvest was reasonably good (bad weather was the cause of difficulties in the North Caucasus).[11] The effect was not only to create acute problems in supplying the cities with bread, but also to threaten supplies of industrial crops. Thus the maintenance of the cotton acreage in Uzbekistan was threatened by grain shortage there. Archive information shows a flow of complaints about this from local party organizations to the Central Committee.[12] Some of the reasons for this critical situation have already been given: the low price of grain, shortage of manufactured goods, the gap between official and free prices. With grain procurement prices so low, peasants naturally concentrated on other commodities. For example, in the area of the Urals grain sales to the state were only 63 per cent of the previous year, but meat sales increased by 50 per cent, while sales of eggs doubled, bacon quadrupled, and eleven times as much bacon fat was delivered as in the previous year.[13] Naturally the peasants waited for the rise in official grain prices, the necessity for which seemed quite obvious. However, Stalin and his colleagues drew very different conclusions.

Ignoring the proposals of Bukharin and others to increase grain prices, Stalin decided instead to launch a direct attack, which revived memories of the excesses of war communism. There had been a good harvest in the Urals and West Siberia. There went Stalin with a task force of officials and police. Free markets were closed, private traders thrown out, peasants ordered to deliver grain and punished as criminals if they failed to do so. Stalin made speeches, which were published only twenty years later, denouncing laggard officials, requiring them to seize kulak grain, demanding that they invoke a hitherto unused article of the criminal code (Article 107) against 'speculation', to legalize the seizures. He mocked the 'prosecuting and judicial authorities [who] are not prepared for such a step'. He used extreme language to the party officials, who were slow to understand that a basic change of attitude was expected of them: 'Can it be that you are afraid to disturb the tranquillity of the kulak gentry? You say that enforcement of Article 107 against the kulaks would be an emergency measure, that it would not be productive of good results, that it would worsen the relations in the countryside. Suppose it would be an emergency measure. What of it? ... As for your prosecuting and judicial officials, they should be dismissed.'[14] At the same time, scarce industrial goods were directed to the grain-surplus regions.

Rumours spread that the government 'will pay all foreign debts with grain and is therefore reinstating prodrazverstka, taking all grain away'. Reporting this and citing contemporary archives, a Soviet writer gave the following typical instance: 'In the village of Pankrushino the kulaks spread the rumour that all grain was being collected in one vast storehouse in the town of Kamensk, that the peasants would be given a bread ration, that armed detachments were scouring the villages for bread and that they would soon arrive.'[15] The same source quotes numerous reports in the press about alleged kulak opposition, though it seems more than probable that this was simply strong peasant reaction to the seizure of their produce.

The kulaks undertook large-scale agitation, asserting that Soviet power impoverished the peasant, did not allow him to improve his income, that NEP was being abolished. Kulaks, priests, former white-guardists endeavoured to utilize in their counter-revolutionary agitation certain cases of distortion [sic] of the Party line in credit and tax policy. ... In the village of Troitskoye in the Don area there was unmasked a priest who hid grain and organized in the cemetery a kulak meeting, where he made a report on 'grain procurements and the international situation'.[16]

This source also admits that there were indeed grounds for agitation.

Not infrequently measures were taken which hit not only the kulaks but the middle peasant. Such measures were: the confiscation of grain surpluses without the judgement of a court under Article 107, administrative pressure on the middle peasant, the use of barrier detachments (i.e. forcible prevention of private transport of grain), the forcible issuing of bond certificates in payment for grain and as a condition for the sale of scarce commodities to the peasants, and so on.[17]

Indeed arbitrary confiscation was a common phenomenon, and by strictly interpreting Article 107 the authorities could choose to regard the mere possession of grain stocks as illegal hoarding with a speculative purpose and, therefore, a fit subject for confiscation without payment. Newspapers of the time were full of reports about evil kulak hoarders of grain, and also of reports concerning peasant meetings in which the delivery of grain to the state was extolled and the kulaks condemned. Such reports must be taken with a pinch of salt. Thus the same source which referred to the payment of peasants with bond certificates instead of money, and referred to it as an impermissible excess, cites approvingly an allegedly spontaneous decision by peasants 'to refuse to accept money and to request payment in bonds for the entire amount of delivered grain', and a resolution worded thus: 'Not a single pound to the private speculator.'[18]

Stalin himself concentrated on West Siberia and the Urals, and other senior officials pursued the procurement campaign in other areas: for example, Zhdanov in the Volga region, Kossior in the Ukraine and the Urals, Andreyev in the North Caucasus. The chief coordinator of the entire operation was said to have been Mikoyan.[19] All the above were devoted members of the Stalin faction.

These arbitrary procedures, later 'disguised' as locally initiated, became known as the 'Urals-Siberian method'. In retrospect this must be regarded as a great turning-point in Russian history. It upset once and for all the delicate psychological balance upon which the relations between party and peasants rested, and it was also the first time that a major policy departure was undertaken by Stalin personally, without even the pretence of a central committee or politbureau decision.

Bukharin, Rykov and Tomsky, three of the politbureau of the party, protested vehemently. In April 1928, at the plenum of the central committee, Stalin beat an apparent retreat, and accepted a resolution condemning excesses, reasserting legality and promising that nothing similar would be repeated. But events showed that Stalin's compliance was a mere manoeuvre. Forcible procurements were repeated in many areas in 1928–9. Stalin soon made it clear that the 'Urals-Siberian method' would be used whenever necessary. Yet surely it was obvious that peasants would not increase marketed production if the state would seize the produce at whatever price it chose to pay, and imprison anyone who concealed grain. The outbreak of argument among the leaders was carefully concealed from the public, and from the party membership at large. Only later it became known that Bukharin, at last realizing what his erstwhile ally was up to, began to speak of Genghis Khan, of 'military-feudal exploitation' of the peasantry, of 'tribute' (*dan'*) levied on the village. It was in the context of these fears and feelings that we must read his plea for balanced growth, mentioned on page 147 above. In the already suffocating political atmosphere, he was unable to voice his real fears in public. These concerned most of all the consequences of the coming clash with the peasants.

COLLECTIVIZATION ON THE AGENDA

Stalin and his faction did not yet show any sign that they had decided on an all-out collectivization campaign. Indeed, there is no evidence at all that such a decision was taken until the early autumn of 1929, and the Soviet public first heard of it on 7 November 1929. Some may consider that Stalin had a secret plan all ready for the propitious moment, but this seems unlikely. What he clearly wished to do after the 'Urals-Siberian' episode of February 1928 was to free the regime from over-dependence on the peasantry. This view he had already expressed during his Urals-Siberian tour: 'In order to put grain procurements on a satisfactory basis other measures are required.... I have in mind the formation of collective farms and state farms.'[20] An immediate consequence of this was the decision, in April 1928, to set up a 'grain trust' (*Zernotrest*), aiming to create new state farms covering 14 million hectares (36 million acres). However, this was still not to be regarded as a step towards more collectivization, since these farms were to be set up on unused land. No one was to be expropriated or forced. Consequently this decision was not a challenge to Bukharin and his friends, and may even have been accepted by them. It proved unworkable and inadequate.

The July 1928 plenum of the central committee was still, so far as its official statements were concerned, dominated by the need to reassure the peasants (or to keep the Bukharin group quiet). True, the resolution spoke of 'voluntary union of peasants into collectives on the basis of new techniques', but no one could object to this, the more so as the absence of new equipment was a principal argument of the 'go-slow' school. The resolution admitted that low grain prices contributed to existing difficulties and resolved to increase them, but this was a case of much too little and too late. The resolution also spoke of 'the further raising of the level of small- and medium-size individual (peasant) economies'. So *at the time* there seemed to be a return to moderation. But here again the historian faces the difficulty that some key policy statements were made only behind closed doors. If they were to be included in a chronological account, the surprise and shock which greeted subsequent announcements of policy changes would be incomprehensible. At the same July plenum, Stalin admitted both the need for 'tribute' from the peasants, and that they did and had to overpay for manufactured goods and were underpaid for farm produce. This followed from the need to 'industrialize the country with the help of *internal* accumulation'.[21] In other words, Preobrazhensky (and Trotsky) were right all along. But this speech too was published only in 1949; when it was made, Trotsky was in Alma Ata under effective house-arrest, and Preobrazhensky had been deported. If anyone chose to remind Stalin of the source of his ideas, this was omitted from the belatedly published record.

In a public speech after the July plenum, Stalin took up an apparently moderate position: 'We need neither detractors nor eulogizers of individual peasant farming,' but he did urge once again the gradual development of collective and state farms. His tone sharpened in his first *public* attack on 'right-wing deviationists' (naming, as yet, no names) on 19 October 1928, and at the November 1928 plenum of the central committee one heard more of the inefficiency of agriculture holding back industry, of the encouragement of collective and state farms, measures to limit the kulaks. In the winter the 'Urals-Siberian method' was quietly reapplied to the peasants, and under cover of anti-kulak measures the method once again hit the middle peasants hard, since, after all, most of the grain came from them. Stalin attacked Bukharin vigorously in the April 1929 plenum of the central committee, but most of his words remained unpublished at the time.

Confusion in the public mind over peasant policies was heightened by the resolution of this plenum, when the word 'tribute', accepted by Stalin in the previous year in an *unpublished* speech, was now treated as a slanderous and lying accusation directed at Stalin by Bukharin, who was now openly attacked. It was admitted that peasants did overpay for some industrial goods, but this would speedily be put right.

The sixteenth party conference, meeting in the same month, approved the 'optimal' version of the five-year plan, as we have seen. This included a section on agriculture. There was to be a marked advance in collectivization, and by the end of the plan it was hoped to have twenty-six million hectares cultivated by the state and all kinds of collectives (including TOZs); these would provide over 15 per cent of the total agricultural output. It was not at all clear by what means this expansion in agricultural collectivization was to be achieved. However, given five years and the necessary resources, it was not an unrealistic perspective. While in the years 1921–7 there had been no move by the peasants towards any type of collective farming, this could be blamed on the lack of inducement, indeed on the neglect of the few collectives that did exist. Such a programme, if carried out, would still have left the vast majority of the peasantry in the private sector, producing the bulk of every crop and owning most of the livestock.

THE DECISION TO ATTACK

Now that Bukharin was at last openly denounced as a right-wing deviationist (he was not expelled from the politbureau until November 1929), Stalin must have felt free to launch the campaign that was maturing in his mind. Yet by not a word or gesture did he prepare the party, the people, the peasants, for the great turn, the 'revolution from above' which was to shake Russia to its foundations. In fact, even as late as 27 June 1929 a decree on agricultural marketing cooperation still assumed the predominance of the private sector in agriculture for an indefinite period, and we shall see that it was not until the campaign had begun that there was an amendment of the plan to achieve a mere 15 per cent collectivization by 1933.

No document exists which can tell us exactly when Stalin made up his mind. During 1929 the strains of the investment programme of the five-year plan began to affect all sectors of the economy. Rationing of consumers in cities was introduced gradually during 1928 in some areas and became general early in 1929, perhaps the first and only recorded instance of the *introduction* of rationing in time of peace. The goods famine increased in intensity. The gap between free and official prices widened, as the following figures demonstrate:

(1913 = 100)

	Food		Manufactures	
	Private	Official	Private	Official
1926 (December)	198	181	251	208
1927 (December)	222	175	240	188
1928 (December)	293	184	253	190
1929 (June)	450	200	279	192

(SOURCE: Malafeyev, *Istoriya tsenoobrazovaniya v SSSR* (Moscow, 1964), pp. 384, 385.)

By 1930 the difference increased very rapidly.

Voices from the right urged slowdown, higher farm prices, a modification of the investment programme. Rykov proposed a 'two-year plan', with emphasis on agriculture. Grain procurement prices were in fact raised, far too late, in 1929, by 14–19 per cent,[22] but the market situation was such that private traders were buying in that year at prices which rose by over 100 per cent. In the Ukraine, for instance, private traders were buying at 170 per cent above state procurement prices.[23] However, there were a number of arguments for an all-out drive forward. In the first plan, it must have already seemed impossible to continue on the basis of a combination of private agriculture and periodic coercion. Secondly, Stalin's faction wanted to prove the right to be wrong, and would benefit from stealing the clothes of the left opposition now that it had been defeated and its leader, Trotsky, exiled. Thirdly, many party activists had all along hated NEP and were willing to throw their energy and enthusiasm into the great tasks of 'socialist construction'. *Pravda* described how the delegates to the fifth congress of Soviets, which approved the five-year plan in May 1929, gazed upon a vast map which showed the various construction projects. 'Before our eyes we saw our country, as it will be in five years' time. Exciting prospect! As if by some magic hand the curtains which conceal the future have been parted. The enthusiasm of the congress found expression in a potent rendering of the Internationale.'[24] There were doubtless a few cynics present, yet the enthusiasm must have been genuine. Of course the majority of ordinary people may have felt very differently, but this was hardly considered relevant in a country in which nearly 80 per cent of the public consisted of peasants, most of whom could not be expected to feel the dynamism of a socialist transformation and who in any case, when they were not dangerous enemies of the regime, did not know what was good for them. Sacrifices? Well, many young communists found great satisfaction in living hard in tents and huts, building the great factories which would change Russia and make a happy future for generations to come. Stalin himself was no romantic, but saw advantages in harnessing such feelings as these.

So the decision was taken: to force up still further the tempos of industrial construction, and launch the campaign to collectivize the peasantry; this meant the majority of the peasantry, not 15 per cent, and immediately, not by 1933. The relevant data are examined in the next chapter.

Why, then, did the 'great turn' happen? Why the revolution from above, why collectivization? Much ink has been expended in discussing these questions. Some of the answers have been indicated in the previous pages. To recapitulate, the following factors were of evident importance:

1. The desire of many party members, and notably Stalin himself, to eliminate an individual peasantry which, as Lenin had said and Stalin repeated, 'produces capitalists from its midst, and cannot help producing them, constantly and continuously'.[25] True, Lenin advised caution, persuasion, example. True, the brutal methods which will be described in the next chapter were quite unjustified by doctrine and ideology, a fact which explains the secrecy and plain lies which were characteristic of the entire operation. But what if adherence to the voluntary principle meant the indefinite dominance of individualist agriculture?

2. The problem of industrial development, with priority of heavy industry, and the linked issues of capital accumulation and farm surpluses. Stalin did not deny that there was an alternative road, that of 'making agriculture large-scale by implanting

capitalism in agriculture'. He rejected this as he rejected the kulaks.[26] He left himself little choice thereafter. (After all, even kulaks were very modest farmers by Western standards.)

3. The price, policies, in industry and agriculture, which developed in 1926 and were obstinately continued, and which could *of themselves* have destroyed NEP, even if no other complications had ensued.

4. The political atmosphere, the prejudices against the market and Nepmen generally, the rise of monolithism and of Stalin, the 'leap forward' psychology. Fears of internal class enemies, and also of the hostile environment, affected both the social policies of the regime and the degree of priority accorded to heavy industry, as the basis of military capacity.

Years later, a Menshevik wrote of Stalin's methods as 'primitive socialist accumulation by the methods of Tamerlane'. He added: 'The financial basis of the first five-year plan, until Stalin found it in levying tribute on the peasants, was extremely precarious. ... [It seemed that] everything would go to the devil. ...'[27]

All this in no way justifies what actually occurred. It did occur, and it was not an accident or a consequence of private whims. To understand is not to forgive. It is simply better than the alternative, which is not to understand.

7. The Soviet Great Leap Forward:
I. Collectivization

SUDDENLY AND WITHOUT WARNING

The events of 1929–34 constitute one of the great dramas of history. They need much more space than they can possibly receive here, and a more eloquent pen than the author's to describe them. They need also a sounder base in reliable data than is available at present to any historian, in East or West. For we are now entering a period in which the lines dividing propaganda from reported fact tend to disappear, and statistics too often become an adjunct of the party's publicity office. Official statements and pronouncements by leaders can no longer be checked against counter-arguments made by contemporary critics, since criticism is silenced, or is confined to minor local detail. The whole flavour of intellectual life underwent a drastic change. Anyone who knows Russian can observe the change for himself, just by reading articles in learned journals on social-economic issues published in 1928 and comparing them with what was published in, say, 1932. Between these dates not only was serious criticism rendered impossible, but articles became increasingly the vehicle for strident assertions of brilliant successes and denunciations of real or alleged deviationists as agents of foreign powers. Therefore the historian must, so to speak, change gear, and use his source-material differently when he gets into the thirties. He has only very limited help from Soviet archival materials. It is true that, since 1956, more has been published, but still very selectively. Besides, the prevailing atmosphere affected the quality and content even of confidential reporting.

The dramatic events to be described affected virtually every aspect of Soviet life, and to treat them chronologically would, on balance, be more confusing than to tackle each sector separately. So we shall begin with collectivization and its consequences, and go on to industry, construction, transport, finance and trade, labour and living standards and the reorganization of planning.

On 1 June 1929 the total number of peasant members of collectives of all kinds was barely one million, and of these 60 per cent were in the T O Z (loose) type of producers' cooperatives. By 1 October the number had risen to 1·9 million (62 per cent TOZ).[1] It was this increase which gave Stalin the basis for his statement, in his famous article of 7 November 1929, that 'the middle peasant is joining collectives' and that the great turn was under way. That these figures were due at least in part to illegitimate pressures is now admitted by Soviet historians, who also now deny that the peasants were in process of 'going collective' *en masse*.[2] It seems that, silently and secretly, Stalin and his friends ordered local officials in a few selected areas to try out mass collectivization by whatever means were handy. When the result showed that victory was possible, Stalin, with Molotov and Kaganovich as his closest associates in the matter, decided to launch the collectivization campaign, using for the purpose the activists already mobilized to enforce grain collection by the well-tried 'Urals-Siberian method'. This, at least, is the reasonable conclusion of M. Lewin, in his admirable study of these events. Readers may be confidently referred to his book for details.

No doubt the final defeat of the right opposition facilitated the opening of the offensive. This, indeed, is a point specifically made by one of the ablest recent Soviet analysts of the period, Moshkov: 'The condemnation of the rightists enabled the central committee to operate more consistently the line of the offensive against the kulaks. . . .' And not only the anti-kulak policy was affected. Moshkov refers also to instructions of the central committee to party organs in selected grain regions, issued in August 1929, urging them to reach high collectivization percentages in that very year. 'In party circles the view was hardening to the effect that only by collectivization could the problem of grain production be solved.' Moshkov laid considerable stress on the effect on the peasants of the 'new system of procurements', which he identifies as having been enforced by the decrees of 28 June 1929 (RSFSR) and 3 July 1929 (Ukraine).

These have not, as a rule, been noted as important by other analysts, and yet Moshkov treats them as in effect signalling the end of NEP in the village.[3]

There is much evidence to support this. Until this date, the forcible collections of grain, which had begun early in 1928, were officially described and viewed as emergency measures. However, these decrees provided for the imposition of procurement plans on particular areas by the government, and empowered the authorities to fine (and in some cases, imprison) recalcitrant households who failed to deliver the quantities specified by the delivery plan as it affected them, and to sell up their property if need be. This power, it is true, was to be exercised by local Soviets, which were obliged to call a general meeting. However, whole villages were now receiving procurement quotas, and were encouraged to place the maximum burden on the kulak or other prosperous elements. But all were doomed indefinitely to deliver grain surpluses to the state at low prices. Moshkov very properly makes two further comments. Firstly, this decree, as applied by the government, served as the judicial foundation of the first wave of 'dekulakization', which, as we shall note, had begun already in the second half of 1929, without any declaration or decree specifically to that effect. That is to say, in selected grain-growing areas the kulaks were deliberately over-assessed for grain deliveries and they were then expropriated for failing to obey. Secondly, and more fundamentally, further great changes were bound to follow, since 'as the experience of the civil war showed, the [imposed] planned delivery of grain to the state at prices which were unfavourable to the peasants inevitably led to the reduction in production of grain to subsistence level'.[4] In other words, the peasants in general (not just the kulaks) were bound to reduce sowings, once the fundamental basis of NEP was subverted by a return to a kind of prodrazverstka. This method of procurement was successful, at least in the short run. The sub-division of the total procurement plan by regions, the mobilization of party personnel, led to a 49 per cent increase in state procurements of grain over the previous year. This could well have increased Stalin's confidence in the effectiveness of political pressure in general, and so 'procurements went parallel with the process of the wholesale collectivization of whole regions . . . and were closely linked with it'.[5] There is much in favour of such an interpretation of events.

Be this as it may, after Stalin's article on the 'great turn', published on 7 November 1929, a plenum of the central committee was held on 10–17 November. It decided that there existed 'a move of the broadest mass of poor- and middle-peasant households towards collective forms of agriculture', which was described as 'spontaneous' (*stikhiinaya*). Given that no such spontaneous move existed in nature, while the entire campaign was conducted on the supposition that it did, and given also that there was no kind of inquiry or prior warning, the events that followed were both confused and, above all, ill-prepared. There is not the slightest evidence that there had been a party or state sub-committee engaged in assessing how best to change the way of life of most of the population of a vast country. Since in fact it was to be decided that the loose TOZ was not 'collective' enough, that the *artel'* with its more advanced degree of collectivism was to be preferred, it is truly extraordinary that nothing was done before December to clarify what kind of *artel'* was intended, for there were many variants: some paid members 'by eaters' (*po edokam*, i.e. in relation to mouths to feed), some in some rough proportion to work done, some in accordance with the land and implements contributed; in some farms a good deal of livestock was collectivized, in others not. Indeed the party cadres were not too clear whether the fully-fledged commune, with total collectivization, was not in the minds of the leaders. We shall see that these confusions had considerable influence on events.

As a Soviet writer on this theme has pointed out, 'Excesses . . . were due in part to the fact that there was no clear explanation of the nature of the methods and forms of wholesale collectivization, or of the criteria for its completion. . . . Many officials interpreted it . . . as the immediate incorporation of all toiling peasants in kolkhozy.' 'Stalin and his closer co-workers did not consider it essential to discuss the party's new policy for the villages in a broad party forum, such as a congress or conference.' If proper discussion had taken place, many mistakes would have been avoided, asserted another writer.[6]

An all-union collective-farm centre (*Kolkhoztsentr*) was

created, as well as an all-union *Narkomzem* (People's Commissariat of Agriculture), under Yakovlev. The same Yakovlev headed a special politbureau commission set up on 8 December 1929, a month *after* Stalin announced the great turn, to discuss how to collectivize. It sprouted a whole number of sub-committees, among them one on tempos, another on the organizational structure of collectives, yet another on kulaks, etc. On 16–17 December they met to argue various proposals. On 22 December the commission presented proposals to the politbureau, which became the basis of a decree passed on 5 January 1930. It might be proper to conclude that it had no time to consider the colossally complex issues involved. Ahead of any report, orders were already going out to the localities, urging instant action. Thus a telegram from Kolkhoztsentr on 10 December 1929 read: 'To all local organizations in the areas of total collectivization: to achieve 100 per cent collectivization of working animals and cows, 80 per cent of pigs, 60 per cent of sheep and also poultry, and 25 per cent of the collectives to be communes.'[7]

Meanwhile the commission proposed the following timetable 'for total (*sploshnaya*) collectivization': the lower Volga by the autumn of 1930, the central black-earth area and the Ukrainian steppes by the autumn of 1931, the 'left bank Ukraine' by the spring of 1932, the North and Siberia by 1933.

According to evidence published in 1965,[8] Stalin and Molotov pressed for more rapid tempos. By contrast, others – such as Andreyev (party secretary of North Caucasus) and Shlikhter (Ukrainian commissar for agriculture) – argued for delay. They were overruled. The same source, which had access to archives and quotes them, tells that the unfortunate Yakovlev's draft included the provision that collectivization should take place 'with the preservation of private peasant ownership of small tools, small livestock, milch cows, etc., where they serve the needs of the peasant family', also that 'any step towards communes must be cautious and must depend on persuasion'. Both these limits on arbitrary excesses were crossed out by Stalin himself. It was Stalin's fault, therefore, that the decree of 5 January 1930 contained nothing to suggest to ill-prepared and confused local cadres that they were not to go ahead and collectivize all peasant property down to chickens, rabbits, hoes and buckets. To make their confusion worse, and to ensure the wildest excesses, the head of the party's agitation and propaganda department, G. Kaminsky, declared in January 1930: 'If in some matters you commit excesses and you are arrested, remember that you have been arrested for your revolutionary deeds.'[9] Stalin and Molotov urged all possible speed. The local cadres appear to have understood their task as – full steam ahead. It was hardly surprising that there was 'unjustified forcing of the pace'. Yakovlev warned in vain: avoid 'administrative enthusiasm, jumping ahead, excessive haste'. The party cadres were to 'lead the spontaneous growth' (*vozglavlyat' stikhiinyi rost*) of collectivization.[10] He and the recipients of his warning were victims of the myth and the lie. How could they lead a non-existent spontaneous movement? How could they achieve voluntarily what they knew from what they saw in front of them was a coercive operation in its very essence? A Soviet researcher found a report in the archives which stated the following: 'Excesses are to a considerable extent explained by the fact that regional and local organizations, fearful of right-wing deviation, preferred to overdo rather than underdo (*predpochli peregnut' chem nedognut'*).' Similarly, Kalinin reported that collectivization of all livestock was being undertaken by officials 'not of their own free will, but owing to fear of being accused of right-wing deviation'.[11]

Local officials announced: 'He who does not join a kolkhoz is an enemy of Soviet power.' They had 'either to achieve 100 per cent (*sploshnaya*) in two days, or hand in your party card'. The assault was launched, regardless of lack of preparation, regardless of local conditions, of opinion, of everything except the great campaign. There was, one can see, some logic against going slow: peasants who knew what was coming would react by cutting down production, perhaps destroying their tools and livestock. Better get it over, and before the spring sowing.

THE LIQUIDATION OF THE KULAKS AS A CLASS

But if whole regions were to be 100 per cent collectivized, what was to be done with the kulaks? During the second half of 1929 a debate on this question went on. It was at this point not yet clear what kind of collectivization campaign there would be, but already the issue of possible expulsion or expropriation of kulaks was posed. The majority view was against such drastic solutions. In June 1929 *Pravda* headed an article with the words: 'Neither terror nor dekulakization, but a socialist offensive on NEP lines.' Others believed in a grave danger to Soviet power in the kulaks,[12] though one might have thought that their opposition was due in large part to the measures which were being taken against them. The debate ceased when Stalin, in his statement to the 'agrarian Marxists' at the end of December 1929, asserted and justified the principle of their 'liquidation as a class'. They were not allowed to enter the collectives, presumably in case they dominated them from within, as they had dominated many a village assembly (*skhod*) in the twenties.

Stalin's justification of these drastic measures showed how, once the opposition was silenced, he became contemptuous of serious argument. Millions were to be uprooted, a mountain of human misery created, because the grain produced and marketed by kulaks could now be replaced by collective and state farms. In consequence,

Now we are able to carry on a determined offensive against the kulaks, eliminate them as a class. ... Now dekulakization is being carried out by the masses of poor and middle peasants themselves. ... Now it is an integral part of the formation and development of collective farms. Consequently it is ridiculous and foolish to discourse at length on dekulakization. When the head is off, one does not mourn for the hair. There is another question no less ridiculous: whether kulaks should be permitted to join collective farms. Of course not, for they are sworn enemies of the collective farm movement.[14]

These harsh phrases put a stop to a painful and serious discussion of the kulaks' fate. But in fact, by a mixture of local party cadres' improvisations and semi-spontaneous quasi-looting, the process of dekulakization had begun before Stalin's words had seen the light of day.

At first there was no clear line. Local officials, acting 'at their own risk and peril', began deportations, these being linked at first not with collectivization but with measures to enforce grain deliveries, as mentioned above. Only on 4 February 1930 was there an instruction issued from the central committee about how to treat the kulaks. Deportation did in fact begin in some regions by the end of 1929, and reached its peak in 1930-31. According to Ivnitsky a total of about 300,000 kulak households were deported (roughly 1.5 million people); he quotes other Soviet estimates ranging from 240,757 (in the official party history) to 381,026, though he thinks that this could have been due to some families being counted twice.[14] What happened to the other kulaks? Ivnitsky refers to a 'complex of measures' other than deportation, and their fate is left unclear.

What is quite clear is that collectivization went hand in hand with dekulakization, and dekulakization with half-disguised robbery. Poorer peasants seized their neighbours' goods in the name of the class struggle, or with no excuse at all, and the officials found themselves instructed to 'win the support of poor peasants' and were then blamed for 'allowing the distribution of kulak property among the poor and landless, in contravention of party directives'.[15] In fact Stalin intervened to prohibit the dispersal of kulak property among poor peasants, since this would make their subsequent collectivization more difficult by giving them something to lose. His conclusion (in February 1930) was: since dekulakization only made sense in relation to collectivization, 'Work harder for collectivization in areas in which it is incomplete'.[16]

Details of just who was or should have been dekulakized are still inadequately documented. Even the text of the decree of February 1930 must be reconstructed from indirect evidence. However, several sources confirm that kulaks were divided by this decree into three categories. The first, described as 'actively hostile', were to be handed over to the OGPU (political police) and sent to concentration camps, while 'their families were subject to deportation to distant regions of the north, Siberia and the far east'.[17] The second category was described as 'the most economically potent kulak households'. These were to be deported outside the region of their residence. Finally, the third group, regarded as least noxious, were to be allowed to remain in the region but were to be given land of the worst kind. The property of the first two categories was virtually all to be confiscated. Those in the third category were to be allowed to keep essential equipment, which implied partial confiscation. On their inferior land they were to grow enough crops to meet the very large demands of the state for compulsory deliveries. The same source specifically mentions extremely high procurement quotas, and taxes rising to 70 per cent of their income. Failure to deliver produce or to pay taxes was considered as anti-Soviet activity, and was often followed by deportation. It is clear from the evidence that many of these deportations took place after 1 July 1930, so it is quite probable that in the end all the persons described as kulaks were in fact deported. Some details of the procedures used may be found in the archives of the Smolensk party committee. Others will be cited in succeeding pages.

It is also clear that persons who were not kulaks at all were arrested and deported. How else can one interpret a warning, to be found in the Smolensk archives, against continuing to deport so-called 'ideological' kulaks, these being plainly opponents of collectivization, rich or poor? In the archives may also be found references to kulaks being robbed of their clothes and boots, and those engaged in the process of dekulakization were known to requisition and drink any vodka found in the kulak house.[18] Orders were issued to stop such behaviour. But what could the government expect? There were few reliable party members in the villages and they had to utilize and encourage any ragged ruffians who could be prevailed upon to expropriate and chase out their better-off neighbours (in the name of the class struggle, of course). The party and police officials found themselves vying with each other in their dekulakizing zeal. If families were separated, children left uncared for, thousands sent on journeys with little food and water to Siberia in railway wagons, then this seems to have been accepted as an inevitable part of the struggle to extirpate the last exploiting class. There were far more warnings against 'rotten liberalism' and sentimentality than there were against so-called excesses. Soviet sources insist to this day that the excesses of this class struggle were due in the main to the strong anti-kulak feelings in the countryside among the ordinary people. This point is made by Trifonov, though he does say that numerous errors of policy also occurred. One would like to see more evidence of the extent of spontaneous action. Some of the resolutions cited in Trifonov's book look suspiciously as if they were adopted by a party activist and rammed down the peasants' throats.

COERCION AND TEMPORARY RETREAT

The great assault was launched amid indescribable confusion. It may be, as has been argued by Olga Narkiewicz,[19] that some or much of collectivization remained on paper, or was confined to reports by perplexed, confused or over-enthusiastic comrades. The fact remains that it was announced by 20 February 1930 that 50 per cent of the peasants had joined collective farms, of which most were either arteli or 'communes'. The TOZ was largely discarded. Half of the peasant population in seven weeks!

Of course the threat of being labelled a kulak was widely used as a means of cajoling peasants to join. Those strongly opposed could be, and were, deported as kulaks, whatever their economic status. This was a vast exercise in coercion, and the bewildered peasants wondered what had hit them. No doubt, in the absence of adequate briefing or preparation, there were great variations in different localities. Until much more is published, we simply cannot tell. But this was indeed a 'revolution from above'.

Large numbers of conflicting instructions have been cited by Soviet analysts, which help to explain the variety of policies followed on the spot. Occasional warnings were published in the central press in January–February 1930, particularly on the undesirability of forcing collectivization in the more backward national republics. However, the warnings were sometimes ambiguously worded, and the regional party committees issued equally ambiguous orders. Thus Bogdenko quoted from the archives of the Siberian party resolutions warning severely against excesses, but demanding at the same time the completion of collectivization by that very spring. Since at the date of the 'warning' (2 February 1930) only 12 per cent of Siberia's peasants had been collectivized, the campaign inevitably continued, or even intensified. In Georgia, Armenia, Kazakhstan and Uzbekistan there were said to be a few areas (ill-defined) suitable for wholesale collectivization.[20] Not very surprisingly, all these measures produced a sharp reaction from the peasants. Thus in Central Asia alone in the first five days of March 1930 the archives record forty-five open demonstrations (*vystupleniya*) involving 17,400 persons.[21] Another source refers to 'rebellions and agitations' (*myatezhi i volneniya*), provoked by 'kulaks and anti-Soviet elements in some places'.[22]

Why deport so many real or alleged kulaks? Did this not, at a blow, deprive Soviet agriculture of its most energetic and knowledgeable husbandmen? Lewin has suggested the most probable reason: to drive the middle peasants into the collectives, not only by scaring them but also by finally slamming in their faces the door to their future advance *qua* individual peasants; that door, it was demonstrated, led to kulak status and that was a fairly sure ticket to Siberia. As well as kulaks, the terminology of the time identified an even less definable category, *podkulachnik*, or kulak-supporter (or 'sub-kulak'), to whom repressive measures were also applied as and when necessary.

A Soviet writer has stated quite frankly that 'most party officials' thought that the whole point of dekulakization was its value as an 'administrative measure, speeding up tempos of collectivization',[23] which clearly means that it had great value as a weapon of coercion in relation to the peasantry as a whole. (Kulaks were not eligible to join the collectives!)

But chaos, despair and coercion would not get the spring sowing done. After encouraging excesses of every kind, Stalin called a halt. With a rare effrontery, he blamed the local officials. They were 'dizzy with success'. He wrote: 'The successes of our collective farm policy are due, among other things, to the fact that it rests on the *voluntary character* of the collective-farm movement' (his emphasis). He warned against ignoring regional and national differences. He admitted that there was some 'bureaucratic decreeing' of collectivization, which lacked reality, and threats, such as depriving some peasants in Turkestan of irrigation water and manufactured goods unless they joined. In the same article, Stalin advocated the artel' form of collectives and said that within the artel' 'small vegetable gardens, small orchards, the dwelling houses, some of the dairy cattle, small livestock, poultry, etc. are *not socialized*'. He denounced the collectivization of poultry, of dwelling houses, of all cows, the removal of church bells, the 'over-zealous socializers'.[24]

This seemed to imply a renunciation of the coercion principle, a condemnation of what the party cadres in the villages had been so feverishly seeking to accomplish, and from the (very) highest level.

Within weeks the proportion of the peasantry collectivized fell from 55 per cent (1 March) to 23 per cent (1 June). Perplexed and demoralized officials were made scapegoats and fools. The letter of one such to Stalin has been published; Khataevich (a prominent party secretary) wrote on 6 April 1930: 'We have to listen to many complaints [from party cadres] that we have been wrongly declared to be dunderheads [*golovotyapy*]. Really, instructions should have been given to the central press so that, in criticizing the deviations and excesses which took place, they should attack and mock not only local officials. Many directives on collectivizing all livestock, including the smallest types, came from Kolkhoztsentr, from the agricultural commissariat.'[25] He might have been trying to shame Stalin. (No prize is offered for guessing whether Khataevich survived the great purge.)

Others 'went so far as to forbid people to read Stalin's article, removed the issues of the newspapers containing the article, and so on'. Archives show that some local officials treated the new policy as a surrender to the peasants.[26] In fact the confusion was increased because Stalin's article was ambiguous. He called, it is true, for the end of excesses and of coercion. But he also called the party to 'make firm' (*zakrepit'*) the existing level of collectivization. It was not too clear whether, and if so on what terms, peasants could be allowed to leave the farms. It took many weeks of clarification before it was finally forced upon party officials in some regions that Stalin's directive, and the resolution of the central committee which followed it, really did mean that one could walk out. The very great regional variations are shown by the following extract from a much longer table:

Percentage of peasant households collectivized, 1930

	1 March	10 March	1 April	1 May	1 June
U.S.S.R. *Total*	55·0	57·6	37·3	?	23·6
North Caucasus	76·8	79·3	64·0	61·2	58·1
Middle Volga	56·4	57·2	41·0	25·2	25·2
Ukraine	62·8	64·4	46·2	41·3	38·2
Central black-earth region	81·8	81·5	38·0	18·5	15·7
Urals	68·8	70·6	52·6	29·0	26·6
Siberia	46·8	50·8	42·1	25·4	19·8
Kazakhstan	37·1	47·9	56·6	44·4	28·5
Uzbekistan	27·9	45·5	30·8	?	27·5
Moscow province	73·0	58·1	12·3	7·5	7·2
Western region	39·4	37·4	15·0	7·7	6·7
Belorussia	57·9	55·8	44·7	?	11·5

(SOURCE: Bogdenko (citing archive and other materials), p. 31.)

Several conclusions follow. One is the fantastic ups-and-downs in the lives of the large majority of the population of the Soviet Union within a few short months. Another is the variation in the extent to which the peasants could (or were allowed to, or wanted to) leave collectives. Thus a large number were retained, no doubt by appropriate pressures, in such key grain-surplus areas as the North Caucasus and Ukraine, whereas in some other areas collectivization was almost abandoned (see figures for Moscow, the West and Belorussia). Finally, the pressure to collectivize in some Asian republics started late and was continued well after 'Dizzy with success', as the Kazakhstan figures show – and this despite particularly emphatic warning to go carefully and slowly in the complex circumstances of these backward areas. But by the end of April there was an outflow of peasants from the half-baked kolkhozes in all areas, though at different rates, while, in the words of a Soviet scholar, 'conditions in the villages, created by excesses, were strained in the highest degree'. In many areas, a very large proportion even of poor peasants and landless labourers walked out.[27] It is interesting that many of them formed what were described as 'cooperatives of the simplest type' and tried to work together.[28] It is one of the tragedies of this period that this and other kinds of genuine cooperation were so quickly wiped out.

Yet, amid all this chaos, the heavens chose to smile. The weather was excellent, somehow most of the sowing did get done, and the 1930 harvest was better than that of 1929, and notably better than the harvest that succeeded it (see table on page 186).

The official Soviet explanation suffers to this day from an inbuilt defect. Thus the authoritative article published in 1965 takes the following line. It asked if it was wrong to press on with collectivization, and answered: 'No. Under conditions of capitalist encirclement and constant threat of intervention, it was impossible to delay for long the reconstruction of agriculture, the liquidation of counter-revolutionary kulaks.' It was admitted that in November–December 1929 Stalin exaggerated the peasants' desire to be collectivized, that he pushed officials into excessive haste and harshness; warnings that 'the Leninist voluntary principle' was being disregarded were ignored by him. In discussing whether heavy losses in livestock could have been avoided, the authors declared: they were avoidable 'if the Leninist principle of the voluntary entry of peasants into kolkhozes were undeviatingly observed'.[29] But this (if the authors will forgive me) is simply not a tenable position. How can one assert the necessity of collectivization (and defend 'dekulakization' too, thirty years after the event), and solemnly assert that collectivization should have been voluntary? It could not have been done without mass coercion, and they must have known it perfectly well. Privately, Soviet scholars are willing to admit this. But this whole area remains thickly strewn with myths.

The old village community organizations (*obshchina, mir*) were formally dissolved, in areas subject to collectivization, by a decree of 30 June 1930. Their functions were taken over by the collective farms and by rural Soviets.

THE OFFENSIVE RESUMED

Gradually, the peasants were forced, persuaded, cajoled, taxed, ordered, back into collective farms. The total figures for the U.S.S.R. (for July) are as follows:

	1930	1931	1932	1933	1934	1935	1936
Percentage of peasant households collectivized	23·6	52·7	61·5	64·4	71·4	83·2	89·6
Percentage of crop area collectivized	33·6	67·8	77·6	83·1	87·4	94·1	–

(SOURCE: *Sotsialisticheskoe stroitel'stvo SSSR* (1936), p. 278. State farm area and households included.)

The full story of how it was done has yet to be told. Only some of the facts are as yet available. Peasants outside the kolkhoz were given inferior land, were loaded with extra taxes or delivery obligations, or both. There were repeated instances in 1931–2 of compulsory purchase of peasant livestock.[30] More areas were declared as due for all-round collectivization. Thus a decree of 2 August 1931 specified the 'cotton-growing area of Central Asia, Kazakhstan and Transcaucasia and beet-growing areas of the Ukraine and central black-earth regions' as being due for collectivization during 1931. A long and bitter struggle raged. Peasants slaughtered livestock. Sholokhov has left a vivid picture of what happened:

Stock was slaughtered every night in Gremyachy Log. Hardly had dusk fallen when the muffled, short bleats of sheep, the death-squeals of pigs, or the lowing of calves could be heard. Both those who had joined the kolkhoz and individual farmers killed their stock. Bulls, sheep, pigs, even cows were slaughtered, as well as cattle for breeding. The horned stock of Gremyachy was halved in two nights. The dogs began to drag entrails about the village; cellars and barns were filled with meat. The cooperative sold about two hundred poods of salt in two days, that had been lying in stock for eighteen months. 'Kill, it's not ours any more. . . .' 'Kill, they'll take it for meat anyway . . .' 'Kill, you won't get meat in the kolkhoz . . .' crept the insidious rumours. And they killed. They ate till they could eat no more. Young and old suffered from stomach-ache. At dinner-time tables groaned under boiled and roasted meat. At dinner-time everyone had a greasy mouth, everyone hiccoughed as if at a wake. Everyone blinked like an owl, as if drunk from eating.[31]

The new farms lacked all experience in handling the collectivized livestock. Many died of neglect. The party activists from the towns sent to supervise the peasants were ignorant of agriculture, suspicious of advice. The already-cited authoritative article admits to something of a crisis in 1932, owing to bad planning, low pay, crude coercion within kolkhozes, poor organization of work, and unfavourable weather ('subjective and objective factors'). With remarkable restraint, the authors comment: 'The kolkhozes could not immediately show the superiority of socialized over individual production.'[32]

Collectivization spread into primitive, pastoral Kazakhstan, with catastrophic results. Livestock losses were disastrous everywhere, but in Kazakhstan they virtually wiped out the sheep population (and many of the Kazakhs too, since this nationality declined by over 20 per cent between the 1926 and 1939 censuses).

Kazakhstan, sheep and goats		
1928	1935	1940
	(millions)	
19·2	2·6	7·0

(SOURCE: *Nar. khoz. Kazakh. SSR*, 1957, p. 141.)

Shortages of fodder were a major cause of the reduction in livestock in some areas, notably in the Ukraine, where the state's exactions left very little on which to feed animals. In 1931 sowing suffered acutely from the appalling state of the hungry horses.[33]

Among methods used to force peasants back into collectives were arbitrary exactions known as 'hard obligations' (*tvyordye zadaniya*) to deliver vast quantities of grain to the state. Thus, to take one example, in September–October 1930 in the Crimea 77 per cent of all those assessed for special obligatory deliveries failed to deliver the required amount, despite what the source called 'the toughest struggle', and they were punished by sale of their property, fines, imprisonment, etc., the exact figures being cited from the archives by the source.[34]

Similar measures were taken in other regions. Kulaks had been largely liquidated in 1930, so the attack was now on 'kulak and better-off' peasants, and was quite clearly intended, in the winter of 1930–31, to drive the peasants back into the collectives. To cite the same source again, 'This struggle grew into another wave of liquidations of kulaks as a class, which in its turn was directly linked with the new wave of collectivization in the winter and spring of 1931.' This was repeated in 1931-2, and there were also many cases reported where obstinate individual peasants' privately-owned horses were compulsorily used on the collective farms.[35] Some victims of these measures were deported, others evaded ever-growing delivery obligations by joining collectives 'voluntarily'. Moshkov commented: 'The [exceptional] delivery obligations affected not only kulaks but also the upper strata of the middle peasants. However, in practice, they were treated differently to kulaks, being given the chance [*sic*] to enter the kolkhozes.'[36] Percentages rose, though detailed evidence shows that some peasants left the kolkhozes, many fleeing to work in towns and on construction sites.

THE 1932-3 CRISIS

In 1932, faced with mass pillage of 'socialist' property by the demoralized and often hungry peasantry, the following draconian legislation was adopted, as an amendment to Article 58 of the Criminal Code: pilfering on the railways and of kolkhoz property (including the harvest in the fields, stocks, animals, etc.) was to be punished 'by the maximum means of social defence, shooting, or, in case of extenuating circumstances, deprivation of freedom [i.e. prison or camp] for not less than ten years, with confiscation of all property'.[37] Even Stalin did not do such things without good reason. The fact that such laws were passed in peacetime shows that he, at least, knew he was at war. His letter to Sholokhov, which Khrushchev cited thirty years later, showed what he thought. Sholokhov had protested against excesses in the area of the Don in 1933, which had included mass arrests (also of communists), illegal seizures, excessive grain procurements; Stalin in his reply admitted that some officials, in working against 'the enemy', also hit friendly persons 'and even commit sadism'. 'But ... the honourable cultivators of your region, and not only your region, committed sabotage and were quite willing to leave the workers and the Red Army without grain. The fact that the sabotage was silent and apparently gentle (no blood was spilt) does not change the fact that the honourable cultivators in reality were making a "silent" war against Soviet power. War by starvation, my dear comrade Sholokhov.'[38]

This, of course, was the point made by Stalin in his famous talk with Churchill, reported in Churchill's War Memoirs. Stalin it was who compared his struggle against the peasants with the terrible experience of the war against the Germans.

The essential problem was all too simple. Harvests were poor. The peasants were demoralized. Collective farms were inefficient, the horses slaughtered or starving, tractors as yet too few and poorly maintained, transport facilities inadequate, the retail distribution system (especially in rural areas) utterly disorganized by an over-precipitate abolition of private trade. Soviet sources speak of appallingly low standards of husbandry, with 13 per cent of the crop remaining unharvested as late as mid-September in the Ukraine, and some of the sowing being delayed till after 1 June.[39] Very high exports in 1930 and 1931 (see p. 180, below) depleted reserves, and the rapid growth of the urban population led to a sharp increase in food requirements in towns, while livestock products declined precipitately with the disappearance of so high a proportion of the animals. The government tried to take more out of a smaller grain crop. We now have food and fodder balances for the years 1928–32, and also *per capita* consumption figures.

	Kilograms per capita							
	Bread grains		*Potatoes*		*Meat & lard*		*Butter*	
	A	B	A	B	A	B	A	B
1928	174·4	250·4	87·6	141·1	51·7	24·8	2·97	1·55
1932	211·3	214·6	110·0	125·0	16·9	11·2	1·75	0·70
			A = Urban	B = Rural				

(SOURCE: Moshkov, *Zernovaya problema v gody sploshnoi kollektivizatsii* (Moscow University, 1966), p. 136, quoting archives.)

These figures show that urban citizens ate more bread and potatoes, in the place of meat and butter. But the peasants ate less of everything. That was the result of deliberate policy. A Soviet scholar commented that the vast increase 'in state procurements during the years of wholesale collectivization, with low levels of grain production, cannot be explained merely by errors, imperfections of planning or . . . by the ignoring of the interests of agriculture and of the rural population, as is alleged by bourgeois writers in the West. The country was laying the foundation of a mighty industrial base.'[40] Yes, but primarily at the peasants' expense. Procurements in 1931 left many peasants and their animals with too little to eat. The Ukraine and North Caucasus suffered particularly severely. Collectivized peasants relied almost exclusively on grain distribution by kolkhozes for their bread, since money was virtually useless in this period; bread was rationed in towns and unobtainable in the country save at astronomical 'free' prices (see next chapter). These excessive procurements threatened the very existence of the peasantry in some areas. In fact, according to Moshkov, exactions were so severe that the state had to return grain which had already been collected (21 per cent of the total in West Siberia, for instance) so that there would be some seed, food and fodder. There were tremendous variations between areas and between farms in the same area, owing to the almost incredible arbitrariness of the procurement organs.[41]

All this led in 1932 to trouble, pilfering, indiscipline, concealment of crops. As a result, Stalin evidently decided to relax the procurements pressure somewhat, and the procurement plan for 1932, which had originally been fixed at an impossible 29·5 million tons, was reduced to 18·1 millions, while greater freedom was offered to kolkhozes and remaining individual peasants to sell on the free market, provided the reduced delivery plan was fulfilled first.[42]

However, conditions grew even more chaotic. Procurement organs relaxed their pressure, and, because of the vast disparity between the low state buying prices and the very high free market prices, grain flowed into unofficial channels, and in particular into the peasants' own storehouses, since the harvest was not a good one and the food shortages of the previous winter were vividly recalled. Discipline collapsed in some areas. The reduced state procurement plan was threatened. Telegrams from Moscow had no result. In the North Caucasus the harvest was particularly poor, a mere 4·4–5·9 quintals per hectare, a miserable crop on the best land in the U.S.S.R. In this area, and in the Ukraine, evil-intentioned persons 'succeeded in awakening private-property feelings, in diverting many kolkhoz peasants from the correct path and poisoning them with individualism. Some kolkhozes in the North Caucasus and the Ukraine ceased to come under the organizing influence of the party and the state.'[43] (These are very strong words indeed for a Soviet author, indicating a kind of rebellion.)

This led to state counter-measures, which in turn led to the great tragedy: the famine of 1933. 'All forces were directed to procurements.' The law of 7 April 1932, which, as we have seen, provided for the death penalty for pilfering foodstuffs in kolkhozes, was used against those who 'with evil intent refused to deliver grain for [state] procurements. This particularly affected socially alien groups. Organizers of sabotage in kolkhozy were handed over to the courts, including degenerate communists and kulak-supporters among the kolkhoz leadership. In accordance with the central committee directives, regions which did not satisfactorily fulfil procurement plans ceased to be supplied with commodities. . . . Illegally distributed or pilfered grain was confiscated. Several thousands of counter-revolutionaries, kulaks and saboteurs were deported. . . .'[44] The party was purged. In the North Caucasus 43 per cent of all investigated party members were expelled. There were some appalling excesses. Stalin declared, in a speech to the politbureau on 27 November 1932, that coercion was justified against 'certain groups of kolkhozes and peasants', that they had to be dealt a 'devastating blow'. Kaganovich announced that rural communists were guilty of being 'pro-kulak, of bourgeois degeneration'.[45] Mass arrests went beyond all bounds; half of local party secretaries in the North Caucasus were expelled on orders of Kaganovich. 'All grain without exception was removed, including seed and fodder, and even that already issued to peasants as an advance [payment for workdays].'[46] The result was 'an extremely grave food shortage in many southern areas', and a 'heavy loss of livestock', which took a long time to repair. Much the same happened in the Ukraine. A local party secretary commented: 'Without administrative pressure on the peasant we will not get the grain, so it does not matter if we overdo things a little.'[47] In January 1933 a more orderly system of compulsory procurements was decreed, based on acreage sown, replacing the purely arbitrary (though nominally voluntary) system of *kontraktatsiya*. But the damage had already been done. The famine, part and consequence of the struggle described above, was terrible.

Grain procurements did indeed increase, as the following figures demonstrate:

State grain procurements					
(millions of tons)					
1928	*1929*	*1930*	*1931*	*1932*	*1933*
10·8	16·1	22·1	22·8	18·5	22·6

(SOURCE: Malafeyev, *Istoriya tsenoobrazovaniya v SSSR* (Moscow, 1964), pp. 175, 177.)

Grain exports					
(millions of tons)					
1927–8	*1929*	*1930*	*1931*	*1932*	*1933*
·029	0·18	4·76	5·06	1·73	1·69

(SOURCE: Soviet trade returns.)

The Soviet population in 1926 was 142 millions, and for 1932 it was officially estimated at 165·7,[48] since it had been increasing at the rate of about 3 millions a year. In 1939, seven years later, it was only 170 million. Somewhere along the way well over 10 million people had 'demographically' disappeared. (Some, of course, were never born.) Many died in the terrible early thirties. Eye-witnesses saw starving peasants, and I myself spoke to Ukrainians who remembered these horrors. Yet neither the local nor the national press ever mentioned a famine.

There have been, as far as can be discovered, only two references in Soviet print to the famine, even in recent years (the official histories mention only a 'shortage of food', at most). One was in a novel: Stadnyuk's *Lyudi ne angely*. The other was in a work by Zelenin, which quoted archives concerning 'mass instances of swelling from hunger, and death' as occurring in the Central black-earth region, an area which Western observers did not regard as seriously affected by the famine.[49]

In his autobiography Koestler described a visit to Kharkov at this period. As well as hunger there was a breakdown of electricity. Newspapers failed to appear. When they were eventually printed, they mentioned neither food shortages nor the power breakdown. Clearly, historians who believe that there is no fact without documentary proof would be hard put to it to describe the events of the period.

Finally to wind up this deplorable story, the nine million peasants left outside collectives in 1934 were duly attacked. They were, it seems, cold-shouldered and treated as hostile elements, but allowed to survive. This toleration was treated as a 'right-wing deviation'. On 2 June 1934, at a conference of officials on collectivization, Stalin demanded – and this is quoted from the archives – that 'in order to ensure the uninterrupted growth of collectivization, there should be a tightening of the tax screw (*nalogovyi press*) on the individual peasants'.[50] Yet this article ends with the still-compulsory myth (this in 1964!): 'The multi-million peasantry became even more convinced of the incontrovertible superiority of socialist agriculture, of the mighty kolkhoz system.'

The organization structure of kolkhozes was at first quite confused. Stalin laid down that the artel' was to be predominant, and in 1931 91·7 per cent of collectivized land was within arteli (4·7 per cent TOZ, 3·6 per cent communes). However, internal arrangements were exceedingly haphazard, peasants' rights were ill-defined, their incomes uncertain not only in quantity (they remained that until 1966) but also in their nature. How was payment to be made? The June 1931 plenum of the party decided that payment must be in accordance with work done, and not per head or per 'mouth'. A rough-and-ready system of piece-rates was to be devised. This gradually became the *trudoden'* (work-day unit), which was 'legalized' by decree of 5 July 1932, and more closely defined in January 1933. These and other rules became ultimately embodied in the model charter of kolkhozes, adopted in 1935, of which more in Chapter 9.

PARTY CONTROL AND THE MTS

Kolkhozes were under the close supervision and tutelage of the party. The party sent out 25,000 urban activists to act as supervisors, farm chairmen, political officers. Their ignorance of rural questions and misunderstanding of the peasant mind contributed to the errors and excesses of the period. A key element in the control mechanism was provided by the procurement organs (*Zagotzerno,* and others), but perhaps the most important were the Machine Tractor Stations (MTS), which require more detailed examination.

The 'ancestor' of MTS was a 'tractor column', a state-run tractor service, rendered to individual peasants as well as to the few collective or state farms, in the Odessa province. The MTS were organized after a decree of 5 June 1929. At first they were run as a kind of joint enterprise, with peasants buying shares in *Traktortsentr,*[51] but they became fully-fledged state-controlled organizations. It was decided during the process of collectivization to give the MTS such tractors and other power-driven machines as were available, and to make of them a kind of compulsory service agency, while simultaneously stressing their role as supervisors (decree of 1 February 1930). In January 1933 the party plenum decided to create political departments in the MTS and state farms, of which more in a moment. So from the first the MTS developed into a unique combination of providing both tractor-power and political-economic guidance. Their contractual relationships with kolkhozes had been based, since February 1933, on payments in kind, usually in the form of a percentage of the harvest. Perhaps for this reason 1933 saw the birth of a statistical device, 'biological yield', which, as will be shown later, overstated the harvest. The state's share, received via payments in kind for the work of the MTS as well as by direct procurements, was increased by this device. Tractor production rose substantially in these years, but at first the net effect was merely to replace the haulage power of horses slaughtered during collectivization.

The political departments of the MTS were, on the face of it, another means of exerting pressure on the peasants. Yet in a well-documented paper on the subject, the Soviet historian Zelenin shows that things did not always work out that way. The political departments were responsible to the party's central committee, and were not under the party secretary of the district which they operated, a circumstance which caused much friction. Each political department included a representative of the OGPU (political police). The head of the department was, *ex officio,* deputy director of the MTS and charged with vast powers over production plans and procurement activities. These heads were specially selected, largely volunteers. Zelenin's evidence shows that, when they reached the villages early in 1933, they saw with their own eyes the dreadful effects of the excesses described above. They talked to the peasants, they argued, they learned. Being told to bring some order into the situation in agriculture, they quickly realized that excessive procurements must be cut down, that peasants must be allowed adequate incentives. They found themselves instructed instead to purge the kolkhozes of subversive elements – for Stalin's line was that the enemy, disguised as storemen, bookkeepers, agronomists, was engaged in 'silent sapping'. So it was reported by the political department that, in 1933, in twenty-four provinces of the U.S.S.R. 34·4 per cent of storemen, 25 per cent of bookkeepers, etc. were dismissed, and 'many were accused of wrecking'.[52] Many political officers came into conflict with their OGPU colleagues, who were too apt to arrest and dismiss, as archive material quoted by Zelenin shows. In the end, many political departments began to defend peasant interests, and in particular to protest against excessive grain procurements, especially when the authorities sought to increase delivery plans over and above the norms supposedly laid down by the compulsory procurement decree of January 1933 (a practice which continued). In the June 1934 plenum of the central committee, the head of the grain procurement organization accused local officials, including the heads of political departments, of 'anti-state tendencies' in seeking to diminish the state's exactions. Such prominent party leaders as S. Kossior, P. Postyshev, I. Vareikis, also accused the political departments of this. Some political officers had the audacity to draw up food-and-fodder balances to prove that the state's exactions were excessive, and were sharply condemned: such balances were, it seems, 'kulak tendencies [*sic*], directed to the breach of the law on grain deliveries'.[53] In November 1934 the political departments of the MTS were abolished. Though there remained a deputy-director (political) of the MTS, he no longer had a department, or any special powers *vis-à-vis* the local party organization.

State farms (*sovkhozy*) were, at first, greatly favoured by the regime. However, their high cost and inefficiency led to a change of policy. This is easy to understand if one bears in mind the principal reason for collectivization, which was procurement of produce *at minimum cost*. In the case of kolkhozes, high cost and inefficiency meant simply that the peasant members were very poorly paid, since they divided among themselves whatever was available, with no guaranteed minimum of any kind. But a state farm worker was a wage-earner, and losses made by such farms had to be met out of the budget. The 'ideological' superiority of state farms none the less led to a sharp rise in their numbers, the area sown increasing from 1·7 million hectares in 1928 to 13·4 million in 1932 and 16·1 million in 1935. It declined thereafter, and state farms did not play a major role in Soviet agriculture until after Stalin's death. (More about state farms in Chapter 9.)

THE FREE MARKET AND PRIVATE PLOTS

How did the peasants survive the confusion and hardship of the 'revolution from above'? They could not have survived without the toleration, in and after 1930, of some private food-growing, and, after the initial excesses of super-collectivization, they were allowed some domestic animals. Great bitterness was caused by the compulsory acquisition by kolkhozes of livestock, especially cows, under conditions in which the collectives had neither the buildings nor the knowledge or experience for looking after big herds (which have to be kept indoors during the winter), and when milk for peasant children could only be provided from their own cows, in the absence of any alternative source of supply (this remains the case, in most of Russia, even in 1969, let alone in 1931). Gradually a sort of *modus vivendi* emerged, and Stalin himself began to make promises to help peasants acquire cows.[54] But the drastic decline in the livestock population made this a rather distant project, and many could manage to keep only goats, which some bitterly described (in whispers) as 'Stalin's little cows'. However, peasant rights became more clearly defined, and gradually there developed an understanding as to the permissible upper limits of collective peasants' private holdings, which emerged finally in the model statute, described in Chapter 9.

The question arose of the right to sell freely after meeting the state's procurement quota, the latter having the legal status of a tax levied on the collective, on peasant members and (more heavily) on the surviving individual peasants. There was sporadic interference with the functioning of any free market, while private traders were being driven out and the process of collectivization completed, and many cases of closing all markets were reported. On 6 May 1932 a decree allowed free sales of grain by kolkhozes and collectivized peasants after the state's procurement plan had been fulfilled. Four days later the same rights of selling in 'markets and bazaars' were extended also to livestock products. On 20 May the tax levied on such sales at markets (this trade never wholly ceased) was lowered, and the right to sell at free prices reasserted. In this decree it was stated that the opening of private shops, and private dealers, were to be barred. On 22 August of the same year, 'speculators and dealers' were to be sent to 'a concentration camp for from five to ten years', to cite the words of the decree.[55] These were the final nails in the coffin of the NEP concept of free trade. Peasant trade was different, in so far as it consisted of sales by the producers of their own surplus produce, and so did not constitute earning a living by trade. This remains the legal position today.

In 1933, which was a very difficult year, the right to market grain was more strictly defined: only 'after fulfilling the procurement plan for the whole republic, *krai*, province, and making full provision for seed'.[56] In these years kolkhoz trade was still on the edge of semi-illegality, since arbitrary exactions of all kinds for the needs of the state could happen at any time, with accusations of speculation and 'kulak' behaviour. This, as well as the acute shortages prevailing, caused an extremely steep rise in free-market prices, which will be documented later in this chapter.

Agriculture reached its lowest point in 1933, and then began a painful recovery, the story of which can be left aside for the present.

SOME STATISTICS

The harvest and livestock statistics of the period were as follows (the 'biological yield' figures, which distorted Soviet data from 1933 until after Stalin's death, are also given, since they were used to falsify reality and to facilitate excessive procurements):

	1928	1929	1930	1931	1932	1933	1934	1935
Grain harvest, real (million tons)	73·3	71·7	83·5	69·5	69·6	68·4	67·6	75·0
Grain harvest, biological (million tons)	–	–	–	–	–	89·8	89·4	90·1
Cattle (million head)	70·5	67·1	52·5	47·9	40·7	38·4	42·4	49·3
Pigs	26·0	20·4	13·6	14·4	11·6	12·1	17·4	22·6
Sheep and goats	146·7	147·0	108·8	77·7	52·1	50·2	51·9	61·1

(SOURCES: *Sotsialisticheskoe stroitel'stvo, 1936*, pp. 342–3, 354; Moshkov, *Zernovaya problema v gody sploshnoi kollektivizasii* (Moscow University, 1966), p. 226.)

Did collectivization in fact contribute to capital accumulation? This has been the subject of controversy, and the reader may be referred particularly to works by James R. Millar and Michael Ellman.[57] It has been correctly pointed out that the real agricultural surplus cannot be measured simply by the volume of sales off the farm; industrial inputs into agriculture and the village must be deducted, and a surplus so defined then depends decisively on price. Evidence is then quoted, especially that presented by the Soviet economic historian Barsov, to show that prices did not move adversely to the village in the years following 1928, when allowance is made for the high free-market prices. Barsov also purports to prove that when valued in 1913 or 1928 prices, or in labour-values, the flow of goods in the two directions did not show any significant rise in the surplus extracted from the village; the contrary is true in some years. From this it may seem to follow that the increased accumulation of capital in the period 1928–37 came not from agriculture but from the urban sector (including former peasants who had migrated to the towns).

This is a most complex subject, which deserves more thorough examination than it can receive here. It is certainly true that the state procured less than it originally intended, out of a much lower output, while having to supply extra inputs to offset losses in draught animals. It is also true that the urban sector also bore sacrifices, that workers' living standards declined. However, apart from serious doubts as to the validity of Barsov's figures, there is an important sense in which the data are misleading in principle; thus the prices of 1913 and 1928, let alone labour-values, bear absolutely no relationship to the relative scarcities of the early thirties. What is the 'value' of bread-grain to a starving peasant? The price which ruled in 1928, when there was plenty of bread? The labour-time required to produce it? Another way of expressing the same point is to say that massive use of coercion was needed to take away grain and other products which, as a result of collectivization, had become acutely scarce, and were consequently valued much higher (by peasants and by consumers) than the price paid for them by the state. Secondly, prices in the early thirties had little meaning when goods were so often unobtainable. A ration card, a pass to a special shop, were more important than money in a hard year such as 1933 in which, it should be noted, millions of *peasants* starved. Finally, the evidence on the volume of sales of industrial goods in these years is of very doubtful validity. An intelligent observer, whose reports are in the British Foreign Office archives, noted that when, following a 1932 decree on sending more consumers' goods to villages, he visited rural shops, he found that deliveries had consisted largely of vodka and cosmetics.[58]

So was collectivization actually counterproductive, i.e. did it impede rather than facilitate industrialization in these years? Millar takes the first view, Ellman the second. It seems unnecessary for us to take sides in this debate, important though the issue undoubtedly is, especially as one is conscious that more research is needed.

In looking back at the impact of those years on agriculture and the peasants, critical comment is superfluous. The events described cast a deep shadow over the life of the countryside, of the whole country, for many years thereafter. Far too many works on the period say far too little about what occurred. Of course, much more evidence has recently become available, and this chapter is no more than a bare summary of such evidence. It is very much to the credit of Soviet scholarship that so much has been made available, after so prolonged a silence (for which the scholars cannot be blamed) about what by common consent must be a painful period, of which many men in high places must feel ashamed in their hearts.

8. The Soviet Great Leap Forward: II. Industry, Labour and Finance

An adequate history of the first five-year plan has yet to be written. Official Soviet accounts overstress the achievements, dwell endlessly on the 'pathos of construction'. The positive features seem also to be overstated in novels of the period. As for anti-Soviet writers, for them the years 1929–33 are composed exclusively of coercion, hunger, shortages and inefficiency, and the achievements are mentioned only as a kind of apologetic afterthought. Here it will be necessary to dwell on many negative features, which are an integral part of the story. Yet so are the achievements which must be seen against a background of appalling difficulties.

OPTIMISM RUNS RIOT

In the previous chapter we noted the adoption of a high 'optimal variant' of the five-year plan. This was speedily followed by super-optimal variants of the most fantastic kinds. The upward revision of the 'optimal' targets began very soon after their adoption. The year 1928–9 proved quite successful in industry, and this caused, in the decree of 1 December 1929, an upward amendment of the plan for the economic year 1929–30. On 5–10 December 1929 a congress of 'shock brigades' adopted a call to fulfil the five-year plan in four years. This became official policy, and in the end the five-year plan was deemed to have run its course on 31 December 1932 instead of 30 September 1933, nine months ahead of schedule, it having been decided (in 1930) to make the economic and the calendar year coincide. Of course, this of itself was an upward revision. Others followed. The sixteenth party congress resolved to review the machinery plans in order 'decisively to free industry and the national economy from dependence on foreign countries'. This last point may be said to be inspired by military-strategic considerations, but it would be incorrect to assert that military expenditure as such was responsible for the acceleration of the heavy-industry plans: military spending remained modest, the Red Army's re-equipment lay still some distance ahead.

The net effect was that during 1929 and 1930 the five-year plan (now a four-year plan) was altered, as the following selection of figures shows:

	1927–8	1932–3 ('*optimal*')	1932 (*amended*)	1932 (*actual*)
Coal (million tons)	35·0	75·0	95–105	(64·0)
Oil (million tons)	11·7	21·7	40–55	(21·4)
Iron ore (million tons)	6·7	20·2	24–32	(12·1)
Pig iron (million tons)	3·2	10·0	15–16	(6·2)

(SOURCES: S. Bessonov, *Problemy ekonomiki,* No. 10–11 (1929), p. 27 and plan-fulfilment report.)

Still wilder figures were encountered in 1930–31. Stalin himself spoke on 4 February 1931 of fulfilling the plan 'in three years in all the basic, decisive branches of industry'.[1] In the same speech he declared: 'It is sometimes asked whether it is possible to slow down the tempo somewhat, to put a check on the movement. No, comrades, it is not possible. The tempo must not be reduced! On the contrary, we must increase it. . . .'[2] It was then that he made the justly famous prophecy: 'We are fifty or a hundred years behind the advanced countries. We must make good this distance in ten years. Either we do so, or we shall go under.' 1941 was ten years away.

But a sense of coming danger is no excuse for attempting the impossible. The extent of the upward amendments in annual plans may be illustrated by the table on page 189, which also shows the downward amendment of the textiles targets, i.e. the shift in mid-plan in the relative priorities of heavy as against light industry.

Needless to say the new targets were far beyond practical possibility. The rush, strain, shortages, pressures, became intolerable, and caused great disorganization. Naturally, supplies of materials, fuels, goods wagons, fell short of requirements.

Index numbers 1930–31 (1927–8 = 100)

	Original Plan*	Amended Plan
Producers' goods, all	196·3	349·9
Coal	155	202·5
Oil	166	266
Ferrous metallurgy	176·8	207·7
Machinery, total	198·8	482·1
Agricultural machinery	222·8	552·6
Electro-technical ind.	235·8	590·5
Basic chemicals	252·3	390·0
Consumers' goods, all	161·9	163·1
Textiles	148·3	121·2
Food	149·8	166·2
All industry, total	176·7	244·5

(SOURCE: A. Koldovsky, *Problemy ekonomiki*, Nos. 4–5 (1930) p. 109.)

* According to five-year plan, optimal variant.

It was then that the government, by stages, imposed upon the economy its own priorities, by ever-tightening control over resource allocation, physical output, credit. The 'Stalin' model was created in the process of trying to do the impossible, and therefore by facing every day the necessity of assuring supplies to the key projects or 'shock-constructions' (*udarnye stroiki*), at the expense of others regarded as of lesser importance.

Plans were born of a conflict between specialists and keen comrades convinced that revolutionary *élan* would perform miracles. One old oil expert, given what he regarded as an absurd order to increase production, is said to have written to the central committee as follows: 'I cease to be responsible for the planning department. The [plan] figure of 40 million tons I consider to be purely arbitrary. Over a third of the oil must come from un-explored areas, which is like cutting up the skin of a bear before it is caught or even located. Furthermore, the three cracking plants which now exist are to be turned into 120 plants by the end of the five-year plan. This despite the acute shortage of metal and the fact that the highly complex cracking technique has not been mastered by us. ... I stand for high tempos, but duty demands ...' and so on. To this, according to the writer, a young woman replied: 'We do not doubt the knowledge or good-will of the professor ... but we reject the fetishism of figures which holds him in thrall. .... We reject the multiplication table as a basis for policy.'[3]

This possibly imaginative reconstruction described a common enough situation. Much waste of effort inevitably resulted, but it could be argued that more was achieved in the end than if 'sound' advice had been taken. The resultant chaos has been described by many writers. Here is the present Minister of Tractors and Agricultural Machinery remembering the birth of the Stalingrad tractor works:

Even one who saw those days with his own eyes finds it hard to picture today how things then looked. .... One chapter of a book of the period is headed: 'Yes, we smashed lathes.' This chapter was written ... by a worker who came to the Volga from a Moscow factory. Even he was full of wonder at the American lathes without belt transmission, with their own motors. He could not handle them. What is one to say of the peasants fresh from the villages? They were sometimes illiterate – it was a problem for them to read and write. Everything was a problem in those days. There were no spoons in the canteen – and this matter was dealt with when Ordzhonikidze arrived at the factory and demanded that things be put right. There was the problem of bugs in the workers' huts – and the secretary of the young communist central committee, Kosarev, made us get rid of them. The first director of the factory, Ivanov, wrote as follows: 'In the assembly shop I talked to a young man who was grinding sockets. I asked him how he measured, and he showed me how he used his fingers. We had no measuring instruments!' Now, after fifty glorious years, we must remember all this in detail, remember how this industry was created, which now produces the largest number of tractors in the world, how and in what conditions the first great tractor-works in the country was built in a year and working to full capacity a year later. All this was done in a country where as late as 1910 over two thirds of the ploughs were wooden.[4]

Such facts, quotations, examples, could be multiplied many times.

Vast projects got under way. While some were the work of forced labour (the Volga–White Sea canal, for instance), there is no doubt that there was enthusiasm too. There was, for instance, the story of Magnitogorsk. A great new metallurgical centre was created in the wilderness, and the workers and technicians worked under the most primitive conditions, yet many seemed to have been fired by a real faith in the future and in their own and their children's part in it. There were, especially in later years, all too many examples of phoney official superlatives, which gave rise to widespread cynicism. So it is all the more necessary to stress that thousands (of young people in particular) participated in the 'great construction projects of socialism' with a will to self-sacrifice, accepting hardship with a real sense of comradeship. Statistics will also be cited to show that others had very different attitudes to their work, not only prisoners or deportees but also peasants fleeing the collectives.

ACHIEVEMENTS OF THE FIRST FIVE-YEAR PLAN

The key figures of plan fulfilment were as follows:

	1927–8 (actual)	1932–3 (plan)	1932 (actual)
National income (milliard 1926–7 roubles)	24·4	49·7	45·5
Gross industrial production (milliard 1926–7 roubles)	18·3	43·2	43·3
Producers' goods (milliard 1926–7 roubles)	6·0	18·1	23·1
Consumers' goods (milliard 1926–7 roubles)	12·3	25·1	20·2
Gross agricultural production (milliard 1926–7 roubles)	13·1	25·8	16·6
Electricity (milliard Kwhs)	5·05	22·0	13·4
Hard coal (million tons)	35·4	75	64·3
Oil (million tons)	11·7	22	21·4
Iron ore (million tons)	5·7	19	12·1
Pig iron (million tons)	3·3	10	6·2
Steel (million tons)	4·0	10·4	5·9
Machinery (million 1926–7 roubles)	1822	4688	7362
Superphosphates (million tons)	0·15	3·4	0·61
Wool cloth (million metres)	97	270	93·3
Total employed labour force (millions)	11·3	15·8	22·8

(SOURCES: 1932 figures from *Sotsialisticheskoe stroitel'stvo* (1934) and the fulfilment report of first five-year plan. For sources of other figures see table on page 146.)

Before proceeding to interpret the table, it is most important to stress the limitations of the statistics. In brief, these limitations are as follows:

(a) All aggregates (indices of national income and industrial output in particular) are liable to statistical 'inflation'. For reasons to be examined, costs rose sharply during the plan period, and the nominal 'constant 1926–7 prices', in which the indices were computed, showed an upward drift.

(b) This is particularly noticeable in the machinery sector. Thus it is surely decidedly odd that the plan for machinery was supposed to have been very greatly over-fulfilled, whereas both metal and fuel output lagged far behind plan. No major metal-using sector under-fulfilled its plan, thereby making more metal available for machine-building, while construction (another big metal-user) exceeded its plan too. Imports rose; thus 1·3 million tons of steel sheet was imported in 1931; but shortages persisted. It is not, as some imagine, that the critics claim that metal and machinery production ought to expand at the same rate. The point is that, if much less metal is produced than was planned, then major metal-using industries must be cut back. Yet there is no sign of this in the figures. The most probable explanation is that many machines were priced high, and were entered in the '1926–7' price index at a large sum in roubles. In part this is a weighting problem; the few machines which were made in Russia in 1926–7 were indeed expensive, and so the index would look much less favourable if a later Soviet price pattern, or Western prices, were used. But few types of machines made in, say, 1932 were being made in 1926–7, or only existed as prototypes. Since plan fulfilment was measured in '1926–7' roubles, it paid all managers to strive for the acceptance of the highest possible '1926–7' price. Official statisticians, under pressure to show good results and high growth rates, would hardly be in a strong position at this time to resist.

(c) There was the fate of handicrafts, small workshops, domestic production of many kinds. Ample evidence exists for the proposition that these categories were in a decline, squeezed out as part of the drive against the Nepmen, and also by the prevailing shortages of materials and fuel. Private workshops were closed, private craftsmen forced into producers' cooperatives, but the latter were frequently suspected of being mere disguises for Nepmen and there were numerous instances reported of their being dissolved for this reason. The net effect was, despite the contrary provision of the five-year plan draft, a marked decline in the numbers of artisans of all kinds. There was also some replacement of domestic by factory production: less bread-baking in the home and more bakeries, less domestic spinning, and so on. There had been a tendency to understate the output of the artisans and omit the purely domestic production, and so this decline did not find its proper reflection in statistics. Only thus can one explain the paradox that is shown in the above table: output of consumers' goods was supposed to be rising rapidly at a time of most acute privation.

(d) The effects of rapid urbanization, in Russia as elsewhere, tend to cause an increase in measurable output much greater than any real increase in consumers' welfare.

But though the claims in their totality are dubious, there is no doubt at all that a mighty engineering industry was in the making, and output of machine-tools, turbines, tractors, metallurgical equipment, etc. rose by genuinely impressive percentages. The fuel group did well, though falling short of the targets (very short of the absurd super-targets referred to on page 188). The metallurgical group was far short of its goals, and the report on plan fulfilment frankly admitted that the possibilities of so speedily constructing and equipping iron and steel-works had been greatly overestimated. However, a great deal was achieved in the Urals, the Kuzbas, the Volga, and the Ukraine. Engineering works in the Moscow and Leningrad areas were expanded and modernized. The gigantic Dnieper dam was built speedily, and provided power for more industries. Apart from developing new production in 'traditional' industrial areas, such as Moscow, Leningrad and the Donbas, and in the Urals area, industrialization began to touch the more backward national republics: a textile mill in Central Asia, mining in Kazakhstan, engineering in Georgia. Achievements were sometimes very great, sometimes very patchy. Chemicals targets were, with the exception of synthetic rubber, not fulfilled. Stalin began to listen to the voices of advisers who told him that agronomic panaceas (the *travopolye* rotation scheme advocated by Vil'yams) would enable great savings to be made; mineral fertilizer would be unnecessary.

Textile output tended to fall rather than rise. This was due partly to the low priority of consumers' goods in the investment programme, but most of all to the balance of payments situation. These were years of depression in the 'capitalist' world, and prices of raw materials and food were falling much faster than prices of machinery. Thus the Soviet government found itself having to export more in order to obtain from the 'capitalists' the machinery and equipment needed for the new factories, and this made the strain worse. Imports of cotton and especially of wool had to be cut. At the same time wool production in the U.S.S.R. itself was adversely affected by the drastic reduction in the number of sheep. The cotton-growing areas were pressed to increase acreage, but there was resistance, at a time when staple foodstuffs were so short and cotton prices were unfavourable (they were greatly raised in 1934). So a shortage of raw materials was the principal bottleneck in the textile industries. Therefore the combination of three factors caused the low living standards of these years: excessive investment in heavy industry, the consequence of collectivization in agriculture, and the worsening terms of trade. One should also mention the transport difficulties. Much had to be done to enable the railways to carry greatly increased freight, since there were few roads or vehicles, and rivers and canals froze in winter. The Turksib line, linking Siberia with Alma Ata, and the important line to the Karaganda coalfields, were built in these years, and many single-track lines converted to double-track. But the plan envisaged the completion of 16,000 kilometres of new railway, while in fact only 5,500 were built. There was grave shortage of railway equipment of all kinds.[5]

Because the global industrial output figures looked satisfactory the government was able proudly to announce success. The five-year plan had been fulfilled by the end of 1932, ahead of schedule. A careful look at the table (page 191) shows that only in the (statistically suspect) machinery and metal-working group was the plan over-fulfilled, and in most important industrial sectors, not to speak of agriculture, the shortfalls were significant. This is not to deny that a leap forward had been achieved, and that the successes which were to be reported in the years 1934–6 were due in a large part to the completion of projects begun in the first five-year plan period. However, 1932, and still more 1933, were years in which strain and disorganization reached their highest point, and the hardships of the population were severe even if they were not publicly admitted.

How can one measure these strains, which were the cost of what had been achieved? The increase in forced savings is indicated by the rise of the share of capital accumulation in the national income, from 19·4 per cent to 30·3 per cent in 1932.[6]

There are several other highly relevant indicators. Let us begin by looking at productivity and labour plans.

LABOUR PROBLEMS

According to the plan-fulfilment report, the increase in numbers actually employed was as follows:

	1927–8	1932–3 (plan)†	1932 (actual)
All branches of the national economy (thousands)	11,350	15,764	22,804
of which: 'Census' industry* (thousands)	3,086	3,858	6,411
Building (thousands)	625	1,880	3,126

* Excludes small-scale industry and handicrafts
† Optimal variant
  (SOURCE: Calculated from *Pyatiletnii plan, nar.—khoz. Stroitel'stva, SSSR* (Moscow, 1930), Vol. 2, pp. 165–71, and *Fulfilment of first five-year plan,* p. 186.)

The urban population in 1932 was estimated by the plan to be 32·5 million. It was actually 38·7 million.

On the credit side, the original plan had expected that substantial unemployment would continue, whereas it had been eliminated from the towns by 1932, and labour shortage was a fairly widespread phenomenon. Also on the credit side was the introduction in these years of a seven-hour day, though the citizens' enjoyment of leisure was complicated by the introduction, beginning in the autumn of 1929, of the 'uninterrupted' *pyatidnevka,* or five-day week. This is not to be confused with a 'real' five-day week, i.e. a free long weekend. The idea was that factories should work every day, with a fifth of the personnel off on any one day (four days' work, one day free). This led to problems. Thus maintenance was neglected, key personnel were liable to be away when wanted and members of families seldom had the same day off. On 21 November 1931 the work 'week' was somewhat lengthened, being turned into five days on, one day off, with some fixed dates at which the works would be closed. All this eliminated Sunday as a regular day of rest, which, at a time of anti-religious excesses, was doubtless one of the objects of the exercise. We shall see that a seven-day week (with Sunday the normal free day) was restored in 1940.

However, the main point was that the productivity estimates of the plan were over-optimistic. Outside the (suspect) machinery and metal-working sector, the productivity gains were very far below expectation and on occasion even negative. Vast numbers of peasants coming in from the country, sometimes as refugees from collectivization, immensely complicated the problems of elementary labour discipline, time-keeping, training. The planned expansion of the labour force called for great efforts to teach new skills, to increase the inadequate numbers of engineers and technologists, to expand educational establishments. This proved extremely difficult to achieve in practice, and mere statistics cannot measure the hasty mass-production of semi-qualified personnel which was rushed into the breach. For this purpose normal secondary education was totally disrupted for a few years (through mass conversion into emergency 'technicums'), as the following figures show:

Numbers in secondary schools

1928–9	977,787
1929–30	1,117,824
1930–31	383,658
1931–2	4,234
1932–3	1,243,272
1933–4	2,011,798

(SOURCE: *Kulturnoe stroitel'stvo SSSR* (Moscow, 1956), p. 81.)

But of course training schemes could barely touch the millions who were recruited, or fled, from the countryside. Soviet leaders and novelists now freely admit that much damage was done by sheer clumsiness: expensive imported machines were smashed by inexperienced labourers or unqualified substitute-engineers. Too often in the prevailing atmosphere there was suspicion of sabotage.

A considerable number of foreign specialists and skilled workers came to the U.S.S.R. in these years; some were under contract with foreign firms to help erect the new factories and teach their Russian colleagues. Others went to Russia as idealistic volunteers, or because of the growth of mass unemployment in the West as the depression deepened. In some cases, these men too fell under suspicion. In 1933 there took place the 'Metro-Vickers trial', in which British specialists employed in Russia by this firm were found guilty of sabotage.

The troubles of Soviet industry and construction were intensified by a phenomenally high labour turnover. The peasant-workers, bewildered by their new surroundings, often short of food and adequate lodging, rootless and unsettled, wandered about in search of better things. Not for them the 'pathos' of the great construction of socialism. The table given below tells its own tale.

Labour turnover, all large-scale industry (per 100 employees)

	1929	1930	1931	1932
Entered employment	122·4	176·4	151·2	121·1
Left employment	115·2	152·4	136·8	135·3

Labour turnover, coal industry (per 100 employees)

	1928	1929	1930	1931	1932	1933	1934
Entered employment	140·4	201·6	307·2	232·8	185·4	129·2	90·7
Left employment	132·0	192·0	295·2	205·2	187·9	120·7	95·4

(SOURCES: *Sotsialisticheskoe stroitel'stvo SSSR* (1936), p. 530; *Trud v SSSR* (1935), p. 109.)

These figures mean that the average worker in the coal industry, to take the worst example, left his employment almost three times during 1930. No wonder Stalin saw the need to discourage free movement and encourage the acquisition of skills.

Many workers already in jobs had retained close family links with their villages, and one effect of collectivization was to cause many of them, miners in particular, to rush home to see to their families' situation. This caused some short-term disruption in these years, and explains why 1930 shows the worst figures.

The great influx of extra labour into industry and building had a number of consequences.

Firstly, housing and amenities in towns were greatly overtaxed. Trams (there were few alternative means of urban transport then) were packed to suffocation. There was a shortage of water, shops, catering facilities. The plan had envisaged an increase of the housing space in cities from 160 million to 213 million square metres (optimal variant). It was recognized that this would provide only a small increase in the very inadequate living space, from 5·7 to 6·3 square metres per head. But housing was cut to make room for an increased factory-building programme, and/or because of shortage of materials. The total housing space in 1932 proved to be not 213 but only 185 million square metres, an increase of 16 per cent not 33 per cent. Thus the urban population rose by much more than planned, housing space by much less, and overcrowding became worse than ever. Neglect of maintenance made conditions even less bearable. No Soviet citizen is likely to deny that lack of space, shared kitchens, the crowding of several families per apartment, often divided rooms, were the lot of the majority of the urban population for over a generation, and that this was a source of a great deal of human misery.

Secondly, there was a particularly large increase in the employment of women, with far-reaching social consequences. Certain professions, notably medicine and teaching, became almost wholly feminine preserves, while tough ex-peasant women provided a large part of the unskilled labour force.

Thirdly, the great increase in employment had, along with other factors, the effect of making nonsense of the plan's cost estimates, and so contributed to inflation (whose manifestations will be described in the following pages).

It must also be mentioned that in this period prisoners and deportees, especially the latter, emerged as significant factors in the economic life of the country. For example, only a small portion of the inhabitants of the new town of Karaganda went there of their own volition.

INFLATIONARY PRESSURES

A considerable contribution to inflation was also made by the very large increase in the volume of investments. Due partly to expensive errors and partly to underestimation of costs, as well as to upward amendment of earlier plans, total state investments amounted in four and a half years to 112 per cent of the five-year plan, while heavy industry's were 145 per cent of the plan. Of course, these figures too were probably affected by statistical 'inflation', especially of the value of installed machinery. However, the very large excess number of building workers, shown in the table on page 195, together with the known fact that new projects were added to the plan and others speeded up, would be consistent with an 'over-fulfilment', though a downward allowance must be made for disinvestment (unplanned) in the private agriculture and handicrafts section. 'Heavy industry' investments have the effect of increasing incomes without any corresponding rise in output of consumers' goods and services.

Another major cause of inflationary pressure (and partly caused by this pressure) was the rise in wages. Demand for labour rocketed, as planners and managers tried to achieve the impossible. They drew in extra labour to make up for inefficiency or planners' over-optimism, they bid against each other for whatever labour was available. In the face of a drive to fulfil plans in time 'at any cost', financial controls were ineffective. Wage bills rose high above expectations. Costs, naturally, rose too. The original plan envisaged, as was shown on page 146, a reduction in costs, also in wholesale and retail prices. The average wage of workers in large-scale industry was planned to rise from 66·90 (1927–8) to 98·28 roubles per month (1932–3, optimal variant), i.e. by 46·9 per cent, but prices were to fall by 10 per cent, so real wages would increase by about 52 per cent. The actual real wages will be examined later on. It is enough at present to note what happened to money wages: in 'census industry' they reached 123 roubles per month per worker in 1932.

Over the entire economy, average wages exceeded plan by 44 per cent, and the total payroll exceeded expectations by much more than this (since the numbers in employment increased very rapidly). True, many peasants suffered a loss in income because of the consequences of collectivization. However, money incomes in total far exceeded plan, but goods and services available were less than had been hoped for.

More will be said later about how the plan was in fact financed. For the present, let us briefly examine the government's price policies, and the effect of these on our task of assessing the impact of the events of the time.

It has already been pointed out that the government, in launching the plan, made much of the need to reduce prices, or at least to oppose price increases. Rationing had been introduced in cities in the winter of 1928–9, and ration prices were kept at low levels. It extended to ever more commodities, including textiles and clothing, as shortages grew more acute. As for industrial materials and fuels, these were strictly price-controlled, and, as costs rose, subsidies were paid out of the budget. Almost every industrial producer's goods came under ever stricter control on the part of the planners, as part of the effort to give priority to the heavy-industry projects which were regarded as the keys to the success of the entire plan.

Demand for materials and fuel could, in principle, be kept down by administrative allocation, though over-taut planning did cause frequent supply and transport difficulties. However, the people had increasing sums of money in their pockets, and less to spend it on. A black (or 'grey') market duly flourished, despite efforts to suppress it, and decrees (e.g. on 22 August 1932) speak of the 'greater frequency of cases of speculation'. Peasant bazaars were at times closed (or their opening made conditional upon meeting the state's procurement quotas), at other times tolerated. As food shortages became acute, off-ration prices in this limited free market became exceedingly high, varying greatly from place to place because of the elimination of a (legal) professional trading class, and shortages of transport. The invaluable Malafeyev has given us important data on this period, and the next pages are based largely on his evidence (he quotes liberally from archive material which is not otherwise accessible).

PRICES, TAXES AND PURCHASING POWER

We have already noted that by 1929 a wide gap had opened between official prices and those charged by private traders. It should be added that until 1933 at least cooperatives in fact were able to evade price control, and even state enterprises found that they could make up for some of their excess wages bills by overcharging.

In early 1929 bread rationing had been introduced, and by 1932 40 million persons were supplied with bread from 'centralized sources', another 10 million were rationed out of local resources. Rationing spread by the end of 1929 to almost all foodstuffs, and then gradually to the rapidly disappearing manufactured consumers' goods. On 11 October 1931 a price committee attached to STO (on 1 April 1932 renamed 'committee for commodity funds and the regulation of trade') was to liquidate the remnants of 'private traders' speculation' and fix prices in the state and cooperative sector, 'aiming at a gradual price reduction'.[7] So the myth of falling prices was still alive, and in fact a few prices of rationed or scarce commodities were cut on 16 February 1930. But prices generally had already then begun to rise, and the state's urgent need for revenue led, on 2 September 1930, to the adoption of a decree on tax reform. Turnover tax came into operation on 1 October of that year, replacing a multitude of excise and *ad hoc* taxes. Price increases thereafter were to a great extent a function of increases in this tax (of which more will be said later on). It became a major means of mopping up surplus purchasing power, arising out of the large increase in incomes and the consequences of high investment expenditure and the disasters of agriculture. Turnover tax was included in the supply price (*otpusknaya tsena*) at which state industry 'released' products to non-official consumers.

However, the government at first tried to cushion the shock, and so ration prices were held at low levels for a time. By the middle of 1931, the resultant acute shortages even of rationed commodities led to the supply of some goods by *ordery* (authorization). The spread of rationing and administrative allocation (even of trousers to stated individuals) led, as in the period of war communism, to the spread of what Malafeyev calls 'liquidationist attitudes to trade, money, and finance. . . . There arose the absurd and harmful theory of the need to abolish money and to shift to direct products exchange. The supporters of this theory considered that our money is already close to becoming work tokens and really has become a merely nominal unit of account, that within the socialized sector it is already not money, while within the private sector it just "seems to be money".' There were cases of unofficial exchange of products, as when a Moscow factory 'exchanged iron and wire for clothing and furniture' with other enterprises.[8]

It is worth briefly dwelling on this odd resurgence of extreme-leftism which swept the country during the great leap forward. It affected much more than monetary theory. It led to a neglect of cost considerations generally, and to idealization of communal living even when, as so often happened, this was the inevitable consequence of overcrowding. These attitudes affected the view taken of economics in general, and even of statistics. After all, statistics is the study of random, unplanned, uncontrolled magnitudes. Such a word, so it was held, was unsuitable for the new circumstances of all-round, all-inclusive planning. So the central statistical administration was placed under Gosplan in 1930, and in December 1931 solemnly renamed the 'Central administration of national-economic accounting' (TsUNKhU was the abbreviation in common use). Note the stress on accounting, in line with Lenin's thoughts of 1917 vintage. It was not renamed 'statistical' until 1941. In line with the same philosophy, when the Commissariat of Trade was divided in 1930 its internal activities were placed under a Commissariat of 'Supplies', avoiding the tainted word 'trade'.[9]

To return to trade and prices, the government, in and after 1931, began to take strong measures against evasion of price controls and against unofficial product exchange, insisting on its own powers over resource allocation. On 10 May 1931 the cessation of rationing of manufactured consumers' goods was decreed, with, however, the continued use of *ordery* (authorizations) for clothing and footwear. By April 1932 food rationing was confined to bread, grains, meat, herrings, sugar, and fats. The share of state as distinct from cooperative trade rose sharply.

The practice of selling some state goods at high prices, which was originally unauthorized or illegal, became a means of increasing the state's own revenue. While the practice grew particularly rapidly in 1931 and 1932, it began as early as 1929, the first instance being the sale at high prices of 16 tons of sugar in July and October 1929.[10] Such sales became known as 'commercial', and in the next few years they grew to substantial dimensions. Special 'commercial' shops were opened on a large scale in 1932, and these sold rationed and scarce unrationed goods of many kinds at prices far above official prices, the difference going to the budget in the form of a special 'budgetary addition' over and above turnover tax.

There were then a number of different prices in existence for the same commodities, to wit:

(1) Retail prices of the 'normal urban fund' of goods mainly sold on ration coupons.

(2) Prices of the 'commercial fund', supposedly freely sold to all buyers, much higher than the 'normal'. These in turn were divided into 'average commercial' and 'higher commercial'. They were supposed to be available to any buyer who could afford them, without coupons, but 'in 1930–32 ... some of the scarcest commodities which were available for sale at commercial prices were frequently sold only on special authorizations (ordery), thus becoming in effect rationed. In the case of such exceptionally scarce items as good-quality cotton and wool fabrics, footwear, etc. consumers employed by various enterprises were temporarily attached to particular commercial shops, and they were issued with a kind of entry permit into this shop, with a limit on purchases.'[11] This refers to 'closed' shops, available only to specially favoured groups of the population. Such groups included workers of factories deemed to be important for the economy. So-called 'closed workers' cooperatives' were set up to supply them, and a contemporary writer emphasized the desirability of linking supplies by this route with the fulfilment of production plans and 'the struggle against absenteeism and flitting'.[12]

(3) There were also sales of both foodstuffs and manufactured goods 'in working-class areas' at what were described as 'average increased prices' (*srednepovyshenye tseny*), below commercial

and above ration prices. (One can imagine the length of the queues!)

(4) From 1933 there were also 'model general stores' (*univermagi*), with prices higher than commercial prices.

(5) *Torgsin* shops sold goods only for precious metals and foreign currency, both urgently needed for balance-of-payments reasons.

(6) Free market prices, either quasi-legal (open bazaars for peasant foodstuffs) or semi-illegal or black-market. The kolkhoz markets were at first (1931) supposed to observe 'the Soviet price policy', but this attempt to control prices did not work, and was abandoned in 1932.

All these categories had local variations, since, as can well be imagined, distribution was poorly organized.

Prices rose rapidly in 'commercial' trade, more rapidly still in the free market. It was not until 27 January 1932 that there was a (very sharp) general increase in retail prices of all (including rationed) goods. But the government, in its anxiety to appease the workers and in its evident indifference to peasant interests, fixed most prices in rural areas at 'commercial' levels. Thus at the beginning of 1932, i.e. before the big rise decreed in January, official retail prices had risen by 7·5 per cent in towns, 42 per cent in the country, since 1928.[13] By the first half of 1932 the prices charged in state and cooperatives had reached a level 76 per cent above 1927–8 levels, while 'private market' prices were 769 per cent of 1927–8 and rising very rapidly, particularly for foodstuffs, reflecting the general scarcity of practically everything. In the three months March to June 1932 rye flour increased in price by 45·7 per cent, rye bread by 35 per cent, meat by 125 per cent, potatoes by 66·7 per cent. The following table shows the general rise:[14]

| | Private market prices | | | | | 1932 Official price | Free price |
	1928	1929	1930	1931	June 1932	(kopeks per kilogram)	
Rye flour	100	225	350	525	2303	12·6	89·5
Potatoes	100	160	280	520	1552	–	–
Beef	100	125	359	663	1264	111	414
Butter	100	201	602	979	1078	502	1146
Eggs	100	134	330	572	868	–	–
Rye bread	–	–	–	–	–	10·5	111·0

In 1933 matters grew worse. 'Normal-fund' (i.e. ration) prices were sharply increased again on 1 January, and in many instances the rise again hit the villages hardest (e.g. kerosene, which in 1932 cost 18 kopeks a litre in towns and 30 kopeks in villages, was further increased in price by 27 per cent, for villages only).[15] Among the largest price increases were 116 per cent for vegetable oil (51 per cent in rural areas), 95 per cent for sugar (163 per cent in rural areas), 20 per cent for bread, and so on. There were increases also in state 'commercial' prices, and particularly severe increases in those prices of manufactured goods sold in rural areas, which had fallen behind the rise in commercial prices (e.g. wool cloth by 77 per cent, thread by 210 per cent). There were further substantial increases in official prices of foodstuffs in 1933: bread by 80 per cent (less in rural areas), butter by 55 per cent, eggs by 80 per cent, etc. Also the general cost of living rose because of an increased share of 'commercial' trade, at higher prices, in the total, from 12·9 per cent of the total 'planned commodities' in 1932 to 26·9 per cent in 1934. In 1933 a number of basic foodstuffs were sold at greatly increased 'commercial' prices, at which supply and demand did balance. In May 1933 the 'commercial' price of rye bread was 20 times above the official ration price, and was thus close to the real free-market price. This was the so-called 'fund of free sales', i.e. sales unencumbered by informal rationing restrictions. At the same date, the commercial price of sugar was 6 times the ration price, sunflower oil 14 times.[16]

Free-market prices continued to rise, and in 1933 were 48·2 per cent above the average of 1932 (bread and grains, potatoes and vegetables by over 60 per cent).[17] But the very substantial increases in 1933 in prices of rationed and other scarce commodities, a policy continued in 1934 when the rye bread price was doubled, reduced the gap between the various price categories and so prepared the way for the ultimate abolition of rationing and of the endlessly confusing multi-price system.

The proper assessment of living standards at this time is rendered almost impossible not only by the existence of rationing, price differences, and shortages, but also of queues, decline in quality, neglect of consumer requirements: 'It is well known that the closed shop system and rationing was frequently associated with a marked worsening in service to consumers, the widespread tendency to impose "compulsory" products, etc. The attitude of the sales staff . . . to the customer could be characterized by the slogan "take what you are given, do not hold up others in the queue, do not prevaricate and fuss".[18]

Therefore any figures comparing wages and prices are bound greatly to understate the decline in living standards. The party's official history prefers to overlook any decline, and for the statistically worst year, 1933–4, one looks in vain for any overall price index in any Soviet source. Malafeyev was allowed to publish only certain facts. He did calculate an index for 1932, for state and cooperative trade only. This came to 255 (1928 = 100), while average wages rose to 226. Thus the real wage index based on the above would come to 88·6.[19] But in 1932, as we have seen, free market prices had risen very much more rapidly. The correct real wage index, if we only knew it, would therefore be well below 88·6. Malafeyev here faced a problem with which one can sympathize. On the one hand, there is a clear indication that real wages had fallen, and this called for explanation, which he duly gave: 'This is understandable. In the period of industrialization of our country . . . the working class, the entire Soviet people, spared neither effort nor resources, and consciously made sacrifices, to drag the country out of its backwardness.'[20] On the other hand, he felt compelled to assert that there were no sacrifices, since 'in the period 1928–32 national income rose by 82 per cent, and the consumption fund in 1928–31 rose by 75·5 per cent'. This, it seems, proves the 'baselessness' of the assertion 'that socialist industrialization subordinates consumption to investment, or that [the party] applied in practice the Trotskyist slogan of primitive socialist accumulation which it rejected in theory'.[21]

In order to facilitate the mobilization of the working class for the 'great tasks of building socialism', and so as to avoid any organized protest about living standards or working conditions, the trade unions were turned 'face to production', i.e. were instructed to act primarily as organizers and mobilizers in the interests of plan fulfilment. Their old leaders, Tomsky and his friends, were dismissed as 'right-wing deviationists'. The protective role of the unions was greatly reduced. If their face was to production, their back was to their members. For many years there was not even the pretence of elections, and the all-union Trades Union Congress failed to meet at all in the period 1932–49. When, with the disappearance of unemployment, it was decided to wind up the People's Commissariat of Labour (in 1934) the trade union centre took over some of its functions and administered social insurance. It thus became for all practical purposes a branch of government. We shall see that not until after Stalin's death was there a revival of its more representative and protective functions. In the thirties the nominal powers of the unions to enforce protective legislation were little used.

What was the contemporary reaction to the harsh conditions of the time? The press and speeches made virtually ignored them altogether. Stalin, speaking in January 1933, calmly claimed that (money) wages had risen by 67 per cent since 1928, omitted all mention of prices, and asserted: 'But we have unquestionably attained a position where the material conditions of the workers and peasants are improving from year to year. The only ones who can have doubts on this score are the sworn enemies of the Soviet regime.'[22] To put it mildly, such language did not encourage independent research on standards of living!

The inclusion in the picture of the peasants would certainly make it worse, in particular in the period 1928–34. Nothing can better illustrate the meaninglessness, in any real terms, of the claim to an increase in the 'consumption fund' by 75 and more per cent. Admittedly, correction must be made for the elimination of unemployment, and for the fact that many of the poorer peasants were earning more as unskilled workers than they had as peasants, so that a falling real wage-level does not necessarily imply that all wage-earners were earning less than before. Social services did grow, too. Factory canteens provided inexpensive meals on a large scale. There was a big expansion of private cultivation on the outskirts of towns. Rents stayed very low. But many corrections require to be made in the opposite direction also. The fact still seems to be clear: 1933 was the culmination of the most precipitous peacetime decline in living standards known in recorded history. Mass misery and hunger reached dimensions whose demographic consequences have already been remarked upon.

Agricultural procurement prices remained very low. Thus prices paid by the state for wheat in the Ukraine reached 8·05 roubles per quintal (100 kilograms) in 1928–9 and remained unchanged through to 1934. In 1935 there was a 10 per cent increase. Prices paid for beef in 1931–2 were actually below the 1928–9 level, pork prices were a little above. Yet all goods purchased by peasants had their prices very greatly increased. However, some peasants were able to benefit from very high free-market prices. This was indeed 'primitive socialist accumulation'. In 1934, however, a notably different policy for industrial crops was introduced: thus cotton prices were raised from 30 to 115 roubles per centner. This was doubtless a necessary consequence of the rise in food prices, since the Central Asian peasants could not be expected to specialize on cotton unless they could eat.

INCENTIVES AND INEQUALITY

To get results a settled labour force was indispensable. A campaign was launched to penalize materially 'flitters' and absentees.

A series of rules and decrees between 1930 and 1933 punished absenteeism (common enough as a result of the habit of drink) by dismissal, eviction from factory housing and loss of various benefits. This led to a marked decline in absenteeism.[23] In 1931 Stalin made his famous critique of 'egalitarianism' (*uravnilovka*) in wages. The object was to reward those who chose to stay put and acquire skills. The result was a new wage-scale which made the difference in tariff rates between least skilled and highly skilled workers as high as 3·7:1. This had some economic justification at a time when skilled labour was exceedingly scarce. Stalin also encouraged a policy of higher pay and privileges for industrial cadres, and abandoned the old rule, established by Lenin, that party members should not earn more than a skilled worker did. Success would be achieved, in his view, by much greater material incentives. This would bring nearer the day of communist abundance. Until then, the slogan of egalitarianism was, in his view, 'petty-bourgeois'.

But non-material incentives were needed too. So there was greater emphasis than ever before on 'shock workers', 'shock brigades', the Red Banner, honours, 'socialist emulation'. Thus, for example, on 28 April 1930 there was a party declaration on 'shock workers', approving the 'shock' (*udarnyi*) movement, but warning against the tendency, which proved chronic, to 'bureaucratize' the movement by such things as 'chasing the formal fulfilment of plan figures, declaratory showmanship', and so on.[24] Workers and factories were thus urged ever onwards to new efforts, and rewarded both financially and morally.

Privileges, whether for selected categories of workers or for officials, tended at this period to take the form of 'perks': access to 'closed' shops, allocations of tolerable housing, a permit to buy a good suit, and so forth. Under conditions of universal shortages, money alone could not do much, without that something extra which Authority could provide or allow. In this sort of situation abuses were quite unavoidable, and the entire relationship between officialdom and the citizenry was adversely affected. It was therefore entirely right to endeavour to abolish rationing and to bring prices up to a proper supply-and-demand balancing level, as was done in 1934–5, as, apart from greatly simplifying trade and price control, the opportunities for rackets of all kinds greatly diminished in scope. However, as shortages of necessities and luxuries alike were very severe, prices had to be very high in relation to the (inflated) wages. But this will be examined in the next chapter.

FINANCING OF GROWTH

It is time to turn to finance. How was the vast investment plan financed? In part by the sales of bonds, which became by this time almost compulsory, in that a recalcitrant citizen who failed to volunteer to buy bonds might well suffer unfortunate consequences. And then there was the printing press. Money in circulation increased as follows:

1928	1929	1930	1931	1932	1933	1934
		(milliard roubles, 1 January)				
1·7	2·0	2·8	4·3	5·7	8·4	6·9

(SOURCE: Malafeyev *Istoriya tsenoobrazovaniya v SSSR* (Moscow, 1964), p. 404.)

The fall in 1934 is surprising, and receives no explanation. The rise was resumed in subsequent years.

Last but not least, there were taxes. Little was obtained in these years by direct taxation of workers and employees, and while there was some revenue from penal taxes designed to drive individual peasants into the kolkhozes, and individual craftsmen into cooperatives, once there they were taxed fairly lightly. The bulk of revenue, after the tax reform of 1930, came from turnover tax, and to a lesser extent from the profits of state enterprises.

'Turnover tax' is a term which covered (and covers) some very different species of taxation, including in 1930–35, the following:

(a) Increased excise duties, especially on vodka, but also on salt, matches, etc.

(b) Taxes included in the industry's selling (disposal) price, on a wide range of industrial consumers' goods, from textiles to sewing machines. These taxes were particularly high in these years on goods destined for 'commercial' sales. There were also 'budget additions' of various categories, so-called *natsenki*, which were added to the price of some goods when allocated for retail sales.

(c) The tax which arose out of the difference between the low procurement price of agricultural produce and the much higher price at which raw or processed foodstuffs were ultimately sold to the consumers (after allowing, of course, for transport and handling charges). Thus in 1933 the grain procurement organization (Zagotzerno) paid roughly 5·70 roubles for a centner of rye, and sold this rye to state flour mills at 22·20 roubles, the budget benefiting from the difference. After a series of retail price increases the situation at the end of 1934 was as follows: the Zagotzerno selling price of rye was 84 roubles per centner, of which 66 roubles was turnover tax. The price cited for wheat was 104 roubles, with turnover tax taking an even higher proportion, 89 roubles. The price of coarse wheat flour was 216 roubles per centner, of which turnover tax took 195·50 roubles.[25] Anti-Soviet propagandists have made much of the very high percentage of the tax ('tax on necessities' of astronomic dimensions), but this is to miss the point, which is the very low level of payments to the peasant producers. The burden of tax was primarily on them: compulsory procurements at low prices contained a substantial element of tax, which showed up in the budget as part of turnover tax revenue. Its importance may be demonstrated by the fact that in 1935 procurement organizations contributed 24 milliard roubles to the budget,[26] within a turnover tax yield of 52·2 milliards, out of a total revenue in that year of 75 milliards.

Agriculture thus made a decisive contribution to the financing of the plan.

Very serious difficulties faced the government in finding the means to finance purchases from abroad. As already mentioned, terms of trade moved sharply against the Soviet Union, in common with other primary producing countries, as the Great Depression developed in the West. Exports of wheat were resumed, and other foodstuffs were exported, including butter (31,000 tons in 1932), despite acute shortages at home. Accusations of dumping, boycotts, protests against Soviet timber because of forced labour, obstruction of Soviet oil sales, all handicapped the efforts of Soviet salesmen. Every attempt was made to obtain credits from abroad, and, with the depression in full swing, Russian orders were very welcome. So a certain amount of aid was rendered from the 'capitalist' West, in the form of loan finance as well as know-how, though of course aid was not the motive of these transactions. In his interview with Walter Duranty, Stalin stated that in 1931 total Soviet indebtedness on various credit accounts amounted to 1,400 million roubles. Assuming that this was expressed in the then official exchange rate, this amounted to 721 million dollars. These were short-term credits, and every effort was made to repay them quickly. By the end of 1933 the sum of only 450 million roubles was outstanding.[27] But these repayments required sacrifices. Apart from exporting essentials badly needed within Russia, it proved necessary to sell art treasures, to encourage the people to disgorge any stocks of gold and foreign currency (via the *Torgsin* shops, see page 204) and so on.

REORGANIZATION OF PLANNING

It is now necessary to take a brief look at the *organizational changes* which accompanied the first five-year plan.

Firstly, the nature and situation of the enterprises were fundamentally altered. On 5 September 1929 the principle of one-man management was asserted in a decision of the central committee. The factory party organization was told not to interfere with the director's responsibility, the trade unions were to be 'energetic organizers of production activity and of the initiative of the labouring masses', even while upholding 'the everyday cultural, amenity and economic needs of the workers'. The director was, in principle, in sole charge. On 5 December 1929 a further decision (still by the Central Committee, a *de facto* legislator in economic affairs, now and later) laid down that the enterprise was to be 'the basic unit of administering industry', and the principle of *khozraschyot* (economic or commercial accounting) was to be based on the enterprise, which was to have appropriate financial autonomy and a legal personality. This followed the logic of a decision taken on 29 June 1927 to increase the powers of the enterprise, a decision which was successfully resisted by the trusts. Now the trusts lost their direct powers over enterprises, and were told to concentrate on 'technical direction, rationalization and reconstruction'. VSNKH's industrial glavki were abolished, and their functions in controlling industry were transferred to so-called *obyedineniya* (associations) which were supposed to be based on the wholesaling 'syndicates'. A similar change had already been made in the textile industry in 1927 (see page 100).

In the years that followed, the immense new burden of work which fell upon the central planning organs caused repeated further reorganization, some 'legal', some *ad hoc*. A very large number of decisions emanated from the central committee and its officials, or were taken by its plenipotentiaries on the spot. Control over resources allocation became ever more complicated, and the need to systematize it led to a re-examination of the relative functions of VSNKH and Gosplan. The latter was no longer engaged only in prognosis and the drafting of 'control figures', since plans had become orders to act. On 23 January

1930 the central statistical administration was placed under Gosplan. Gosplan came directly under the Council of People's Commissars (not the STO) by decree of 3 February 1931. Gosplan increasingly took over the functions of planning from VSNKH.

VSNKH meanwhile was undergoing a painful internal transformation. The thirty-five associations set up in December 1929 were supposed to 'plan production, plan and control investment, technical policy, supplies and disposals, commercial and fiancial activities, labour, training and distribution of cadres, appointment and dismissals of managerial staffs'.[28] Some of these associations had as their sole task the full control of enterprises and trusts of all-union significance, but some of them controlled enterprises which were within the competence of the republics or local soviets. Confusion arose over the question of responsibility of the associations both with regard to republican organs and in respect of the presidium of VSNKH itself, which retained a wide range of powers. It exercised them with the advice of a reorganized and strengthened version of Promplan, the clumsy title of which was: 'Central planning-technical-economic administration.' As the extent of the planning grew, and industry became more complex and difficulties accumulated, this structure became unworkable. Some of the associations controlled far too many enterprises: thus *Soyuzlesprom,* the association responsible for timber, had under it one thousand production units. So a process of sub-division started. The number of associations grew. VSNKH's coordinating and controlling function was rendered ever more difficult. By the end of 1930 the newly-appointed head of VSNKH, Ordzhonikidze, decided to reorganize it yet again. This time there grew up a series of 'sectors' (e.g. Metallurgical, Chemical, Fuel, etc.), within each of which there were several associations. The 'Central planning-technical-economic administration' was abolished, and in its place there were created 'functional sectors': for plans, accounting and auditing, finance, supplies and disposals, cadres, labour, investments. The functional departments interfered directly with the activities of enterprises and trusts, thereby causing much crossing of wires; this was later denounced as the disease of 'functionalism' (*funktsionalka*). In 1931, industrial 'sectors' of VSNKн began once more to be called glavki. But all this re-labelling of the same officials did not save VSNKн. On 5 January 1932 its work was divided. The largest part was to be handled by the People's Commissariat of Heavy Industry. Other industrial commissariats took over light industry and timber and woodworking. Food and sugar production had already been handed over to the Commissariat of Internal and External Trade on 17 June 1930.[29] By 1932 this had been split, and the food and sugar industry found itself within the Supplies Commissariat (Narkomsnab), which also ran internal trade. Later on in the thirties, as we shall see, the number of these People's Commissariats greatly increased by sub-division.

Chief departments (glavki) within the commissariats acquired direct powers over planning and administering 'their' enterprises, either directly or through republican organs. By the end of 1932 the *obyedineniya* (associations) had become superfluous and had almost wholly vanished from the scene. So was born the 'ministerial' system of industrial administration (People's Commissariats were renamed Ministries in 1946), which survived with few changes, other than growth by sub-division, until 1957. As a result of the fact that no one body at this level now covered all industry, Gosplan's role as coordinator naturally increased, and became the key element in the planning system under the authority of the party and government. A more detailed analysis of the planning system in the thirties must be left to the next chapter.

There was also a reform of credit and banking. As we have seen, in the NEP period state trusts could make agreements and grant credits with one another on their own initiative. In 1930–33 there were cases of 'products exchange' between state enterprises. Evidently, no effective control over the activities of sub-units of the state economy would be possible so long as they could obtain and spend money in unplanned ways. So on 30 January 1930 a decree introduced 'direct bank credit' through the State Bank, and abolished inter-enterprise and inter-trust credits. On 23 May 1930 came a decree on 'a unified financial plan' incorporating the finances of 'all financial resources of the socialized sector of the economy'. On the same day it was decided that investment allocations from the state budget to socialized enterprises were to be non-returnable, these being distinguished from short-term credits granted by the Bank to cover temporary needs for funds (e.g. to cover the gap between outlays and receipts of payment for goods). A series of further measures during 1931[30] tightened up control over bank credits, linking them more closely with both the output plan and with contracts signed with the customers designated by the plan. A more thorough distinction was made between 'own' working capital and that financed by bank credits. This was 'control by the rouble', but in the peculiar circumstances of 1929–33, both the expansion of short-term credits and budgetary grants (in the form of subsidies to cover operating losses) helped to finance inflationary increases in wages.

THE PARTY'S RULE AND METHODS

A word is necessary about the rule of the party, its plenipotentiaries, its direct interventions. An example may be taken from the railways and agriculture. Here, for instance, is Postyshev at the seventeenth party congress:

Repressions were, in these decisive years, the effective method of 'leadership' in many party organizations of the Ukraine. To characterize these ... I will give the following example, typical of very many regions of the Ukraine. In the Novograd-Volynsk region, as in other regions, there was set up the operational *troika* for sowing, consisting of the secretary of the regional [party] committee, the plenipotentiary from the central committee and the head of the regional OGPU, with the participation at the troika meetings of the regional procurator, the head of militia and of the control commission. Here is an extract of the minutes of a meeting of the troika: 'Remove and send for trial the chairman of the retail cooperative. Kondratiev to select a new chairman, and the militia commander to arrange [*oformit'* – in proper form] for the trial. Dismiss the chairman of the village soviet and put him on trial. Members of the village soviet and all rural organizations to be severely reprimanded through the press. Deprive Kolkhoz No. 2 of all favourable treatment. Remove the bureau of the [party] cell.' ... This is the undisguised arbitrariness which in these decisive years became the predominant form. ...[31]

Voroshilov, in his speech, referred to similar arbitrariness by the heads of the political departments (*politotdely*) on the railways: 'When we read that [a polittotdel man] arrived at the depot, sacked some, appointed others, moved wagons, chased out locomotives, it becomes both funny and sad. If this is to be the normal work of politotdely, if they will replace the men in charge, then we will not put transport right, we will further disorganize it.'[32] An example of the (positive) effect of central committee intervention was mentioned at the congress by Zimin: 'Before the decision of the C.C., . . . there were a mere fourteen engineers and 141 technicians to work on repairs to all the rail wagons of our country. After the C.C. decisions there were thrown (*brosheno*) there 450 engineers and 1,550 technicians.'[33] And Voroshilov ended *his* speech by declaring: 'Since Comrade Stalin has now taken up transport questions, then, comrades, joking is superfluous.'[34]

No account of economic affairs in these years would be complete without a reference to the changed 'tone'. The general coarsening of life went far beyond merely bad language, table-thumping by boorish commissars, overcrowded communal dwellings and the like. There was a decline not only of living standards but also in standards of discussion, and in the substance as well as the form of orders given. In the twenties too, needless to say, politics and economics were interlinked and strong words were used in argument. But by the early thirties real argument ceased, to give way to abuse. Thus two unfortunates called Vol'f and Kovarski expressed views in 1933 favouring greater autonomy for farms, thus 'underestimating the role of the proletariat in collectivization and the leading role of the state in kolkhozes'. Perhaps they were wrong. But in a leading academic journal they were accused of being enemy agents, wreckers, of 'infecting horses with meningitis'.[35] Another example is worth citing at length. It concerns rail transport, which, as already stated, proved unable to cope with the vast increase in traffic, in 1933 in particular. This led to a major reorganization and the appointment as commissar of Kaganovich, the very embodiment of the 'trouble-shooter', and 'shooter' is to be taken literally in this instance. 'Political departments' were set up to 'unmask and smash sabotage'. The crisis on the railways was real enough. In the first half of 1933 only 85·5 per cent of the traffic plan had been fulfilled and the tonnage carried had fallen compared with the analogous period of 1932. Oil and coal could not be moved, contributing to an acute fuel crisis. Among the causes were shortages of spare parts, lack of time for maintenance, lack of skilled men.[36] But above all, sabotage was suspected. The trials in 1930–31 of an alleged 'Industrial party', a 'Peasant party', a 'Menshevik bureau', all spread a feeling of insecurity and fear, particularly of sabotage, and placed almost all specialists under suspicion. Stalin asked for the rapid training of reliable specialists of unimpeachable origins, urged communists to 'master technique'. Meanwhile the following appeared in a leading article in the organ of Gosplan – hardly a rabble-rousing publication:

Until now 'many party cells show class blindness and do not un-mask class enemies who creep into the railways and attempt to wreck and disrupt'. As a result, class enemies, white-guardists, kulaks, still have the opportunity here and there on the railways to creep into 'modest' and 'insignificant' jobs, like those of greasers, and, 'silently sapping', they carry on their wrecking, becoming organizers of crashes and accidents, destroying essential parts of the railway and disorgani-zing its work. Those who openly disrupt the carrying through of essential party and government decisions are direct collaborators and agents of the class enemy. . . . Direct trickery by saboteurs of party and government [measures] reaches the proportions of a pillar of Hercules, as is quite clearly shown by the case of degenerative and criminal eye-wash by trade-union falsifiers at the Osnova station of the Southern railway line, who in one night [wait for it!] fabricated twenty-four false reports of workers' meetings, supposedly organized to carry through the decisions [of the party] on transport. To raise the level of watchful-ness over the class enemy which has crawled into transport, to show up and unmask every kind of open and hidden saboteurs so that, with every available means which only the proletarian dictatorship pos-sesses, their criminal activities can be ended, that is the duty of every communist . . .[37] [and so on, *ad nauseam*].

This purple passage is not reproduced here as a piece of anti-communist propaganda, or for a cheap laugh. It was typical of the time, and this pattern of thought and action deeply affected economic administration at all levels. To omit this is to leave out a significant part of history, an essential part of the political-sociological background to the great purge.

The same language extended to economics, and, as already pointed out, measures against alleged saboteurs extended also to the economists. The same issue of the journal from which the above passage was quoted contains an abusive 'review' of a work by Strumilin, who was a leading party economist. He was accused, *inter alia*, of apologetics for Trotskyism and Bukharinism, and it seems lucky that he was not swept away with so many others, and has in fact lived to celebrate in honour his ninetieth birthday while these pages are being written in 1968. The critics accused him of 'failing to unmask the bourgeois wrecking nature of the [modest] first drafts of the five-year plan'. These were 'deliberate minimalism of the wrecker-planners'. That they could have made honest mistakes is out of the question: 'The class nature and tactics of the wreckers are analysed superficially by the author, and do not in any way arm the party and workers in its battle against wreckers.' Even exceedingly ambitious plans were deliberate wrecking, according to the reviewer, the object being 'to create deep and artificial disproportions'. And so on. Therefore, since Strumilin allowed for the possibility of error and real difference of opinion, his book was 'harmful . . . vulgar, opportunistic from beginning to end', the author is a 'purveyor of anti-Marxist theories', etc.[38] Many less eminent and well-connected economists found themselves pilloried as enemy agents and were never seen again.

It is perhaps clear by now that serious discussion was hardly possible. We have already seen Stalin's own contribution to its suppression in his totally false statements on living standards. He took time off his numerous duties in 1931 to write a slashing attack on the 'rotten liberalism' of the editors of a leading party journal for allowing discussion of wrong and harmful ideas.[39] The semi-literate name-callers whom I have quoted were following the party line to the best of their doubtless limited ability. As a psychological fact, the lack of ability explains much. After a cultural meeting in Leningrad, in which a literary commissar (Averbakh) had used tough and uncompromising language, the excellent humorist Zoshchenko said to the (then young) novelist Kaverin: 'Such devilish energy. If only there was talent! But there is no talent, and hence the other qualities.'[40]

RETURN TO SOBER PLANNING

In 1932 the sky was still the limit. On 30 January the seventeenth party conference met, and adopted Kuibyshev's proposals as a basis for a second five-year plan. The following were the targets for 1937, with the actual 1937 figures in brackets:

Electricity (milliard Kwhs)	100	(36)
Pig iron (million tons)	22	(14·5)
Coal (million tons	250	(128)
Oil (million tons)	80–90	(28)

The second plan, when finally adopted, was more modest, as we shall see. By then the fever had abated. The terrible events of 1933 may have had their influence, by a kind of shock therapy. 1933 saw a sharp (unplanned) *reduction* in investments; these were 14 per cent below 1932 levels. The slogan 'consolidate' was adopted. Consolidate collective farms, increase productivity and yields, learn, become more efficient, improve quality, pull up the lagging consumers' goods industries. 'Technique decides everything.' The worst was over. In 1934 agricultural output, and the livestock population, began its painful climb upwards. Free market prices began to fall from the astronomical heights of 1933, and some state commercial prices could at last be reduced. In 1935 rationing of foodstuffs was abolished, by stages, though the new prices were very high. Stalin launched the slogan: 'Life has become better, comrades, life has become more joyous.'

In January 1934 the seventeenth congress met, [41] the so-called 'congress of victors'. Over whom, it may be asked? Over the bourgeoisie, petty and less petty. The peasants were nearly all collectivized. A great new industry was being built, and appropriate statistical presentation emphasized the undoubted successes and slurred over failures. Livestock losses were admitted, but were mainly due to kulak ferocity. Living standards? Well, people are living better. The monolithic party greeted Comrade Stalin and his report with tumultuous applause. And yet – as is now becoming clear – there was a secret session at which it was decided to reduce Stalin's powers, to relax the terror, to bring forward S. M. Kirov, himself an earthy and tough party official, as a symbol of a new mood. This cannot be documented yet, but is accepted by most historians. To prove this point is not part of an economic history. But the effect of the harsh years seems to have been to predispose the party, while maintaining a unanimous façade to the world, to change to a different gear.

The seventeenth congress was notable for the appearance of members of various oppositions, who were penitent and appealed to everyone to support the great Comrade Stalin. Among them was Preobrazhensky, the principal theoretician of Trotskyism, the popularizer of the theory of 'primitive socialist accumulation'.

After the compulsory praises of Stalin and criticisms of the defeated oppositionists, he came to the key point of his speech:

> You know that my chief error consisted in ... elaborating the law of primitive socialist accumulation. ... I thought that by exploiting the peasants, by concentrating resources of the peasant economy in the hands of the state, one could build a socialist industry and develop industrialization. This was a crude analogy with the epoch of primitive capitalist accumulation. ... Events totally disproved my ideas, and those prognoses triumphed which Lenin made and which the party made reality under the leadership of Comrade Stalin. Collectivization, that was the point! Did I anticipate collectivization? I did not. ... Collectivization of the peasants is the greatest of our conquests. ... You know that neither Marx nor Engels, who wrote a great deal about problems of socialism in the village, had any definite idea how the transformation would come about. You know that Engels thought that this would be a fairly lengthy evolutionary process. In this question, what was needed was the greater far-sightedness of Comrade Stalin, his great courage in the formulation of new tasks, the greatest hardness in carrying them out, the deepest understanding of the epoch and of the relationship of class forces. ... This was the greatest of the overturns (*perevorotov*) known to history.[42]

Preobrazhensky was surely expecting at least some of his audience to see the point. Stalin had 'exploited the peasants by accumulating the resources of the peasant economy in the hands of the state'. Of course he had! But Preobrazhensky had not seen *forced* collectivization as a way out. He did, as is documented, advocate collectivization, but he took it for granted, as Engels had done, that it would be 'a fairly lengthy evolutionary process'!

Tomsky, speaking for the penitents of the right opposition, made a valid point with a certain amount of dignity. 'We did not see that the revolution should take place other than on the basis of NEP, on the basis of concessions [to kulaks, etc.], on the basis of market relations. . . . We saw only the assimilation, and re-education of the kulak, i.e. that crude but apt phrase which Bukharin had used: the growing of the kulak into socialism.'[43]

The majority of the Congress booed the ex-oppositionists, and affirmed their faith in the achievements of the past years. They may well have felt a great sense of relief that the dreadful job had been done, that now things would have to get better. Kirov in his speech expressed a mood of genuine confidence and exultancy: 'Our successes are really immense. The devil take it, to speak frankly, one so wants to live and live! After all, look and see what is going on around us. It's a fact!'[44]

Before the end of the year, Kirov was to be assassinated, and soon the wave of terror would sweep over Russia; it would engulf most of the delegates to 'the congress of victors'.

Finally, looking back at the years of the great leap, the following are some observations.

The errors were vast, the cost immense. Could it have been done otherwise? In one sense, the answer must be 'yes'. No one can seriously assert that all the mistakes, crimes, excesses of the period were inevitable and unavoidable. But, so far as industrial planning is concerned, one could with some reason resurrect Lenin's argument about 'Port Arthur' (see page 120). Under Stalin's leadership, an assault was launched against the fortress, defended by class enemies, and on objective obstacles to rapid industrial growth. The assault succeeded in part, failed in some sectors, but failures could be said to be inherent in the process of learning. The later improvements in planning technique were based on lessons learnt in the course of storming the heavens.

Stalin could also bring up in his own defence the need to create military-industrial capacity as quickly as possible. Thus the Urals-Kuznetsk combine was very expensive, and required a great deal of capital, but where would the Russian army have been in 1942 without a Urals-Siberian metallurgical base?

However:

(a) The attempt to go much too fast went altogether too far. The sacrifices imposed were on a scale unparalleled in history in time of peace. The resultant bitterness, disloyalty, repression, also involved a heavy cost, including a military cost. Some soldiers had no difficulty in surrendering, some villages greeted the Germans as deliverers from tyranny (until they learnt better).

(b) The excesses in policy and in plan were in large part due to the deliberate suppression of all opposition, and in particular the strong tendency to regard warnings against overdoing it, whether coming from fellow-communists or non-party specialists, as right-wing deviation at best, or maybe wrecking and diversionism. Sound advice from imaginative economists and other specialists was ignored, the advisers too often dismissed or imprisoned.

(c) Collectivization had disastrous effects on agriculture, and on life generally in and out of the village, from which recovery was slow and painful. The deportation of the bulk of skilled and ambitious peasants did great harm, perhaps never undone. No other communist country followed this practice. Justified on the plea of danger from the terrible kulaks, the hostility of the state made the kulaks into enemies, and so this was a kind of self-fulfilling prophecy. As Lewin put it, the kulak who tried to escape his fate by merging himself in the kolkhoz was 'not an inveterate agent of capitalism, but a frightened desperate creature'. Treated as pariahs, deported, robbed, their families left destitute when not deported with them, their children expelled from school, who should be surprised if some of them tried to kill commissars? (A colleague who visited the Ukraine as late as 1936 told me that he saw two children starving, apparently dying, in a village. He asked about this. 'Kulak children', he was told, indifferently.) It was an odd class struggle, rather one-sided.

(d) The administrative methods which were devised, in towns and country alike, were brutal and went hand in hand with increasing police terror.

Yet a great industry was built. It should be added that morale in Russia itself, and the impact of her achievements internationally, were affected by the co-incidence of the Great Depression. Russia was growing, the Western capitalist system was apparently collapsing, with massive unemployment and social disruption, culminating in America with the paralysis of 1932–3, in Germany with six million unemployed and the triumph of Hitler. The worst period of Russia's own crisis coincided with crashes and bankruptcies in the 'capitalist' world, and at least Russia's troubles could be seen as growing pains. It is as well to remember that the West was no sort of model for Russia or anyone else to follow in the years we have been describing and analysing in the present chapter.

9. From Leap Forward to War

THE SECOND FIVE-YEAR PLAN

The present chapter covers the period in which the 'Stalin' economic model established itself. The convulsions which characterized the years of frantic industrialization and collectivization were replaced by more orderly and systematic methods of planning and administration. Sufferings and excesses were followed by a more tolerable life. True, Stalin's slogan, 'Life has become better, comrades, life has become more joyous', became an excuse for evading some disagreeable realities, since living standards were low in town and country alike, compared with 1928. None the less, a period of relaxation did undoubtedly begin in 1934, and a rapid improvement was maintained, in output of all kinds of industrial goods, for three years. Then came a sharp check to economic growth, and we will be discussing how far this can be attributed to a shift of resources into military preparation and to the impact of the police terror which reached its apogee in the years 1937–8.

But first we must return to the second five-year plan. In Chapter 8 it was shown that this plan, covering the period 1933–7, was originally drafted during the period of 'great leap-forward' psychology, and that the original targets were wild and utterly unrealistic. But by the end of 1932 it became clear that the economy was overstrained. 1933 was a year characterized not only by famine but also by the transport crisis, and by disequilibria and shortages so severe that it was necessary to call a halt. We saw that the volume of investment declined by 14·3 per cent in 1933, and gross industrial production, which had risen (according to the official claims) by an average of close to 20 per cent per annum in the years 1929–32, increased by only 5 per cent in 1933. Many key industries made no progress at all. It plainly became essential to redraft the five-year plan, and this was done during 1933, although this was supposed to be the first year of the plan period. The five-year plan was adopted by the seventeenth party congress, at its sessions in February 1934.

So 1933 was, in an important sense, the concluding year of a period, or perhaps a compulsory breathing-space occasioned by 'leap forward' overstrain. This is why many aspects of that year were considered in the previous chapter, despite the fact that it was the first year of the second five-year plan.

The plan as it finally emerged included the following target figures (the actual 1937 achievements are also given):

	1932 (actual)	1937 (plan)	1937 (actual)
National Income (1926–7 prices) (million roubles)	45,500	100,200	96,300
Gross industrial production (1926–7 prices) (million roubles)	43,300	92,712	95,500
of which: Producers' goods	23,100	45,528	55,200
Consumers' goods	20,200	47,184	40,300
Electricity (milliard Kwhs)	13·4	38·0	36·2
Coal (million tons)	64·3	152·5	128·0
Oil (million tons)	22·3	46·8	28·5
Pig iron (million tons)	6·2	16·0	14·5
Steel (million tons)	5·9	17·0	17·7
Rolling mill products (million tons)	4·3	13·0	13·0
Machine tools (thousands)	15·0	40·0	45·5
Cement (million tons)	3·5	7·5	5·5
Cotton fabrics (million metres)	2,720	5,100	3,448
Wool fabrics (million metres)	94·6	226·6	108·3
Leather footwear (million pairs)	82	180	183
Sugar (thousand tons)	828	2,500	2,421
Tractors (thousands) (15 h.p. units)	51·6	166·7	66·5
Fertilizer (million tons gross)	0·9	9·0*	3·2
Gross agricultural production (million roubles)	13,070	36,160	20,123
Grain harvest (million tons)	69·9	104·8	96·0
Employment, total (millions)†	22·94	28·91	26·99
Employment, industry (millions)	7·97	10·20	10·11
Average money wage (roubles per annum)‡	1,427	1,755	3,047§
Retail price index (1933 = 100)	100	65	180§
Volume of retail trade (1933 = 100)	100	250·7	150§

* Plan specified ten-fold increase over 1932.

† Total employed by state institutions and enterprises.

‡ Average pay of all employed persons.

(SOURCES: Five-year plan documents, and *Promyshlennost' SSSR* (Moscow, 1957). Items marked § from Malafeyev, *Istoriya tsenoobrazo-vaniya v SSSR* (Moscow, 1964), p. 208, 407.)

Attention must at once be drawn to certain misleading features of the above table. First of all, the plan fulfilment figures are highly suspect when they refer to any aggregate sum in roubles. For reasons already mentioned, these tend to overstate reality. Therefore the claim to have over-fulfilled the industrial plan in aggregate terms is not to be taken literally. How indeed can one reconcile it with the under-fulfilment of the plans for almost every major item, which emerges clearly from the table? Even if one allows for the large increase in arms production, and the very good showing of the machinery sector, something seems awry.

Secondly, certain individual figures are misleadingly low, others high. Thus the tractor output of 1937 (66,500 in 15 h.p. units) was very far below that of 1936. Presumably, had there not been a shift towards tank production, the original plan estimate would have been substantially over-fulfilled, since the 1936 figure (173,000, expressed in 15 h.p. units) was actually higher than had been intended for 1937. By contrast, the grain harvest in 1937 was abnormally high, due to very favourable weather, so in this instance the use of a single year's results overstates actual achievements.

But above all, the table conceals a very uneven pattern of growth. 1933 was a bad year. This was followed by three exceptionally favourable years, when all the evidence shows that there were great achievements in industry, construction and transport, though agricultural recovery was adversely affected by bad weather in 1936. In 1937 began a period of relative stagnation. This is shown particularly clearly in the key metallurgical sector. The figures were as follows:

	1932	1933	1934	1935	1936	1937	1938	1939
Steel (million tons)	5·93	6·89	9·69	12·59	16·40	17·73	18·06	17·56
Pig iron (million tons)	6·16	7·11	10·43	12·43	14·40	14·49	14·65	14·52

However, let us look at the second five-year plan first of all as a statement of intent. It was a more fully worked-out plan than the first, covering greater detail by industry and by region, and the document includes an impressive list of academic and other consultants involved, in 1932–3, in the work of preparation. The three guiding principles were: firstly, *consolidation*, or bringing into effective operation (*osvoeniye*); secondly, *mastering technique*, and thirdly, a special effort in the direction of *improving living standards*. The plan envisaged a larger increase in output of, and investment in, consumers' goods than producers' goods. Whatever may have been the ritual phrases about the good life, Stalin and his colleagues must have understood and appreciated the urgency of remedial action. In the first plan-period priority went almost exclusively to heavy industry. Now something would be done to redress the balance. Collectivization would be completed, but the collectivized peasants would live much better, with a very much larger number of private (as well as collective) livestock, and urban real wages would double, as a result of an increase in money wages and a simultaneous fall in retail prices, as shown in the table above. Optimistic assumptions were made about productivity: an increase by 63 per cent in large-scale industry. As money wages were to rise on average by only 23 per cent, costs of production were to fall by substantial percentages. Housing conditions would improve through an addition of 64 million square metres of space, which would increase the total space by about 40 per cent.

Reality proved very different. The plan was, evidently, altered drastically as the years went by, with a shift again towards heavy industry. The result was that the consumers' goods, housing and real wages plans were not fulfilled.

The basic reasons for this shift are not so easy to determine. They obviously include, as a major factor, the rise of Hitler. In 1933 the Nazis won in Germany and made no secret of their determination to make war on the Soviet Union. The Soviet armed forces were still poorly equipped, and the young heavy-industrial sector had to be switched over to manufacture not only arms but also the means of making arms of all kinds. The increase in arms expenditure is somewhat overstated in the table below, in that prices were rising. However, the growing share of defence in the budget is clear.

	1933	1934	1935	1936
		(millions of roubles)		
Total budget expenditure	42,080	55,444	73,571	92,480
of which: Defence	1,421	5,019	8,186	14,883
per cent of total	3·4	9·1	11·1	16·1

	1937	*1938*	*1939*	*1940*
		(millions of roubles)		
Total budget expenditure	106,238	124,038	153,299	174,350
of which: Defence	17,481	23,200	39,200	56,800
per cent of total	16·5	18·7	25·6	32·6

(SOURCE: *Sotsialisticheskoe narodnoe khozyaistvo v 1933–40* (Moscow, 1963), pp. 609–11 and K. Plotnikov, *Ocherki istorii byudzheta sovetskovo gosudarstva* (Moscow, 1954), p. 255.)

In the period 1933–8, according to the above source, the output of the defence industry increased by 286 per cent, and in 1934–9 the armed forces doubled in size, taking away many of the most productive men and machines.

However, while accepting this as a major cause of the recasting of the plan, one is forced to remember that this kind of revision occurred also in 1930, when Hitler was still a powerless rabble-rouser; and even under Khrushchev, many years later, one saw the tendency to take less seriously those plan objectives which concern the needs of citizens and to give top priority to heavy industry. In so far as the second five-year plan was over-sanguine in its forecasts, probably the consumers' goods and housing goals would have been sacrificed to some extent even if the Nazis had not come to power.

Be this as it may, the international situation was certainly responsible in some considerable part for the fact, to be further investigated, that real wages rose by 'at least 20 per cent' (Malafeyev) instead of doubling; that only forty-two, not sixty-four million square metres of housing was built, and so on. (Let it be said at once that real wages must in reality have risen by more than 20 per cent, since the index used by Malafeyev does not allow sufficiently for shortages and rationing in 1932.)

INDUSTRIAL GROWTH AND PRODUCTIVITY

Whatever the validity of certain official claims, it remains true beyond question that the second five-year plan period was one of impressive achievement, as is clear from the commodity statistics in the table on page 225. (It may be worth reminding readers that doubts about officially-claimed growth rates are confined, in industry at least, to aggregate totals such as national income or gross industrial output.) The excellent growth rates recorded in the years 1934–6 were in large part due to the completion of big plants begun in the hard years of 1929–33; for example, the Magnitogorsk, Kuznetsk, Zaporozhye, 'Azovstal'', Tula and Lipetsk metallurgical works were wholly or partly brought into operation during these years.[1] There was a most impressive leap forward, too, in the machinery and metal-working sector, which is insufficiently reflected in the above table. Both in volume and in degree of sophistication the advances recorded in these years did help to transform the whole balance of industry and to diminish very substantially the U.S.S.R.'s dependence on foreign countries for its capital goods. This dependence was very great indeed during the first five-year plan. In 1932, 338 million roubles' worth of machine tools were imported, and this represented 78 per cent of all machine tools installed in that year.[2] In 1936 and 1937 less than 10 per cent of machine tools were imported. Total purchases of all kinds of machinery and equipment fell substantially. This greatly relieved the strain on the balance of payments, and debts incurred in the preceding period were paid off.

Of course, the U.S.S.R. continued to import machinery (and still does so on a large scale). However, such imports could be concentrated on specialized equipment, or on items which the U.S.S.R. had not yet succeeded in developing. The basic tools of industrialization, and of arms production, were, by 1937, made in the Soviet Union.

Other branches of industry also showed very significant gains, though without fulfilling the very optimistic forecasts of the planners. Coal production increased substantially in the Kuzbas (Siberia), in Karaganda and in the Urals, and mechanization and training enabled productivity per miner to rise from 16·2 tons in 1932 to 26·9 tons in 1937.[3] A principal cause of the failure to fulfil the plan was delay in completing new mines. Electricity generation expanded particularly rapidly in 1934–6, the average growth rate being 26 per cent per annum. However, as also in metallurgy and in many other branches of industry, there was a marked slowdown in and after 1937. The least satisfactory among the fuel industries was oil. The Baku fields were unable to cope with their plans, and the rich Urals-Volga oil area, which had already been discovered, was developed much too slowly, owing to 'the weakness of the technical equipment in the area, a very large labour turnover due to poor housing conditions, and the backwardness of geological surveying and drilling work'.[4] Another extractive industry, iron ore, also lagged behind, and this was one cause of the subsequent difficulties in the production of ferrous metal.

Industrialization led, as was expected, to a very rapid expansion of demand for non-ferrous metals, and plans were made accordingly. While production did increase, demand grew faster, and imports continued. Copper production and import statistics were as follows:

	1932	1937
Production (thousand tons)	45	99
Imports (thousand tons)	12	65

(SOURCE: *Sots. nar. khoz.*, *1933–40*, pp. 216, 626.)

The supply of many non-ferrous metals continued to be a bottleneck.

The chemical industry grew, but here again the plan was not fulfilled. Branches of special interest to armaments production, and synthetic rubber, did a great deal better than did mineral fertilizer. There were inevitable difficulties in expanding this very backward branch of industry: lack of experience among management and labour alike, delays in construction, and so forth.

Consumers' goods industries also advanced, though, with the exception of footwear, at a pace well below plan. But the modernization of some textile and (especially) food-processing industries made big advances. New bakeries and ice-cream and meat-packing plants opened in many areas. In comparison with the hard years of the early thirties much more was available. More could be spent on imports of raw materials for the textile industry, but still not enough, especially in view of the disastrous losses of sheep in the U.S.S.R. This explains the very great shortfall, in relation to plan, of wool cloth production. At this period, too, a few consumer durables made their appearance, notably gramophones and bicycles.

The regional balance was changing. The Urals-Kuznetsk combine has been referred to already. Non-ferrous metal mining grew rapidly in Central Asia and Kazakhstan; an industrial complex was arising based on the Karaganda coalfield; new factories appeared in Baku and a sizeable modern textile mill was built in Tashkent. There was a conscious effort to develop the more backward national republics, even where it was not economically the most rational way of using scarce resources.

LABOUR

Productivity in the economy as a whole rose substantially in these years, at least up to 1937. As the table on page 225 shows, the number of employed persons in total, and also in industry, were below plan, in striking contrast to the situation in the first five-year plan. We must now examine the reasons for this favourable development.

Firstly, *incentives* were made much more effective, by the recasting of the wage scales and widening monetary differentials, then by the gradual abolition of rationing and the greater availability of goods to buy (see page 248). The very high prices of basic necessities, in and after 1934, stimulated harder work on piece-rates, so as to be able to make ends meet. This same problem also caused many more wives to go out to work.

Secondly, *training schemes* at all levels were gradually transforming the labour force. In 1930–33 unskilled peasant labour caused many breakdowns and was exceedingly inefficient, as well as unpunctual; labour turnover was very high. In 1933 a survey carried out in Moscow showed that only 17 per cent of those recruited to industry in that year had any skill or qualifications. Even in 1935, 60 per cent of workers in the great Urals machinery factory were unskilled.[5] Training schemes of many kinds were brought into being: apprenticeship on the job, courses run by enterprises, technical colleges run by industries, the so-called FZU or factory training schools, and so on. Some errors were committed because of the great shortage of trained instructors, and some courses were so accelerated that they produced poor-quality craftsmen; therefore training was lengthened, numbers were cut. In 1938 and 1939 FZU schools trained only 320,000 skilled workers, though the plan had specified 1,700,000.[6] This was partly due to the rapid expansion of 'academic' secondary education. In 1940 this led, as we shall see, to compulsory call-up of youth for training. But whatever the errors and omissions, a huge task was performed and a more efficient working class was created. Similarly, very great strides were made in higher technological education, with every effort made to educate the abler and more energetic workers. Since many of these had not had a secondary education, the results were not always happy, but by 1936 an expanded secondary school system was able to provide the recruits to the universities and technological institutes, and a decree regulating entry and demanding a more effective system of education dates from 23 June of that year. Stalin's slogan, 'Cadres decide everything' (speech of 4 May 1935), was designed to emphasize the urgency of the need for better-qualified and trained staffs. Their pay and privileges were enhanced. The following figures show (in thousands) the extent of the increase in their numbers in the course of twelve years:

	1928	1941 (1 January)
Total graduates in the national economy	233·0	908·0
of which: Engineers	47·0	289·9
Total of 'secondary specialists'	288·0	1492·2
of which: Technicians	51·0	320·1

But here too there was some reduction in entrants after a great rush forward earlier in the decade, which had affected quality. The number of students accepted for higher education was 245,000 in 1932, and only 158,300 in 1937.[7]

Thirdly, prominent among the reasons for improved productivity was the so-called Stakhanovite movement. This was a complex phenomenon which deserves careful attention. It emerged out of the 'socialist competition' and shock-brigade campaigns, which grew rapidly in scope and intensity during the first five-year plan period. They were linked with efforts to encourage enterprises to offer to achieve more technical progress, output, cost reduction and productivity than was originally proposed. These were the so-called counter-plans (*vstrechnye tekhpromfinplany*), and rank-and-file workers were brought into the task of ensuring that every enterprise and workshop under-took to go well beyond existing norms. It was in this context that a coal miner, Alexei Stakhanov, achieved output fourteen times greater than the norm, in September 1935. He did this not just by working hard, but also by intelligent use of unskilled auxiliaries. The party took this up, and 'Stakhanovism' spread rapidly to other branches of the economy. The December 1935 plenum of the party's central committee decided to urge planners and managers to revise upwards the existing technological and labour norms. Higher productivity 'would make the U.S.S.R. the most prosperous country in the world'. The effect of all this was to use equipment more fully and to intensify and rationalize labour. No one can doubt that there was ample scope for such things in 1935. Norms were low, having been based on the inefficiencies of the early thirties. 'Stakhanovism' was a means of dramatizing and publicizing a necessary change. Following a series of conferences held early in 1936 work norms were very sharply raised: by 30–40 per cent in engineering, 34 per cent in chemicals, 51 per cent in electricity generation, 26 per cent in coal mining, 25–29 per cent in the oil industry, and so on.[8] Despite this, and some reductions in piece-rates, total earnings rose on balance. No doubt the campaign had a positive effect on productivity.

However, some of the undertakings made were clearly exces-sive: e.g. to double the output of sulphuric acid without any new capital equipment or to 'double or treble our so-called projected utilization of productive capacity' (this last was said by the young Khrushchev).[9] Speed-up and higher work norms caused much strain in sectors in which there was no easy way of increas-ing output per man. Many a 'Stakhanovite' team was given favourable treatment by managers, who expected to benefit from the publicity which accompanied record-breaking, at the cost of neglecting the interests of the rest of the workers. Naturally, this was resented, and some 'Stakhanovites' were threatened or even beaten.

Work norms were further increased, especially in 1938 and 1939, but in most cases they were still surpassed, by fairly large percentages.

Discipline improved throughout the economy, due partly to stronger penalties for absenteeism; these were not yet criminal penalties, but took the form of administrative fines, threatened expulsion from the union (and therefore partial loss of social insurance benefits), dismissals with loss of housing space. Even failure to attend training courses outside working hours was held to be absenteeism.[10]

A particularly welcome improvement was recorded in the building industry. In the first five-year plan period the bulk of the labour force was exceedingly unskilled and footloose. There was a steady shift later in the decade towards the creation of permanent, as against *ad hoc*, building enterprises, and this helped to reduce labour turnover. This, and the introduction of more labour-saving equipment, made possible a substantial reduction in numbers employed on construction and installation: these fell from 2,289,000 in 1932 to 1,576,000 in 1937, yet the value of work done rose by almost 45 per cent.[11]

TRANSPORT

Productivity increased impressively in rail transport, which had been a very weak spot in 1933 and 1934. Due to better organization and equipment (bigger locomotives and wagons, better track, etc.), a very marked improvement was achieved, from a total of 239,400 to 356,700 ton-kilometre equivalents per railway employee in the period 1933 to 1937.[12] Yet even as late as 1936 there was a labour turnover of 103·8 per cent, i.e. the average worker left after less than a year.

The railways had to bear a particularly heavy burden. Owing to lack of roads, and the fact that rivers froze in winter, plus the very great distance over which it was necessary to carry such bulky items as coal and ores, the railway system was the key to the entire plan. Lack of modern rolling-stock, the poor quality of rails, inadequate ballast, made things very difficult. In 1932 the average speed of goods trains was less than nine miles per hour. The great efforts made did produce some very creditable results. Responsibility for the freight plan was centralized in 1934, in the People's Commissariat of Transport, which imposed priorities determined by the government. Since very few new lines were built in the old industrial areas, the existing lines were overloaded, and it was frequently necessary to reduce or cancel the movement of passenger trains. The total of freight and passengers carried in 1939 was 91·8 per cent higher than in 1932. But the most rapid rate of growth ended, here as elsewhere, in 1937.

In 1938 the performance of the transport system was very unsatisfactory; freight carried by the railways actually fell. The reason given: 'The majority of the railway managers and heads of political departments, and many of the leading cadres in the People's Commissariat of Transport, were dismissed and arrested.' There were also fuel crises in the winters of 1937–8 and 1938–9, which led to numerous cancellations of trains. Fuel shortages also hit road transport: shortage of oil led to the conversion of many vehicles to use gas-generators.[13]

THE INDUSTRIAL SLOWDOWN

We must now return to the problem of the slowdown of 1937. This is not the same problem as that of the shift of resources away from the consumers' goods sector. We have seen that the metallurgy industry virtually stopped growing, and yet iron and steel was an essential part of the building of Soviet military capacity. What happened?

In the case of iron and steel, priority was given, during 1934–6, to the completion of plant already started. Little was done to press on with new construction in these years, due to 'imperfection in the planning of investment, and organization of fulfilment of the plan', plus 'lack of highly-qualified construction staffs, and breakdown in supply of equipment and building materials'.[14] All these deficiencies seriously held back completion of plant. Furthermore, in 1936 there was a 35 per cent reduction in investments in iron and steel as against 1935, and the lower 1936 level continued through to 1939.[15]

This was followed in 1937 by a drastic cutback in all investments. According to an official Soviet recomputation expressed in 1955 prices, the investment figures (excluding kolkhozy) were as follows:

Investments
(Millions of roubles, prices of 1955)

1932	2,350
1933	1,943
1934	2,552
1935	2,984
1936	4,070
1937	3,621
1938	3,807

(SOURCE: *Nar. khoz.*, *1960*, p. 590.)

There was a particularly severe drop in 1937 in construction work.

One reason for the slowdown, then, was this fall in investments, occasioned probably by a shift of resources into the arms industry and the consequent shortages of skills and materials. But Soviet historians admit that another factor had an adverse effect on output and on planning at this period.

This factor was the great purge. It swept away a high proportion not only of leading party cadres, but also of army officers, civil servants, managers, technicians, statisticians, planners, even foremen. Everywhere there were said to be spies, wreckers, diversionists. There was a grave shortage of qualified personnel, so the deportation of many thousands of engineers and technologists to distant concentration camps represented a severe loss. But perhaps equally serious was the psychological effect of this terror on the survivors. With any error or accident likely to be attributed to treasonable activities, the simplest thing to do was to avoid responsibility, to seek approval from one's superiors for any act, to obey mechanically any order received, regardless of local conditions.

Already in Chapters 7 and 8 we have seen the spread of this hysteria into the administration of the economy, particularly in agriculture and the railways. It spread further in the worst years of the purge, paralysing thought and action. No one can tell how much economic damage was done by all this, but we can surely agree with those Soviet commentators who assert that the damage was serious. 'Breaches of socialist legality, mass repressions, caused an immense loss of economic cadres and disorganized the normal work of industry,' as an authoritative textbook put it.[16] Key planners were among the victims, including the Minister of Finance, Grin'ko, and also Pyatakov, deputy to Ordzhonikidze, the latter being driven to suicide.

It is no part of the present work to analyse the causes of the great wave of terror, about which historians will doubtless still be arguing and debating well into the twenty-first century (if our world survives that long). The terror has, certainly, a largely political explanation, connected with Stalin's desire for absolute power, and other non-economic factors. It is sufficient here to stress only a few of its aspects. One, already mentioned, was its adverse economic effect. Another was the probable connection between the extreme harshness of the policies described in the previous two chapters and the growth of police repression. Obviously, the dreadful struggle with the peasantry, and the sacrifices and strains of the industrialization drive, could not but give rise to resentment and bitterness, not only among ordinary folk but among party members too. We saw in Chapter 7 that local party officials were dismissed for pro-peasant ('pro-kulak') tendencies. No doubt the Stalin group also feared the exploitation of discontent by former oppositionists, which was why they arrested so many of them and compelled the leaders of the 'rights' and the 'lefts' to accuse themselves of treason and wrecking. Finally, the trials provided scapegoats. Shortages of consumers' goods, breakdowns in supply, errors in planning, could be attributed to malevolent plotting by enemies of the people, in the pay of Hitler, the Mikado, and 'Judas-Trotsky'. The published reports of the trials abound in economic self-accusation. Thus the shortage of eggs was supposedly due to the efforts of wreckers, who smashed eggs in transit just to deprive the Soviet people of the fruits of their labour and to create politically exploitable discontent. Or so the readers of *Pravda* were supposed to believe. Bukharin's – and other oppositionists' – doctrines would be publicly associated with treason, and thereby placed outside the area of legitimate discussion or comment.

A census was taken in 1937. Its preliminary results displeased the authorities for reasons unknown, so it was scrapped and its authors arrested. A new census was taken in 1939. One may be pardoned for concluding that these and other similar experiences affected the work of the surviving statisticians, and therefore the quality and reliability of statistics.

In fact, the quantity of published statistics began to decline in the middle thirties, and the material on the last years of peace is markedly deficient compared with the early years of the decade.

The third five-year plan period began in 1938, and was interrupted by the war. Its pattern and progress will be further discussed at the end of this chapter.

AGRICULTURE UP TO 1937

In Chapter 7 it was demonstrated that agriculture reached a state of acute crisis in the harvest year 1932–3. A slow recovery began with the relatively better harvest of 1933. The drive to collectivize the bulk of the remaining individual peasants was launched in 1934 and was practically completed in 1937, when almost 99 per cent of all cultivated land was included within collective or state farms. The proportion of peasants collectivized at this date was slightly smaller (93 per cent), but evidently the majority of the surviving 'individualists' earned a living other than by cultivating the soil, and they ceased to play any significant role in agriculture. By contrast, the private operations of collectivized peasants grew in importance, particularly in livestock. Indeed, the rapid recovery of the livestock population from the disasters of 1930–33 was in part due to the government's willingness to tolerate and even encourage private ownership of livestock, subject to maxima (see page 184). The actual figures, in millions of heads, were as follows:

| | January 1934 | | January 1938 | |
	State and collective	*Private*	*State and collective*	*Private*
Cattle, all	12·3	21·2	18·0	32·9
Cows	4·6	14·4	5·5	17·2
Sheep and goats	16·3	20·2	29·3	37·3
Pigs	6·2	5·3	8·8	16·9

(SOURCE: *Sel'skoe khozyaistvo SSSR* (1960), pp. 263–4.)

NOTE: Some of the private livestock was owned by workers.

While livestock numbers were still below the pre-collectivization level, the great increase during the period 1933–7 benefited everyone: there was more meat and milk in towns and villages, and revenues of both the farms and the peasants rose from sales of livestock products.

Grain output rose exceedingly slowly, though good weather in 1937 resulted in a record harvest. For many years, as we have seen, statistics were given in terms of biological yield, and even these estimates 'were consciously overstated ... so as to include the kolkhoz within a higher harvest category so that it should pay more produce for the services of the MTS'.[17] This motive grew in importance because these payments in kind became the largest single source of grain procurements during this decade, and the payments were related to the (nominal) size of the harvest.

The average grain yield for 1933–7 was said to have been 9·1 quintals per hectare, against 7·5 in 1928–32. But revised figures published in 1960 gave the real figure for 1933–7 as only 7·1. In other words, the yields actually fell. Only one year, 1936, was exceedingly unfavourable climatically, but this was balanced by the very good weather of 1937. The revised figures for total grain harvests are as follows:

	1933	1934	1935	1936	1937
Millions of tons	68·4	67·6	75·0	56·1	97·4

(SOURCES: Moshkov (for 1933–5); *Nar. khoz.*, *1958*, p. 418 (for 1937). Average for 1933–7 given in *Sel'skoe khozyaistvo SSSR* (1960). The 1936 figure is deduced from the above.)

While harvests (except in 1937) were below even the inadequate levels of 1928–32, procurements rose:

State procurements
(millions of tons)
1928–32:   18·2
1933–7:   27·5

In 1928–30 some grain was sold by peasants through private traders. None the less, less grain remained in the villages (except in 1937). The fall in the numbers of livestock, especially horses, reduced demand for grain and this relieved pressure somewhat. (We shall see that after 1937 the revival of the livestock population began to be severely limited by lack of fodder supplies.)

Procurement quotas were supposed to be based on the area sown or livestock owned by the given kolkhoz or peasant, and the authorities were not supposed to vary these quotas; but they did so arbitrarily, the needs of the state taking priority over all other considerations.

In 1933–4 one of the most severe problems was shortage of haulage power. In the five years following 1929 a net total of 17·2 million horses were lost; the amount of tractor horsepower increased by only three million, and there were few lorries. Poor maintenance, lack of skilled labour and of spares, ensured frequent breakdown of the tractors available. However, by 1937 tractor numbers had substantially increased, and there were more horses.

Therefore 1937 was in many respects a good year. We shall see that peasant consumption, of grain in particular, reached the highest point of any year in the period 1932–41.

THE KOLKHOZ CONGRESS AND MODEL STATUTE

By 1935 the shape of the kolkhoz as an institution was more or less settled, and a congress was called to adopt a 'model statute', which remained the basis of kolkhoz organization until well into the sixties.

The kolkhoz was declared to be a voluntary cooperative, whose members had pooled their means of production in order to produce in common. The members ran their own affairs, and elected a chairman and a management committee responsible to a meeting of all members. The kolkhoz was, however, bound to obey instructions from local state and party organs on matters of agricultural production and procurements, so in practice its operational autonomy was severely limited. The raion party organization, the MTS, the raion agricultural department, issued numerous orders to the chairman, and the party secretary in particular had the effective power of appointment and dismissal.

The payment of members was proportionate to the trudodni

(workday units) worked by each man; all collective work was measured in these units, and more highly skilled occupations were worth more trudodni than unskilled ones. The amount paid to a peasant, in cash and in kind, in any given year depended on two things: the number of trudodni he earned, and the worth of each trudoden'. The latter depended on the cash and produce available for distribution. Members shared the cash and produce which was left after other requirements had been met. Cash revenues were subject to deductions for taxes, insurance, the capital ('indivisible') fund, administrative and cultural expenses and production costs (e.g. purchase of fodder, seed, fuel). At the period we are describing this left, on average, 40–50 per cent of the gross cash revenue for distribution to peasants. We shall see that, owing very largely to the low prices paid, the sums in question were very small.

As for produce, payment was generally confined to grain, sometimes potatoes, often also hay (for private livestock). Compulsory deliveries to the state, payment in kind foɪ MTS services, seed, fodder, relief for sick and aged people, all took priority. In spite of this, at this period roughly a third of the kolkhoz grain harvest was consumed by the peasants (or their animals).

The 'residuary legatee' status of the kolkhoz peasant, the lack of any guaranteed minimum income, inevitably led to very great variations between kolkhozes. It rendered impossible any calculation of costs, for what, in such circumstances, is the cost of labour? In fact until well after Stalin's death no such calculations were even attempted. But the system of payment had a logic, given the objective of taking out of agriculture all possible resources for the priority needs of industrialization.

The statute gave formal recognition of the right of the kolkhoz household to a private plot of land, amounting to between ¼ and ½ hectare in most of the country (i.e. an acre or less). Possession of livestock was limited to one cow and calves, one sow and piglets, four sheep, and any number of rabbits and poultry. Livestock was generally pastured on collective land. This private sector – which Soviet statistics for some reason always treated as part of the socialized economy – is to this day by far the most important element of private enterprise in the Soviet Union.

The produce belonged to the peasants and, subject to compulsory delivery obligations to the state, they could eat it, or sell it on the legal free market.

In 1937 and for many years thereafter the collectives produced grain and industrial crops (e.g. cotton, sugar-beet, flax). In other produce the private sector was predominant. We have seen (on page 238) that it outnumbered state farms and collectives in all categories of productive livestock, and therefore in the supply of meat and milk. It was almost the sole producer of eggs, and provided most of the potatoes, many of the vegetables, virtually all the fruit. The peasant household usually relied on its kolkhoz for bread-grains, sometimes for potatoes, but the rest of the food, and the major part of the cash, came from their private holding. Thus in 1935 the average household received 247 roubles a year for collective work (which would buy them one pair of shoes), but must have obtained at least twice as much from sales in the free market.

Thus a kolkhoz household divided its time between collective and private activities, and its income was a combination of cash and produce paid for collective work, plus consumption of its own produce and cash from sales in the market and to the state. As payment for collective work was very poor, pressure had to be exercised to compel work to be done for the collective. This was formalized in 1939 by legislation requiring a compulsory minimum number of trudodni of all adult members.

Many kolkhoz members earned extra money on seasonal work for state enterprises, including state farms and forestry.

At the beginning of 1937 there were nearly 243,000 kolkhozes and 3,992 state farms. The bulk of agriculture was undertaken by or within kolkhozes. Each averaged 76 households, 476 hectares of sown area, 60 cattle, 94 sheep, and 26 pigs. There were, of course, big regional variations, the farms being larger in the south and east. One MTS served at this period roughly 40 kolkhozes on the average, using kolkhoz labour on a large scale alongside its own paid specialists.

There are other sources of agricultural produce. There are allotments cultivated by state-employed persons (state farm workers, railwaymen, rural officials, dwellers in suburbs of big cities, etc.) who own some livestock. There are also very numerous small farms and market-gardens operated by non-agricultural state enterprises, and used to provide foodstuffs for the factory canteen or the enterprise's shop (ORS).

Agricultural production was discouraged by the exceedingly unfavourable price structure. While prices of cotton and some other industrial crops were substantially raised in 1934, prices of livestock products in 1933–7 hardly differed from those of 1928–9, and grain prices were only a little above 1928–9 levels. Yet the prices of industrial goods, and particularly those purchased by peasants, rose very greatly. Thus for instance the price of cotton cloth (*sitets*, calico) on 1 January 1928 was 34 kopeks per metre, while on 1 January 1937 it was 2·30–3·25 roubles, i.e. it increased seven to nine times. Prices for soap multiplied by five, for sugar by six, etc.[18] This was the price 'scissors' with a vengeance. It should be added that agriculture was neglected in other ways; even as late as 1940 only 4·2 per cent of kolkhozes had any electricity.[19] Any available mineral fertilizer was likely to be allocated to industrial crops.

Not very surprisingly, while cotton output rose very satisfactorily – production doubled during the decade – performance was much less satisfactory elsewhere. With prices at exceedingly low levels coercion became an essential element in relations between state and collective farms. Since incentives were lacking, 'there developed excessive centralism in the planning of kolkhoz production, arbitrariness (*administrirovaniye*) developed in the control over kolkhozes. This led to a weakening in the creative initiative of the kolkhoz management. Arbitrary methods eliminated economic motivation. ... The production plans of kolkhozes frequently became a mechanical dividing between them (*razverstka*) of delivery obligations passed down from the centre. The organization of kolkhoz production, from the first processes of farm work down to the final delivery of produce to the state, was strictly regulated and centralized.' The Soviet source from which this is quoted is in no doubt that such practices were intimately connected with the very low prices paid.[20] It should be added that, for reasons of accountancy, state farms were, at that period, paid the same low prices, and so required large subsidies from the state budget.

State farms suffered exceedingly at first from 'gigantomania', the belief that enormous 'grain factories' were conducive to efficiency. Thus in 1931 the 'Gigant' state farm had an area of 239,000 hectares! In 1931–2 harvests were unbelievably low, averaging on all state farms a mere 3·6 quintals per hectare, when individual peasants and kolkhozes were managing 7–8 quintals.[21] One reason was their narrow specialization on grain only, another was total neglect of agronomy.

From 1933 state farms, as well as MTS, were supervised by 'political departments'. These survived until March 1940, whereas, as we have seen, the MTS political departments were abolished in 1934. The MTS and kolkhoz sector was left to the *raikom* (local party) committee and to organs of the People's Commissariat of Agriculture, while state farms came under a separate Commissariat of state farms (from 1932), and 'the raikomy did not interfere in the affairs of state farms'.[22]

In the period after 1934 many excessively large state farms were divided up, and in some cases land was transferred to kolkhozes. As a result of increased investment and better cultivation yields rose in 1937 and subsequent years somewhat above kolkhoz levels, though without significantly exceeding them. Given that state farms tended to have more equipment and power than kolkhozes, as well as regularly paid wage-earners, their achievements must be regarded as moderate, the more so as their raising of livestock was most unsuccessful.

Despite all these defects, agricultural output did recover from the depths to which it had fallen in 1933.

INCOMES AND SOCIAL SERVICES

(a) Peasants

Since we have been discussing agriculture it seems convenient to begin by surveying the change in peasant incomes.

No complete set of figures exists in any published source, and some of the data that have appeared are not entirely consistent. Cash payments undoubtedly increased, as the following figures show:

	1932	1935	1937
Per trudoden' (roubles)	0·42	0·65	0·85
Per household per annum (roubles)	108	247	376

Distribution of grain varied greatly according to the harvest and procurement policies:

	1932	1933	1934	1935	1936	1937
Per trudonen' (kilograms)	2·3	2·9/3·1	2·8	2·4	(1·6)	4·0
Per household per annum (quintals)	6	9·0/9·6	9·9	9·1	(6·2)	17·0

(SOURCES: *Sotsialisticheskoe narodnoe khozyaistvo v 1933–40*, (Moscow, 1963), p. 388; and I. Zelenin, *Istoricheskie zapiski*, No. 76, p. 59. Figures differ in the two sources for the year 1933. The 1936 figures are a rough estimate only.)

The total number of trudodni worked for the collective increased, hence the total income per household rose faster than the payment per trudoden. It must be noted that prices of manufactured consumers' goods rose very substantially in the period 1932–7, so that real purchasing power increased relatively little, though the range of goods available at official prices was, of course, much the greater in 1937.

The above are averages. While in 1932 an average of 600 kilograms per household represented a good deal less than a peasant family's normal consumption of bread, even this would not have represented famine conditions had it been evenly distributed. Yet, as we saw in Chapter 7, famine there certainly was – in some areas. *A fortiori*, 900 kilograms, while not good, was tolerable, and the 1,700 kilograms of 1937 was very satisfactory. But whole provinces were unable to allocate enough bread grains. Even in 1937, the best year of the decade, 28½ per cent of all kolkhozes distributed less than 2 kilograms of grain per trudoden, i.e. less than half of the average, and those households were short of bread by spring. Yet the harvests of 1933–6 had been well below those of 1937, that of 1936 being in fact disastrously poor; a high proportion of the peasant families must have been seriously short of bread, under conditions in which other sources of food were few and ordinary baked bread was, for the less favoured peasant, prohibitively expensive. Thus the retail price of bread in 1935 was one rouble per kilogram, which equalled average cash earnings from one day's work (in a typical day, a peasant could earn a little more than one workday unit).

At this same date one kilogram of vegetable oil cost 1·30

roubles, one kilogram of the cheapest rye flour was 1·60 roubles, wheat flour cost 4·60 roubles, good leather shoes between 466 and 1,000 roubles.[23] This was bad enough. But even as late as 1939, 15,700 kolkhozes (out of a total of about 240,000) gave their members no cash at all for their work, and 46,000 kolkhozes distributed a mere 20 kopecks or less per trudoden. The Soviet source from which the above figures are taken remarked that low productivity was 'due in the main to the lack of material stimuli, lack of personal interest of the mass of the kolkhoz peasants in their work and in the results of the collective economy'.[24]

Yet the conditions of life of the peasants undoubtedly improved very greatly in the period 1933–7 (except in 1936, when the weather was bad), since their private livestock holdings and their income from sales in the free market rose substantially. But the very large amount of grain available in 1937 proved a flash in the pan. We shall see that it declined sharply in subsequent years.

The rate of migration from country to town declined substantially during the second five-year plan. It may be that this contributed to the marked slowdown, already remarked upon, in industrial output, especially construction, which became apparent in 1937. It is possible that one reason was that rural conditions of life improved, while overcrowded housing and the low pay of unskilled labour acted as a disincentive to peasants contemplating a move townwards. A further contributory factor may have been the fact that in 1934–7 1·2 million tractor-drivers and many other 'mechanizers' were trained,[25] thereby providing a rural outlet for the more competent and ambitious village youths.

Technical training for some, better education for most, the almost total elimination of illiteracy; these were significant gains, which ought not to be overlooked. They were greater in the formerly backward areas, such as Central Asia. There were also major improvements in rural health services.

(b) Wages, prices, and living standards, urban sector

We have seen that wages rose by much more than had been planned. Presumably at the time the plan was drafted it had been hoped to reduce costs and to fix retail prices at levels not greatly in excess of the ration prices of 1932–3. We have seen that prices did in fact rise, and in 1934 and 1935 very large increases accompanied the elimination of rationing. This made it essential to increase wage rates of those workers and employees left behind in the course of the wage drift which had already developed in the previous quinquennium and which continued throughout the decade (though at a reduced rate). The government sought so to control wages as to ensure special advantages to priority industries, and we have already noted that piece-rates and bonus schemes, including the high incomes associated with 'Stakhanovism', were designed to provide and dramatize incentives for higher productivity.

Money wages in fact increased as follows:

1932	1933	1934	1935	1937
		(roubles per annum)		
1427	1556	1858	2269	3047

(SOURCE: *Trud v. SSSR* (1936), and fulfilment report on second five-year plan.)

By 1935 the effect of these wage increases was to cause heavy losses in, and subsidies to, basic industries, whose selling prices were deliberately kept from rising. The following figures illustrate this:

	Planned costs	Selling price
Coal (per ton)	19·12	9·65
Iron ore (per ton)	10·20	5·70
Steel (per ton)	120·00	83·00
Cement (per ton)	46·31	27·13

(SOURCE: *Sots. nar. khoz. v. 1933–40*, p. 75.)

In 1936 substantial increases in factory wholesale prices – coal and steel cost more than double – helped to eliminate most of the losses and subsidies.

A more difficult task is to describe the evolution of retail prices and so of real incomes. The basic problem is the existence of a multiple price system and rationing at the beginning of the period we are describing. In 1933 ration prices were low, 'commercial' prices for the same goods exceedingly high (see page 204). During 1934 commercial prices were reduced (bread by 31·2 per cent compared with the very high prices of March 1933), but 'normal-fund' prices, for goods rationed or otherwise subject to controlled distribution, were increased. There was a strong case for the speedy liquidation of multiple prices and of rationing. Firstly, the sheer complexity of the system was costly. Secondly, it gave excellent opportunities for speculative resale, and there were many instances of state enterprises illicitly selling or reselling in higher price categories to cover financial deficiencies or just to make money. Finally, labour incentives were ineffective unless goods could be freely bought.

However, given the shortage of goods to buy, unified prices would have to be substantially in excess of the so-called 'normal-fund' prices.

So when bread rationing was declared abolished as from 1 January 1935, the price was fixed at 1·00 rouble for the cheapest rye loaf, which was 37 per cent below the existing commercial price but 100 per cent above the ration price ruling in the second half of 1934, and 12½ times the 1928 price. After the fairly good harvest of 1935, and following upon the partial recovery of the livestock sector, on 1 October the rationing of meat, fats, fish, sugar, and potatoes was also abandoned, and bread and flour prices were reduced. According to a survey of prices in one industrial town, quoted by Malafeyev, the net effect of the price changes introduced on 1 October 1935 was to reduce prices of bread by 13·4 per cent, of flour and pulses by 10·6 per cent, of herrings by 21 per cent. By contrast, meat prices rose on average by 101·4 per cent, butter by 33 per cent, margarine by 50 per cent, potatoes by 39 per cent, these being the reflection of the sharp rise in prices involved in de-rationing.[26] In 1936 milk, dairy produce, salt and vegetable oils were increased in price continuing the policy adopted in October 1935. Price rises in 1937 affected meat, milk, and potatoes, but the effect on the index of food prices was only 1 per cent.[27] As for manufactured goods, their prices had already been greatly increased by 1933. *De facto* rationing and multiple prices continued, and it was not until the end of 1935 that a series of government decisions narrowed the difference between the levels of commercial prices and those of the 'normal fund'. Though this difference became slight, 'in the period 1936–8 the retail prices for non-food products were not yet in a full sense unified'.[28] Commercial prices were reduced in 1935, but 'normal' prices rose; the exact figures are not available. In 1937 the average price of all non-food items fell by 3·8 per cent. Malafeyev, who used archive materials extensively, has calculated that all state and cooperative prices rose on average from 1932 to 1937 by 110·2 per cent. During this period free-market prices behaved very differently, as the following table shows:

Free-market prices (1932 = 100)

		1933	1934	1935	1936	1937
All goods		148·2	90·8	64·6	55·3	62·3
of which:	Grains	160·5	80·3	46·1	27·8	34·4
	Potatoes	164·0	77·9	50·1	30·5	45·2
	Meat	140·0	126·0	96·4	73·2	83·2

It should be noted that the 1932 prices were also very high (see Chapter 8), and also that the free-market turnover was small and only semi-legal in that year. By constrast, there was little difference between official and free prices of most foodstuffs in 1937.

Taking into account the lower market prices, the retail price index as a whole rose, according to Malafeyev, not by 110·2 per cent but by only 80 per cent in 1932–7. Average wages rose by 113 per cent. Presumably allowing for services, he reaches the conclusion that real wages rose in this period by 'at least 20 per cent'. This surely understates the improvement, because of greater availability of goods and better trading arrangements. In any event, the lowest point in terms of welfare was not in 1932 but in 1933, a year in which both official retail and free-market prices rose very substantially indeed and free-market prices reached their peak. There were also, as we have seen, very large 'official' price increases in 1934, affecting particularly 'normed' or rationed prices of foodstuffs. These increases far outstripped the rise in wages, so any index of purchasing power for these years would look exceptionally unfavourable. This was followed by three very good years for the consumer, and in 1936–7 especially the real-wage index takes a substantial leap forward: in those two years retail prices fell slightly, while wages rose by 30 per cent.

Accepting that in 1937 real wages were, say, 35 per cent higher than in 1935, they were lower than in 1928. Malafeyev clearly implies this; wages rose 4·5 times, official prices 5·3 times. After correcting for prices of services, which rose by less than this, one must conclude that there was a decline. Janet Chapman[29] calculated that real wages in 1937 were 85 per cent of 1928 if 1937 prices were used as weights, but only 58 per cent of the 1928 pattern of consumption. (The Russian émigré experts, Prokopovich and Jasny, had come to similar conclusions earlier, using rougher data.) If 1937 represented perhaps a 35 per cent improvement on 1935, it then follows that 1935 living standards were very grim indeed. Yet, since real living standards depend in the last analysis on the goods and services available, and fewer were available in 1933 and 1934 than in 1935, life was even harder in those years. Of course, one must also take into account the improvement in social services, the elimination of unemployment, and the undeniable fact that many ex-peasants who moved into industry were earning more than they would have earned had they remained on the farms. Finally, those who had been workers in 1928 had very good opportunities of promotion, and the vast majority of them were probably making a good deal more in 1937 than in 1928. The average is pulled down by the dilution of the urban labour force by unskilled 'immigrants' from the villages. Aggregate consumption went up, because *per capita* measurable consumption by peasants was several times lower than that of the urban population. Urbanization, in all countries, tends to lead to an overstatement, in welfare terms, of the statistical expression of total consumption and also of the total national income.

It is necessary to mention also that availability of housing space in urban areas continued to decline. Stalin seemed more concerned with prestige projects, such as the lavishly decorated Moscow underground railway, than with ordinary housebuilding or the maintenance and repair of existing houses. Conditions in big cities became very difficult, and Moscow was among the worst. The following statistics relate to 'old houses' (i.e. nearly the whole of Moscow) and to the year 1935. They come from the statistical annual *Trud v SSSR* published in 1936 and are unique of their kind. They give figures in terms of 'renters' (*syomshchiki*), i.e. units which were mostly families but in many cases also single persons. At this date 6 per cent of 'renters' had more than one room, 40 per cent lived in one room, 23·6 per cent occupied a part of a room, 5 per cent lived in kitchens and corridors, 25 per cent in dormitories.

TRADE

Retail trade was a neglected sector. With a general shortage of trained personnel little was available for such a non-priority activity. The number of shops, after the squeezing out of the private trader, increased much too slowly: thus there were 116,000 shops in urban areas in 1932, and only 133,000 in 1937. Queues were sometimes due to lack of retail outlets, but also to poor distribution. Allocation of 'funds' of planned commodities was both centralized and clumsy, with consumer demand largely neglected. 'Particularly unsatisfactory was service to customers.'[30] The psychology of the seller's market went deep.

It will be recalled that during the period of the great leap the word 'trade' acquired a somewhat negative connotation, as there was a recrudescence of illusions about 'commodity exchange' and the withering away of monetary relations. Trade was linked with the food-processing industry, and both were run by Narkomsnab, the People's Commissariat for supply. It was not until 1934 that it was divided into the People's Commissariats for Food and Industry, as well as for Internal Trade. Under the latter were union-republican commissariats and local trade departments. The transition to unrationed trade at unified prices was carried out by the new Commissariat.

In 1934 retail trade was being conducted by both state and cooperative shops, in town and country. Some of this trade was run by so-called ORSY, the 'departments of workers' supply', set up in 1932; these were shops run by enterprises for their own employees, and such shops continue to exist, with their food supplies often coming from the enterprises' own small farms.

On 29 September 1935 it was decided to concentrate the cooperatives exclusively on trade in rural areas, and state shops (plus ORSY) supplied the towns. The cooperatives were also responsible for sales to kolkhozes of such items as lorries, building materials, and nails, which the kolkhozes had to purchase at retail prices, which were much higher than those charged for the same goods to state enterprises.

The government intended greatly to improve distribution, especially in rural areas, and retail turnover did increase. However, 'the general level of trade turnover in villages was totally inadequate. ... In many village retail shops there were lacking even goods which were in sufficient supply at wholesalers' stores, including such necessary commodities as salt, matches, kerosene, etc.'[31] Despite this criticism, any picture of life in the period 1933–7 must stress the great improvement which characterized these years.

One effect of the abolition of rationing was the sharp reduction in 'communal feeding', i.e. public catering. In the hard years of the early thirties many urban citizens relied greatly on factory and office canteens, which were supplied with rationed foodstuffs at low prices. In 1935 canteen prices were very greatly increased, and, with the end of rationing, more food could be bought (if one had the money) in the shops. Consequently, the number of meals (or more strictly dishes) served by public catering enterprises fell, from a peak of 11,800 million in 1933, to only 4,200 million in 1937. The fact that this represented an increase in roubles of close on 100 per cent gives one some idea of the magnitude of the price increase.[32]

FINANCE

The pattern which developed in the early thirties was maintained with little change. Turnover tax remained by far the largest item of revenue, and this was levied largely but not exclusively on consumers' goods. (Heavy industry paid roughly a tenth of all turnover tax in 1935.) A large proportion of this tax arose out of the wide difference between the low compulsory-procurement prices of farm products and the much higher (and increased) retail prices. The extent of this difference has already been illustrated (page 210). In 1935 agricultural procurements and the food industry were responsible for nearly 60 per cent of all turnover tax receipts. The rates of tax on most manufactured consumers' goods were exceedingly high at this period. Here are some examples, in which the tax rate represents the percentage of the price *inclusive* of tax (i.e. a tax of, say, 70 per cent means that within a price of 100 roubles there is 70 roubles tax. In Britain this would be expressed in relation to the price before tax, and the rate would then be not 70 per cent but 233 per cent.) Bearing this in mind, the following rates were applicable in 1935: kerosene 90 per cent, nails 28 per cent, light bulbs 10 per cent, galoshes 70 per cent, cotton cloth 35·6 per cent, thread 34 per cent, salt 82.9 per cent, kitchen soap 45–57 per cent, toilet soap 69–72 per cent, vegetable oil 74 per cent, sugar 77·2 per cent in towns, 86·2 per cent in rural areas, sausage 54·9 per cent. In addition there were a variety of 'budgetary additions' charged at the retail stage: this was 37 per cent for leather shoes sold in towns, no less than 142 per cent in the country. By contrast, producers' goods paid low rates: thus coal, cement, bricks, and steel paid 2 per cent. There were vast numbers of exceptions and local variations.[33] All this was very complex, with a great multiplicity of rates which varied by region and sometimes even by factory, as well as by model or type of product. In 1939–40 the tax system was overhauled. For many multi-product industries, such as textiles, the tax was henceforth calculated as the difference between the wholesale price at which the enterprise sold the given commodity (which was related to production costs) and the retail price (less trade margins). This made for a more logical structure of retail prices, as well as simplifying the tax. One effect of the revision of both prices and tax rates was to increase enterprise profits somewhat at the expense of turnover tax, and the proportion of turnover tax in total budgetary revenue fell from 73·1 per cent in 1937 to 59·3 per cent in 1940, while the state's revenues from profits rose from 8·9 per cent in 1937 to 12·3 per cent of the budget in 1940.

Direct taxation continued to play only a minor role; income tax and local tax ('cultural and housing levy') were at very low rates, the maximum being 3·5 per cent and 3 per cent respectively. These taxes were combined and increased in 1940, so that the tax paid by a prosperous citizen (earning 3,000 roubles a month) by then amounted to 12·46 per cent. The above rates relate to wages and salaries paid by state enterprises and institutions. They were much higher in the case of private earnings; for instance, individual craftsmen were charged a top rate of 52·3 per cent.[34]

Modest taxes on both kolkhozes and collectivized peasants (the latter only in respect of their private activities) were levied in the middle thirties, but their impact, as we shall see, became much more severe during and after the Second World War.

Purchases of bonds became virtually compulsory and most employed persons found themselves in effect contributing two weeks' wages, these being deducted over the whole year. This was a more important source of revenue than direct taxation: in 1938, for instance, out of a total revenue of 127·4 milliard roubles all direct taxes came to 5 milliards, loan subscriptions to 7·6 milliards.

In the first years of the first plan bond purchases were still regarded as a voluntary act, interest rates were high (8–12 per cent) and the repayment period was short. In 1936 the government converted all bonds previously sold to the citizens to the much lower rate of 4 per cent interest, for twenty years.[35] In fact, there were few repayments.

Budgetary expenditure increased with the expansion of the economy, the more so as the principle was established, in a government decision of 9 March 1934, that capital investments in state enterprises should be financed by outright non-returnable budgetary grants (except where they were financed out of profits or the depreciation fund). Until the 1936 price reform, subsidies were a serious budgetary burden, and they had to be resumed at the end of the decade as wages and costs rose. Defence expenditures, as had already been shown, rose very rapidly, much more rapidly than had been planned. Since the state budget, on both its revenue and its expenditure side, is very closely connected with the operation of the economy, it followed that the financial and economic plans required revision together. Thus, to take an example, if the planned growth in the output of wool textiles was abandoned in favour of producing guns, then, assuming that investments were simply switched from one to the other, expenditures would remain the same, but revenues would be adversely affected by loss of turnover tax, which would have been paid on the wool textiles; there would also open up a gap between the total income of the citizens and the goods available to them at established prices, since wool textiles are consumers' goods and guns are not. An increase of turnover tax, and so of retail prices, would then simultaneously restore budgetary equilibrium and the balance between supply and demand of consumers' goods. An alternative would be to permit the imbalance to continue, to bridge the gap by increasing the currency issue and to allow the emergence of queues and black-market phenomena. We shall see that in fact this is what happened in 1939-40.

Expenditure on social services increased rapidly during these years. The exact extent of the increases is hard to calculate because of the impact of price changes. The big educational effort has already been commented upon. Health statistics show a marked improvement; between 1928 and 1940 medical doctors increased from 70,000 to 155,000, the number of hospital beds from 247,000 to 791,000. A large proportion of the newly trained doctors were women. There was also a big rise in spending on social insurance and social security.

THE LAST PRE-WAR YEARS

(a) Industry

The period 1938-41, until the German attack was launched on 22 June 1941, covers three and a half years of the third five-year plan. This plan, prepared during 1937-8 and adopted formally by the eighteenth party congress in 1939, aimed at impressive increases over the five years: 92 per cent in industrial output, 58 per cent in steel, 129 per cent in machinery and engineering, 63 per cent in wool cloth, and so on. Among other objectives, it was intended to achieve universal secondary education in cities, a minimum of seven-year schooling in the countryside.

But the ever-increasing danger of war led to a drastic recasting of the production programme, which to a large extent explains the shortfalls which developed, particularly in consumers' goods. In addition, as already pointed out, the disorganizing effects of mass arrests were also responsible for failure to carry out plans. Another difficulty was shortage of labour: according to official sources industry, construction, and transport were short of 1·2 million recruits in 1937, 1·3 million in 1938, 1·5 million in 1939.[36] Aggregate industrial output did rise appreciably, by more than was planned, according to official claims. But progress was exceedingly uneven. Thus in the three years 1938–40 the output of machinery and engineering (which include arms) was said to have reached 59 per cent of the total planned for five years, and footwear production even achieved 60 per cent. By contrast, steel output had increased insignificantly, by only 5·8 per cent of the five-year plan increase, rolled metal by a mere 1·4 per cent, cement by 3·6 per cent, while output of sugar, which should have risen by 40 per cent, declined. Output of oil, a vital and strategic item (one would have thought), expanded exceedingly slowly, contributing to a fuel crisis. It is difficult to find an adequate explanation for this pattern. It is not enough to blame the need for military preparation, for the shortfalls include some commodities of vital importance militarily. It may be, as already suggested, that the effects of the purges and the strain due to the rapid switch into armaments production combined to disorganize the entire planning system. In the case of metallurgy a key problem was shortage of iron ore and coke, and this in turn was due in part to the effect of 'mass repressions' on management.[37] The very large number of persons in concentration camps (we still have no idea how many) included many highly qualified managers, technicians, officers. One cannot imagine that they were effectively utilized. One must also mention the impact of the Finnish war of 1939–40. This disrupted transport, and in February 1940 many Leningrad factories stopped production owing to the non-arrival of coal.[38]

Despite all errors and waste, the U.S.S.R. did succeed, in the ten years beginning in 1928, in creating the industrial base for a powerful arms industry. But this base was still too weak to enable the civilian investment and consumers' programmes to survive the effects of a redoubling of arms spending.

The development of the armaments industry, and war preparations generally, will be discussed in the next chapter.

In 1939 the U.S.S.R.'s productive capacity grew significantly by the absorption of new territory: Estonia, Latvia, Lithuania, the eastern provinces of Poland (and in June 1940, also Bessarabia). This must be taken into account in interpreting statistics for 1940. For example, there was a large increase in the number of individual peasants; collectivization in the newly acquired regions was not completed until 1950.

(b) Agriculture

The agricultural policies of the party underwent a change in 1939. The central committee plenum which met in that year decided to reduce the size of private plots. It was found that many families had 'too much' land, and a total of 2·5 million hectares was taken away from the collectivized peasants. It was hard to imagine a less popular act. In the same year stricter discipline was introduced, with compulsory minima of workday units per able-bodied peasant. In the same year a decree (of 8 July 1939) instructed kolkhozes greatly to increase their livestock holdings. Owing to shortage of fodder this led to some fall in the already low productivity of collective livestock: thus both milk yields and average live weight of animals delivered for meat declined after 1938.[39] It had a further consequence: fodder for private livestock became difficult to obtain, and there was a fall in their numbers. Another unpopular act was a marked increase in compulsory delivery quotas for livestock products, levied on both farms and peasant private holdings. Peasants were under pressure to sell for meat (or to their kolkhoz) animals which in many cases they could not feed. There was also an increase in compulsory delivery quotas for many crops, and the basis of the quota was altered from the area sown with that crop (or animals actually owned) to the total arable area. This was used further to screw up the delivery quotas, and the compulsory principle was extended to items hitherto not subject to it, including many vegetables and even sheep's milk cheese. Private livestock ownership fell:

Animals per 100 kolkhoz households

	1938	1940
Cattle	138	102
Pigs	70	46
Sheep and goats	169	169

(SOURCE: *Sotsialisticheskoe narodnoe khozyaistvo v 1933–40* (Moscow, 1963), p. 390.)

There was an increase in the rate of payment in kind for the work of the MTS, and a decrease in the proportion of the grain and potato crop sold at the higher over-quota prices, and this too adversely affected kolkhoz incomes. Furthermore, a decree of

1 August 1940 instructed kolkhozes, after fulfilling their delivery obligations, to set aside seed, food, and fodder reserves, and only thereafter to distribute grain to their members. Although reports on the harvest are now known to have been exaggerated, orders were issued 'to halt the anti-state attempts to understate the harvest yield'. The press reported prison sentences for kolkhoz chairmen who 'understated' the harvest or held up grain deliveries.[40] While total kolkhoz cash revenues rose from 14 milliard roubles in 1937 to 20·7 milliards in 1940, an increase was ordered in the allocation to the 'indivisible' (capital) fund, as well as in taxation, and the average cash distribution per trudoden' (workday unit) rose insignificantly, from 0·85 to 0·92 roubles, certainly less than the increase in prices. Meanwhile the rise in state procurements left much less grain in the villages, and the grain payment per trudoden' fell from the record figure of 4 kilograms in 1937 to only 1·3 kilograms in 1940.[41]

It is evident that, at this period, the peasants had many grievances: less bread, fewer private animals, too little cash, a cutback in the size of their plots. One might suppose that this was no way to ensure loyalty on the eve of war. Of course, higher grain procurements did facilitate the building up of reserves. Incentive schemes were introduced in kolkhozes in 1939–40; these provided bonuses for kolkhoz peasants who carried out plans, but in any case the plans were often set unrealistically high, and the relative decline in cash and, especially, produce available for distribution acted as a counter-incentive. Yields of many crops declined: sugar-beet, flax, sunflower, and potatoes all did worse than before. An exception was cotton, where prices were much more favourable. In 1938–40 yields of this crop were 34 per cent up on 1933–7, and almost reached the 1909–13 average, on a very much greater area of irrigated land.[42] Grain yields were a little better but still very modest, averaging 7·7 quintals per hectare in 1938–40, against 7·5 in 1928–32 and 7·1 in 1933–7. Since the area sown with grain increased between 1932 and 1940 by a mere 1 per cent (on comparable territory) the increase was quite insufficient to meet the growing needs of the state and citizens.

A great deal of harm was done at this period by the imposition by the centre of its own pet ideas, regardless of local advice or circumstances. One such was the so-called *travopolye* crop rotation system, which greatly increased the area under grasses, with the idea that this would increase soil fertility without using fertilizer. In some areas this was nonsense. Another hare-brained scheme was the much publicized introduction of an alleged rubber-bearing plant, *kok-sagyz*. Still another was the futile attempt to grow unirrigated cotton in the Ukraine. Lysenko, the pseudo-scientist, was beginning to influence official decisions, with promises of instant results at low cost. Agriculture does not lend itself to bureaucratically imposed innovation, and such things led to unnecessary losses.

(c) Living standards and trade

The years 1938–41 were, no doubt unavoidably, a period in which the improvement of living standards came to a halt. Yet the index of real wages showed up well. The reason for this apparent paradox was the price policy followed at this period. Wages rose from 3,047 roubles per annum in 1937 to 4,054 roubles in 1940, or by 35 per cent. Official prices rose by only 19 per cent according to Malafeyev (but by 26–32 per cent according to Janet Chapman). However, free-market prices rose by much more than this, reflecting the increasing shortage of food-stuffs in state shops. Official and even unofficial Soviet figures are few and scattered, but one must agree with Janet Chapman that free-market prices roughly doubled between 1937 and 1940, while supplies to this market fell. Some Soviet sources imply that free-market prices were 75 per cent above official prices in 1940. After allowing for higher taxes and bond sales, Mrs Chapman arrives at a fall in real wages, between 1937 and 1940, of 10 per cent or so. The fall was most pronounced in 1940. Goods of all kinds were much more difficult to obtain, there were more queues, there was unofficial rationing of foodstuffs, and a resumption of sales at 'commercial prices in special shops'. Retail prices were increased in the course of the year but supply and demand remained well out of equilibrium.

Foreign trade had diminished in volume in the middle thirties, and the substantial reduction in imports of machinery made possible the emergence of a balance-of-payments surplus, which enabled the Soviet government to repay debts incurred during the first five-year plan period. Trade with Germany fell to a very low level owing to the anti-Soviet attitude of the Nazi regime. In 1939 the outbreak of war cut the U.S.S.R. off from most of its Western trading partners, but the Nazi–Soviet pact caused a large increase in trade with (and especially exports to) Germany during 1940 (and in fact right up to June 1941). The following figures demonstrate this:

	Imports from Germany	*Exports to Germany*
	(millions of roubles)	
1938	67·2	85·9
1939	56·4	61·6
1940	419·1	736·5*

(SOURCE: *Vneshnyaya torgovlya 1918–40*, p. 23.)

* Source mistakenly gives this as 7,365·0.

(d) Labour and social services

Returning to domestic affairs, it is necessary to mention very severe measures designed to strengthen labour discipline, including not only absenteeism but also unpunctuality and slackness at work. At first these followed precedent by merely imposing administrative penalties: fines, dismissals, evictions, reduction of social insurance benefits and the like (decree of 28 December 1938). In the same month, labour books were issued to all employed persons, and these, together with the internal passport, were intended to control movement and to keep a check on discipline (it was a form of industrial conduct sheet and performance record). But this was not enough. The decrees of 1940 (26 June, 2 October, 19 October) went very much further. Their essential features were:

(a) The imposition of direction of labour on specialists of many kinds ('engineers, technicians, foremen, employees, qualified workers').

(b) Compulsory call-up of up to one million young school-leavers for 'labour reserve schools' for training. This was the consequence of the shortfall of trainees, already remarked upon. This in turn was attributed to the rapid growth in the secondary school intake.

(c) The absentee from work was treated as a criminal, subject to criminal law penalties of up to six months' 'forced labour at place of employment' (a kind of hard labour on the job, with up to 25 per cent loss of wages). Sometimes this meant a compulsory shift to another job. Another such offence 'qualified' the offender as a 'flitter', and so for a mandatory jail sentence, if it occurred within the period of the first sentence.

(d) Anyone more than twenty minutes late for work was to be treated as an absentee. This would include returning late from lunch break or going off early. Two such offences, once again, meant prison as a 'flitter'.

(e) No one was to be allowed to leave his or her job without permission. This was only to be granted in special circumstances, some of which were listed (e.g. old age, call-up to the army, move of husband to another town, admission to higher educational establishment, etc.). If anyone disobeyed and left work he would be subject to criminal-law penalties and imprisoned as a 'flitter'. Sentences of four months were quite common.

(f) The working day was lengthened from seven hours to eight hours, the working week from five days out of six to six days out of seven (Sunday was to be the normal day of rest), without additional pay.

This draconian legislation was supposedly introduced on representation from the trade unions! Although in retrospect it seems proper to treat it as due to the imminence of war, it was not justified at the time by such reasoning. It seemed to fit a pattern of thought based on the priority of discipline (and terror). It was the duty of the citizen to work for the common good, and the state knew what the common good was. It would punish anyone who sought to substitute his personal preference for his duty to society. The 1940 decrees were not fully repealed until 1956, although it seems that they fell into disuse in the early fifties.

The impact of these decrees in 1940 can be studied in the legal press. A good summary of the evidence is given by S. Schwarz.[43] Some of this evidence is almost beyond belief, but sceptics have only to read the files of the organ of the Soviet procuracy for the year 1940, from which all the instances cited below are taken.

The idea of punishing workers as criminals in peacetime for

offences against labour discipline was so contrary to tradition that a campaign had to be launched against 'criminal inactivity, direct covering-up, ... slowness, liberalism' by the prosecution officers (*prokurory*), many of whom were dismissed and themselves prosecuted.[44] Judges were accused of 'rotten liberalism', reprimanded, dismissed, tried, and sentenced, for not ruthlessly implementing the decree.[45] Managers were punished, sometimes very severely, for 'covering-up' (not reporting) lateness and absenteeism by their staffs, and it was made clear that such 'crimes' were very common; thus sixty managers were put on trial for this offence in the Ukraine within a couple of months of the promulgation of the decree.[46] Doctors were denounced for being too 'liberal' with medical certificates for alleged malingerers.[47] 'Mild' sentences were made more severe by the higher courts on the 'protest' of prokurory. The courts, compelled to deal with cases within five days, were overloaded with work. By decree of 10 August 1940, these trials were to be handled by one judge (without assessors) and with no preliminary investigation (*sledstvie*).[48]

Faced with such pressures, judges and prokurory began to overdo things, as was only to be expected, and managers reported for prosecution cases of absence or lateness which were quite evidently legitimate, even under this decree. Thus one woman teacher was prosecuted while she was actually in a maternity home;[49] a woman with a sick breast-fed baby at home, and five months pregnant was sentenced to four months' imprisonment for leaving work, and this sentence was actually confirmed by the republican supreme court;[50] another woman with two young children whose baby-minder left was sentenced to two months';[51] still another one was given three months' imprisonment for being absent, after producing a medical certificate proving an attack of malaria,[52] and so on, and so forth. By the end of the year, in cases like these, the chief prokuror in Moscow was intervening to set aside unjustified sentences. To explain such outrages one must remember that the great purge, with its excesses of terror, was fresh in everyone's mind.

This legislation was only very occasionally linked with the danger of war. In the main, it was presented as logical and right in itself, justified by selected quotations from Lenin and Stalin.

This was the time of the Nazi–Soviet pact, and the citizens were being told that the danger of war had receded. These very severe measures seemed to be a bad psychological preparation for the trials to come, when popular support would prove an indispensable source of strength.

Other unpopular measures were taken in 1938–40. Some social insurance benefits were cut: thus maternity leave had been reduced from 112 days to seventy days already in 1938. Full social insurance benefits, under the 1938 'discipline' decree, were paid only to those with a long period of service in the same enterprise. Finally, fees were introduced in 1940 from upper forms of secondary schools and in higher education, despite a provision of the 1936 Constitution guaranteeing free education at all levels for all citizens. (The Constitution was amended, but not until several years later.) This measure, clearly, was linked with the call-up to labour reserve schools: the expansion of academic secondary education was having an adverse effect on recruitment to the ranks of skilled workers.

Thus in the last days of peace the leadership was engaged in retrenchment, strengthening discipline, laying in stocks.

PLANNING AND ORGANIZATION

During the thirties there was developed the system sometimes known as the command economy, or the Stalin planning model, based on stern centralization. We have noted its emergence (Chapter 8). It is necessary briefly to sketch in its further development, since, apart from minor changes principally of designation, the system survived for several decades.

Its essential features were (to some extent still are) the following:

(1) State enterprises (other than those of local interest) were placed under the orders of the appropriate People's Commissariat (Ministry), and the director of each enterprise was subject to orders from his commissariat on all matters. Parts of the economy were under republican or local authorities, but much of their output and activity too was generally allocated and regulated by the centre, directly or indirectly.

(2) Plans had the force of a binding order given by a hier-

archical superior. These plan instructions covered such questions as the quantity and assortment of output, purchases of inputs (from whom, and in what quantities), the delivery obligations of the enterprise, prices, wages, staff establishments, costs, and much else besides, depending on the industry in question. Appropriate output plans were devised for enterprises engaged in retail or wholesale trade (value of turnover), transport (ton-kilometres), and so on. Amid all these planned objectives, one was at this period supreme: the gross-output indicator. In cases of conflict between various goals, this was usually the one to which priority was given.

(3) Plans were devised by People's Commissariats for their own enterprises, subject to the authority of the party and government, which laid down the general policy objectives and key targets. The coordinating and advisory body with a key role in the process was Gosplan, the state planning commission, which by the end of the decade was headed by an able young man, N. Voznesensky. It worked out the logical consequences of policies, reconciled them with proposals and representations received from below, and endeavoured to achieve consistency.

(4) Its chief methodological weapon was the system of 'material balances'. For the plan period (in detail for one year, in lesser detail for a five-year plan), Gosplan drew up a balance sheet in quantitative terms: thus, so much steel, or cement, or wool cloth, would be available next year (production, less exports, plus imports, plus or minus changes in stocks). Utilization estimates would then be made. If, as often happened, demand exceeded supply, utilization plans could be cut, or endeavours made to increase supply, and, in a long-term plan, investment decisions would be made to increase productive capacity. Any change in the plan required hundreds, even thousands, of changes in material balances. Thus a decision to make more tanks calls for more steel; more steel requires more iron ore and coking coal, in quantities which can be estimated from technical norms and past experience; these in turn may require more mining operations and transport; but these need more machinery, power, construction, wagons, rails, and to provide them still more steel is needed, and many other things too.

(5) The overriding criterion at all levels was the plan, embody-

ing the economic will of the party and government, and based
not on considerations of profit or loss but on politically determined priorities. Therefore in production and investment
decisions the role of prices was reduced to a minimum, and rate-
of-return considerations were ignored, or confined strictly to
choice of means to achieve given ends (e.g. electric as against
steam traction), though even in this narrow area they were
largely disregarded. Prices were out of line with costs, changed
at infrequent intervals and not even conceptually related to
scarcities, so the profit motive, had it been allowed, would
have operated extremely irrationally.

(6) It is true that under the system of khozrashchyot the
enterprises had financial autonomy, were encouraged to keep
down costs and to make profits. However, this was a means of
ensuring economy in the use of resources, and not a way of
determining what should be produced, or even how (with what
materials) it should be produced, for such things were specified
in the plan. Planners could and did take away any of the enterprise's assets without compensation if they so desired.

(7) The breakdown of aggregate material balances and output
targets into an operational plan required, in the last analysis,
detailed production and delivery plans for every major item of
output and input, for every enterprise. The operational plan was
made for a year, and for shorter periods. (The five-year plan was
not in this sense operational, since it was not an order to anyone
to do anything, though investment decisions were derived from
it.) The most burdensome task was that of material allocation,
i.e. the planned distribution of all important outputs between
the enterprises which require them (i.e. x tons of sheet steel to
plant A from plant B during the third quarter of 1936). For
key items, the task of allocation generally belonged to a branch
of Gosplan.[53] For less important items, it was the job of the
disposals department of the appropriate People's Commissariat.

The number of People's Commissariats in the economic field
grew rapidly. At the end of 1934 all manufacturing and mining
was covered by four Commissariats: for heavy industry, light
industry, timber, and food. The first-named, the largest, was soon
found to require further sub-division. This could be done by
taking one of the chief departments (glavki) of which the Com-

missariat was composed and promoting it to ministerial status, placing the task of coordination upon the shoulders of Gosplan and the Economic Council (*Ekonomsovet*, the economic committee of the government). Sub-division went further and further. By 1939 there were no less than twenty-one industrial People's Commissariats, with such designations as Textiles, Munitions, General Machine-building, Coal, Chemicals, Non-ferrous Metallurgy, etc. There was further sub-division in 1940–41.

As already mentioned, there was a close link between the plan and public finance. Since direct taxes were few it could be said that the state ran itself, and paid for defence, investment, subsidies, social services, out of the proceeds of the sale of goods made by its own enterprises. These reached the budget mainly via turnover tax and the bulk of the profits of the enterprises. Investment controls operated in large part via the process of financing them. Decisions on investment priorities could be enforced by requiring specific authorizations from the government for any project costing above a relatively modest sum.

The party maintained a grip on the economy at all levels. Not only did it lay down basic policy priorities, but its plenipotentiaries were repeatedly called upon to take decisions on who was to get or do what, ignoring the formal administrative structure. Appointments and dismissals of planners and senior managers were in practice made by the party's personnel departments. A senior party leader in charge of an economic sector – Kaganovich in transport, Ordzhonikidze in heavy industry, to take two examples from the first half of the decade – issued orders personally on all sorts of matters, great and small. It was said that Ordzhonikidze had a direct telephone line to every factory in his bailiwick, and constantly used it arbitrarily to shift around people, resources, equipment. The party's interference in everyday operations was particularly systematic in agriculture, but was frequent anywhere. Especially after 1936, officials of the NKVD (People's Commissariat for Internal Affairs, or police) exercised important supervisory functions through the economy, and they also ran a big economic empire using forced labour, until the break-up of this empire after Stalin's death.

In the everyday working of the system much depended on unofficial links between people at all levels, which helped to overcome many deficiencies and gaps in the plan. Sometimes these were illegal; often they made possible the fulfilment of the plan by improvisation of many kinds. The state tried to prevent this by threats of punishment – for instance, for directors who sold off equipment they did not need. Many stories and memoirs tell of sharp practice and breach of rules, without which it was impossible to survive. Penalties beset the managers; there was a decree (of 10 July 1940) making poor quality production a criminal offence. Persistent shortages of goods inevitably led to intrigue and string-pullings designed to persuade the allocation authorities that this or that project or enterprise was deserving of official priority. Of course, anything in which the central committee, and above all Stalin personally, took a direct interest could rely on getting all that it needed. Others had to make do with what was left.

GROWTH AND URBANIZATION

During the whole period 1928–40 a large part of Soviet growth was a consequence of the shift of labour out of agriculture to urban industry and construction sites, where, despite all the inefficiencies of the period, their productivity was very much higher. It was higher than it would have been in the village, but of course this labour was quite unaccustomed both to factory conditions and to urban living, and all this created acute sociological problems of adjustment, especially in the overcrowded conditions of the cities.

The two censuses of 1926 and 1939 give some numerical idea of what occurred:

(millions)	Total population	Urban	Rural	Per cent urban
1926	147·0	26·3	120·7	18
1939	170·6	56·1	114·4	33

10. *The Great Patriotic War*

PREPARATIONS

By the spring of 1941 the Soviet Union was preparing for war. True, Stalin placed exaggerated hopes on the Nazi–Soviet pact, continued export deliveries to Germany until the eve of the attack, failed to warn Soviet troops of its imminence. But even he well understood that a great trial of strength would come, even if he was tragically wrong as to its date.

Germany's economic power was greater than Russia's and she had at her disposal the industries of occupied Europe. Her armies were well equipped, and the equipment had been tested in the battlefield. Despite the very greatest efforts and sacrifices in the preceding decade, the Soviet Union found itself economically as well as militarily at a disadvantage.

The Soviet Union had vast mineral resources, and even vaster space into which to retreat. However, many of her resources were still undeveloped. Despite the big growth in output in the Urals, Siberia, and northern Kazakhstan, key industries were still situated in the vulnerable west. To cite a Soviet author:

Enterprises making high-quality steels were mainly in the south, in militarily the most vulnerable areas, while few duplicate plants were built in the east. The eastern areas were underdeveloped industrially. ... Ministries and departments, concerned with their own particular interests, emphasized development of industry in the central areas. Thus in the three'and a half years of the third five-year plan period investments in the economies of the Urals, Siberia, and the Far East combined were only 23 per cent, in the Central area 19·4 per cent, of the total. ... In 1940 the investments of state and cooperative organizations (excluding kolkhozes) in the Central area were 2·9 times greater than those of the Urals, 7·7 times those of West Siberia, 7·2 times those of East Siberia.

There was a severe under-development of energy resources in the east.[1]

In a sense these Soviet criticisms do not do justice to Stalin's achievements. It is true that there was a tendency on the part of officials in planning offices to set up new enterprises in already-developed areas of the centre and south, because they were under heavy pressure to achieve results with limited resources, and the easiest way to do this was to save on 'social overhead capital' by utilizing already-existing towns, railways, public utilities. However, the plans did include substantial developments in the east, of which the Urals-Kuznetsk combine was only the most spectacular. More, no doubt, would have been done had it been possible to foresee the extent of the retreat of the Soviet armies. One could hardly base an economic plan on the assumption that there would be hostile armies on the Volga and in the North Caucasus.

Much more serious was the effect of the purges on the military-industrial complex. First-rate generals, staff officers, designers, managers, were imprisoned or executed. Many of their replacements were second-rate, and decisions on technical matters were too often slow and wrong-headed. It was a tragedy that the key Defence Commissariat was headed by the colourful but incompetent crony of Stalin's, Voroshilov, and even more tragic that his three deputies were ignorant nonentities: Kulik, Mekhlis and Shchadenko. The result was that the adoption of modern weapons was so delayed that good-quality tanks and aircraft, which were a real match for the Germans', were not yet fully in mass production when the war began. This applied to the KV and T-34 tanks, to the Yak-1, MiG 3, LaGG-3 fighters, the Yak-2 'stormovik' fighter-bomber, and the Pe-2 bomber, which were all to show their high military worth as the war proceeded. In June 1941 few soldiers or airmen were as yet trained in the use of this modern equipment. The standard Russian fighter, the I-16, well known in the Spanish civil war, was actually a good deal slower than even the Ju-88 German bomber.[2] The Russian TB-3 bomber was so slow and poorly armed that its crews stood no chance, and much the same was true of the crews of the obsolete Russian tanks. There was a grave shortage of automatic weapons and anti-tank guns. In the circumstances, given the weakness of the leadership (other than Stalin himself) in decision-making, and frequent shortages of necessary materials and components, the defence industry commissariats and actual plant managers performed well before the war began,[3] extraordinarily well when war was raging. But the Soviet army was seriously under-equipped for its task, in terms of quality at least, when the Germans struck. There was not sufficient urgency in the last few months of peace, because neither Stalin nor his immediate colleagues believed that action was desperately needed.

The course of the war, and the causes of the initial disasters, are not in themselves our primary concern. But their consequences, as might be expected, had the most profound impact upon the economy.

INDUSTRY AND TRANSPORT

By the end of November 1941 the Soviets in their retreat had lost vast territories, which contained 63 per cent of all coal production, 68 per cent of pig iron, 58 per cent of steel, 60 per cent of aluminium, 41 per cent of railway lines, 84 per cent of sugar, 38 per cent of grain, 60 per cent of pigs.[4] Other major centres, notably Leningrad, were effectively isolated. This was a staggering blow. Not only was there a shortage of basic materials, but indispensable supplies of components, special steels, equipment, spare parts, were suddenly cut off. The Nikopol' manganese deposits were another serious loss. Big efforts were made to dismantle and remove enterprises and workers eastwards. However, apart from the fact that mineral deposits cannot be moved, the sheer speed of the German advance made evacuation impossible in many cases. None the less, in the period July to November 1941, there were evacuated from the threatened regions no fewer than 1,523 industrial enterprises, of which 1,360 are described as large-scale.[5] Everything possible was done to send eastwards truckloads of fuel, equipment, grain, cattle, amid tremendous difficulties and inevitable hardship. This entire operation was conducted in the face of daunting obstacles. The task was supervised by the Committee on Evacuation, set up as early as 24 June 1941 by the central committee and the government. The work had to be carried on in the very greatest haste, day and night. Thus 'in only nineteen days, from 19 August to 5 September 1941, there were removed from the *Zaporozhstal'* steelworks 16,000 wagons of vital machinery, including exceptionally valuable sheet-steel rolling-mill equipment. . . . The generator of the large turbine of the Zuevo power station . . . was dismantled and loaded in eight hours.' Trainloads of equipment and staff were sometimes mis-routed; there was some inevitable confusion, but on the whole the task was accomplished with remarkably few errors. The equipment was re-erected with speed, despite the greatest difficulties and hardships. Thus the source reports that 'the last train carrying the equipment arrived [in a Volga town] on 26 November 1941, and in two weeks' time, 10 December, the first MiG plane was assembled. . . . By the end of December the factory had already produced thirty MiG planes and three IL-2 stormoviks.'[6] The evacuated workers lived in over-crowded premises, and suffered from shortages of every kind.

The equipment was re-erected in any available space, but it often proved necessary to build temporary factories under the most adverse conditions at the beginning of winter. The many stories told of the great evacuation reflect great credit on those concerned. Ten million persons were moved eastwards. Nothing on this scale had ever happened before. Evacuated plants were gradually re-established in the east, especially in the Urals (667 enterprises), West Siberia (244), Central Asia and Kazakhstan (308). Eventually they were able to contribute to the war effort. To supply them, and to replace other sources, it was necessary greatly to expand coal mining in Siberia, the Urals, the Pechora basin in the north, the Moscow basin, and so on.

However, this took time, and meanwhile the Soviet economy suffered terrible blows. By November 1941 over three hundred armament factories were in occupied areas.[7] The Soviet troops abandoned much equipment in their retreat; output fell. In November 1941 industrial production totalled only 51·7 per cent of the output of November 1940. Only by March 1942 did the production curve show a steady upward trend. Matters were to improve very greatly thereafter.

Some of the losses in production were due, as might be expected, to the indirect consequences of enemy occupation, such as the lack of essential materials and components, but also to disruption of rail transport, acute shortage of fuel, the call-up of some of the skilled labour.

In 1942 the Germans occupied the North Caucasus and the Don area, and emerged on the Volga by Stalingrad. This cost the economy the best of the remaining grain lands and the Maikop oilfield, and for a time the transport of Baku oil was halted. This was a further severe blow.

War industry was reorganized under the authority of the State Committee of Defence, of which Stalin was chairman. This committee acted throughout the war as supreme authority on all matters, replacing (to all intents and purposes) the Council of People's Commissars, and its members or plenipotentiaries were repeatedly sent to key sectors to impose order and establish priorities. The various industrial Commissariats continued to function, executing the orders of the State Committee of Defence. The most vigorous efforts were made to create or adapt productive capacity in the unoccupied regions, particularly in the Urals and West Siberia. The result was as follows, in terms of calendar years (the production figures were at their worst, of course, in the winter of 1941–2):

	(1940 = 100)			
	1941	1942	1943	1944
National income	92	66	74	88
Gross industrial output	98	77	90	104
of which: Arms industries	140	186	224	251
Fuel industries	94	53	59	75
Gross agricultural output	62	38	37	54

(SOURCE: *Istoriya Velikoi otechestvennoi voiny, 1941–5*, Vol. VI, p. 45.)

The following table shows how output fell in the first period of the war:

	1940	1942
Pig iron (million tons)	14·9	4·8
Steel (million tons)	18·3	8·1
Rolling-mill products (million tons)	13·1	5·4
Coal (million tons)	165·9	75·5
Oil (million tons)	31·1	22·0
Electricity (milliard Kwhs)	48·3	29·1

(SOURCE: E. Lokshin, *Promyshlennost' SSSR, 1940–63* (Moscow, 1964), p. 52.)

'The production of rolled non-ferrous metals fell by the end of 1941 almost to zero, and that of ball-bearings, so vital for the production of aircraft, tanks and artillery, diminished 21-times.'[8] One can imagine the gigantic problems which were faced in reconstructing the Soviet economy onto an effective war footing. Yet by re-deployment and improvisation, and the imposition of ruthless priorities, even in 1942 the arms industry managed to produce 25,436 aircraft, 60 per cent more than in 1941, and 24,688 tanks, or 3·7 times more than 1941.[9] Mobilization for war was extremely thorough. Control over all resources was very strictly centralized, and both materials and labour were directed to serve the war effort, to a degree unknown elsewhere. In 1940, 15 per cent of the national income was devoted to 'military purposes'. In 1942 the figure had risen to 55 per cent,[10] perhaps the highest ever reached anywhere. No doubt the experience of centralized planning in the previous ten years was a great help. In the process of tightening control over resources the government resorted to quarterly and even monthly plans, in far greater detail than in peacetime. The practice of material balances was used successfully to allocate the materials and fuel available between alternative uses in accordance with the decisions of the all-powerful State Committee on Defence. (It is worth noting that so many of its members – Molotov, Malenkov, Beria, Voroshilov – became politically discredited in later years that few Soviet histories mention names.) An emergency war plan was adopted in August 1941, covering the rest of that year and 1942. There were annual economic-military plans thereafter, as well as some longer-term plans, including one for the Urals region covering the years 1943–7. Needless to say, wartime planning involved many errors, some of them 'to a considerable extent due to the personality cult of Stalin'. However, as in other warring countries, centralization was essential to mobilize resources, and the U.S.S.R., after suffering what could have been crippling losses in the first months of war, carried out centralization very effectively.[11]

Recovery in the second half of 1942 was the result of desperate efforts in the midst of continuing military defeats and retreats. At the end of the year the military tide turned, but little of the great improvement which occurred in 1943 was due to the reoccupation of territory. The Germans were so thorough in their wrecking that in 1943 the gross output of industry in the (Soviet) Ukraine was a mere 1·2 per cent of the total of 1940, though by the second half of that year Soviet troops had occupied Kharkov and by November had also captured Kiev.

A particularly notable achievement was the expansion of production in the Urals area. By 1945 well over half the metallurgical output of the Soviet Union was produced there (as against one fifth in 1940). Steel production in that area rose from 2·7 to 5·1 million tons in this period, coal from 12 to 257 million tons, electricity generation doubled.[12] Altogether 3,500 new industrial enterprises were built during the war, and 7,500 damaged ones restored.[13] There were some remarkable examples of improvisation. Most of the fuel used was wood, which was available locally, as coal, and the means of transporting it, were extremely scarce. Components and spare parts were made locally in any premises and by whatever means were available, since it was impossible to rely on long-distance deliveries. Whole new regional complexes came into existence.

According to the official history, Soviet industry produced the following during the war:

> 489,900 guns
> 136,800 planes
> 102,500 tanks and self-propelled guns

as well as vast quantities of ammunition of all kinds. The history points out that the following were imported from the United States and Great Britain:

> 9,600 guns
> 18,700 planes
> 10,800 tanks (some of them obsolete)[14]

It is true that the U.S.S.R. produced the bulk of what was used. Furthermore, to the great credit of designers and everyone responsible for manufacturing, the quality of a great deal of Soviet equipment was very good, the tanks being especially effective. True, the Red Army was somewhat backward in signalling equipment, and the air force was under-supplied with bombers. But it is quite beyond dispute that the vast majority of the best aircraft, tanks and guns were of Soviet manufacture

It is therefore not only a matter of (understandable) national pride, but also of fact, that Western aid supplied comparatively few of Russia's armaments.

The West contributed much more to road transport. One of the weaknesses of the pre-war Soviet economy was in the production of vehicles. In 1928 hardly any were produced, and several new factories had to be built in the thirties. However, the needs of mobile warfare could not be met by Soviet productive capacity. A large part of the growth in the number of motor vehicles in the armed forces, from 272,000 at the outbreak of war to 665,000 at the end, came from U.S. lend-lease.[15]

Rail transport, however, remained the key, and performed remarkably well in the face of truly formidable handicaps, not least of which were uncertain fuel supplies and inevitable overloading and under-maintenance. The cutting of direct communications by the Germans placed great additional burdens on the railways. Thus in the winter of 1942–3 it was necessary to transport Baku oil to Central Russia via Kazakhstan and Siberia by rail, since both the Volga water route and the North Caucasus pipe-line were cut. Coal also had to be transported for longer distances, until the Donets mines could be reactivated. Then the fact that arms and equipment had to be transported from distant industrial centres in the Urals and Siberia was a further source of great strain. Finally, as the Soviet armies advanced they had to restore lines and bridges wrecked by the Germans in their retreat.

Lend-lease and other imports provided a significant number of machine-tools (44,600), locomotives (1,860), non-ferrous metals (517,500 tons), cable and wire (172,100 tons); these deliveries certainly helped to overcome some bottlenecks in industry, transport and communications.[16]

The sternest priorities were imposed, and the slogan 'Everything for the front' was never more meaningful. With even key sectors of heavy industry handicapped by shortage of fuel, skilled labour and materials, it was natural that consumers' goods industries should be extremely hard hit, and the fact that the Germans occupied the principal food-producing areas caused a very sharp fall in the output of agriculture (in 1942 and 1943 it was under 38 per cent of the level of 1940), and the output of the food industry fell steeply, as did supplies of agricultural raw materials. It was hard enough to feed and clothe the army. Civilians had an exceedingly tough time. In 1942 textile production fell to a third of the 1940 level, meat and dairy produce to half, sugar to a mere 5 per cent.

AGRICULTURE

As already shown, losses were exceedingly severe. The principal source of vegetable oil (sunflower seed) was in occupied regions, and the potato crop diminished to a third of pre-war levels. The peasants who formerly grew industrial crops had to switch to food-growing in order to survive, and the cotton crop in Central Asia fell rapidly to 38 per cent of its 1940 level. Grain harvests were adversely affected by loss of the most fertile lands, by shortages of every kind of labour, of haulage power (tractors and horses alike were mobilized), of fertilizer, of fuel, of spares and equipment of every kind. The following statistics require no comment:

Grain

	1940	1941	1942	1943	1944	1945
Area sown (million hectares)	110·5	81·8	67·4	70·7	81·8	85·1
Yield per hectare	8·6	6·9	4·4	4·2	6·0	5·6
Total harvest	95·5	56·3	30·0	30·0	48·7	46·8
						(47·3)
State procurements	36·4	24·4	12·4	12·4	21·5	20·0

(SOURCE: *Istoriya,* *1941–5*, pp. 67–9. Some figures calculated from index numbers. Several figures exist for 1945.)

Of course, these figures reflect the loss and reoccupation of territory. They exclude the areas under enemy occupation.

In the circumstances the reduced labour force in the villages performed its task very creditably. Since, as will be shown, the black-market or free-market prices of produce were exceedingly high, the temptation to pilfer, or work on one's own account, was very great. There was an increase in 1942 in the compulsory minimum of trudodni to be worked for the collective, from 90 to 100–120 trudodni per annum in most areas (50 trudodni for juveniles between 12 and 16). However, the average worked was well in excess of this; with every family having sons and husbands in the army, the sense of civic duty had its effect. A decree of 13 April 1942 mobilized non-agricultural labour to help bring in the harvest. Here again, an immense job of reconstruction had to be undertaken in reoccupied areas in and after 1943.

Livestock numbers fell rapidly at first. Particularly severe were the losses in horses and pigs. The situation was as follows:

	1940	1942	1943	1945
	(Million heads, end of year)			
Cows	27·8	13·9	16·4	22·9
Horses	21·0	8·2	7·8	10·7
Pigs	27·5	6·1	5·5	10·6

(SOURCES: *Istoriya*, *1941–5*, p. 68; *Selskoe khozyaistvo SSSR* (1960), p. 263.)

There is a first-rate and well-documented account of the position of the peasantry during the war by the Soviet historian Yu. B. Arutunyan.[17] With men mobilized and few tractors and insufficient horses, the burden and the hardships were borne in the main by women. There were, perhaps inevitably, errors of planning, as when orders were issued to increase sown area even when the means to harvest it were lacking. Excessive procurement quotas were sometimes enforced by harsh measures: '(individuals) who had no grain were compelled to sell their house, their belongings, their livestock', while party and farm officials who failed to meet delivery quotas were dismissed and tried as saboteurs.[18] Pay per trudoden (workday unit) fell, in both cash and kind, to extremely low levels:

Average distribution per workday unit, U.S.S.R.

	1940	1941	1942	1943	1944	1945
Grain (kg)	1·60	1·40	0·80	0·63	n.a.	0·70
Potatoes (kg)	0.98	0·33	0·22	0·40	n.a.	0·26
Cash (roubles)	0·98	1·07	1·03	1·24	1·12	0.85

(SOURCE: Arutunyan, p. 339, citing archival materials.)

The money meant little, as there was hardly anything to buy. 'The peasant received from the kolkhoz less than 200 grams a day and about 100 grams of potatoes' (per mouth to feed). These averages include tractor and combine operators, who received much more than the rank and file: 'in 1945 the tractorists received 2·8 kilograms of grain on average, while kolkhoz members got 0·6 kilograms'.[19] No *uravnilovka* (egalitarianism), even in the war years! Private plots were essential for physical survival.

The leadership decided, in November 1941, as an 'emergency measure', to reintroduce political departments (politotdely) in state farms and MTS.[20]

The administration of agriculture in occupied Russia is a major question of its own, and cannot be gone into here. Sufficient to say that many peasants hoped that the German occupation would lead to the abolition of kolkhozes and the return of private peasant cultivation. They were disappointed. Despite the advice of German experts, the occupation authorities preferred to utilize the kolkhoz system to organize compulsory procurements of produce for the Germans. This, and the revulsion of feeling due to the brutalities of the occupying troops, led some peasants who at first met the Germans with indifference, or even welcomed them, to take to the forests and join the partisans. It was only well after the Stalingrad disaster that the Germans decided to bid for support by promising to abolish kolkhozes. It was too late.[21]

LABOUR, WAGES

The disciplinary measures of 1940 were further reinforced. Workers in industries connected with the war effort were mobilized and then placed under military discipline. So were transport workers. Holidays were suspended. By decree of 26 June 1941 there was monetary compensation for holidays lost, but on 9 April 1942 this payment was cancelled. Overtime was compulsory. No one was allowed to leave his job of his own volition. Anyone not actually 'engaged in social work' was liable to mobilization: 'In 1943 alone labour mobilization provided 7,609,000 persons: 1,320,000 for industry and construction, 3,380,000 for agriculture and 1,295,000 for timber.'[22] Pensioners, juveniles and other normally inactive persons volunteered or were drafted. By decree of 28 July 1941 pensioners were allowed to retain their entire pension in addition to their wage.

As in other warring countries, large numbers of women took over industrial tasks normally performed by men. Retraining was undertaken on an immense scale, usually on the job. While compulsion was used, and breaches of discipline were severely punished, the achievements of those engaged in the war economy must receive the highest possible praise. They worked under extremely unfavourable material conditions. Food was short, 'housing' may have consisted of a corner of a room or a bunk-bed in an overcrowded hostel. Thus the average 'housing' per worker in the coal industry fell to 1·3 square metres per head, or 13 square feet![23] Clothing was often unobtainable, and footwear was irreplaceable if it wore out. The army naturally took priority. In 1943, as against 1940, there was available for sale: 14 per cent of the quantity of cloth, 10 per cent of clothing, 16 per cent of knitwear, 7 per cent of footwear.[24] Heating in winter was sometimes adequate, sometimes not. There were long queues, and the rations were sometimes unavailable. The word 'hardship' hardly conveys how people lived in these terrible years. But this, of course, was quite different from the bad years of the early thirties, which were the result of domestic policies. The sacrifices were borne to save Russia from a deadly enemy.

There was a very substantial expansion in small-scale market-gardening on allotments on the outskirts of towns, and factory canteens provided inexpensive meals. This was essential, since, as we shall see, any food in excess of the modest ration was extremely expensive.

Wages were kept in check fairly effectively, and in some sectors of the economy (administration, medical and health services, etc.) there seems to have been very little increase. Wages did rise in industry: 'Monthly average pay of workers in all-union industry rose from 375 roubles in 1940 to 573 roubles in 1944, i.e. by 53 per cent. . . . The pay of engineer-technical staffs in all-union industry rose from 768 roubles a month in 1940 to 1,209 roubles in 1944.'[25] (But there could have been little increase in pay per hour.) Key industries fared best. Statistics of average pay for all categories of workers and employees which have recently been published, which are not strictly comparable with earlier data, give a figure of 434 roubles per month, 5,208 a year for 1945.[26]

Productivity was adversely affected by call-up of skilled labour, and all the other troubles of wartime, particularly in such branches as mining and textiles. But productivity in the arms industry rose rapidly as mass production methods were applied to the new types of weapons and to ammunition. For example, a T-34 tank required 8,000 man-hours to produce in 1941, but only 3,700 in 1943.[27]

TRADE, PRICES

Ration cards were introduced in Moscow and Leningrad in July 1941. In subsequent months all towns were rationed for food as well as for manufactured goods. Some rural residents (over 25 million in 1944) received a bread ration. Supply of the few manufactured goods to peasants was linked with the fulfilment or over-fulfilment of delivery quotas. The ORSY (Departments of workers' supply, see page 251) were very greatly expanded, as were their market-gardening activities. As in the hard days of the early thirties, the significance of catering establishments increased greatly, especially in factories and other institutions.

Rationing was differentiated by categories, with privileges for those engaged in important sectors and on heavy work. Bonuses sometimes took the form of an extra meal in the factory canteen. A significant contribution from America was tinned meat, sales of which in 1945 were forty-six times above the levels of 1940, due partly to increased imports.[28]

The actual rations in December 1943 in Moscow were as follows, according to U.S. official sources quoted by S. Schwarz (in his valuable study to which several references have already been made); all figures are in grams:

Category	Bread (per day)	Groats* (per month)	Meat and Fish (per month)	Fats (per month)	Sugar (per month)
I	650	2000	2200	800	500
II	550	2000	2200	800	500
III	450	1500	1200	400	300
IV	300	1000	600	200	200
V	300	1200	600	400	300

* e.g. oatmeal, barley, etc.

(I = heavy workers; II = ordinary workers; III = office staffs; IV = dependents; V = children under 12.)

Vodka prices were increased five-fold soon after the war began. There were only minor increases in prices of basic foodstuffs, amounting by the end of 1942 to 11·6 per cent. Prices of manufactured goods in state shops (when obtainable) rose by 26·4 per cent. While not negligible, the rise in prices was kept relatively small (vodka apart). None the less, due to the heavy increase of vodka prices, the retail price index for official trade rose in 1942 to 156 per cent of June 1941.[29] 'Commercial' stores, selling rationed goods at very high prices, were opened in 1944. By 1945 the overall official price index was 220 (1940 = 100); by then the 'commercial' prices were somewhat lower than in 1944, though many times above the rationed price (examples will be cited in Chapter 11).

Since the volume of goods available declined, so did the turnover of state and cooperative trade (including catering), as the following figures show:

1940	1941	1942	1943	1944	1945
		(millions of roubles)			
175·1	152·8	77·8	84·0	119·3	160·1

(SOURCE: Malafeyev, *Istoriya tsenoobrazovaniya v SSSR* (Moscow, 1964), p. 222.)

It is true that numbers employed declined too, owing to military call-up and enemy occupation; thus in 1942 the numbers were 59 per cent of 1940, rising again to 87 per cent in 1945.[30] None the less it is evident that total incomes grew much more rapidly than the volume of goods available in shops. The extra purchasing power forced up prices in the limited free market.

Prices there show the acute shortages and hardship of the period. The all-union average figures were as follows (and it should be noted that free-market prices in both 1940 and 1941 were perhaps 75–100 per cent above official prices):

Free-market prices

	July 1941	July 1942	July 1943
Grain and products	100	921	2,321
Potatoes	100	1,121	2,640
Vegetables	100	711	2,138
Meat	100	769	1,278
Dairy produce	100	1,160	1,875

(SOURCE: Malafeyev, op cit., p. 230.)

But these indices conceal immense regional and seasonal variations, rendered inevitable by the extreme shortage of transport.
Even in peacetime the variations were (and are) substantial in
the free market. But in wartime they became exceptionally large.
The following prices, in roubles, were registered on 15 July 1943:

	Lowest price	*Highest price*
Rye flour (kilograms)	45·0	300·0
Potatoes (kilograms)	14·0	300·0 (L)
Beef (kilograms)	40·0	467·0 (M)
Milk (litre)	7·0	162·0 (L)

(SOURCE: Malafeyev, op. cit., p. 232, citing archives.)

(L) = Leningrad, (M) = Moscow. The lowest prices were all in Central
Asia.

The very highest of the prices quoted for milk and potatoes, in
Leningrad, were perhaps exceptional, in that the city was still
partially blocked by the enemy (though no longer suffering the
appalling conditions of the winter of 1941–2, when over
630,000 civilians starved or froze to death). But in Central
Russia the index for all products in 1943 was 2120 (1941 = 100),
while in Kazakhstan, Central Asia and Transcaucasia it was
between 1205 and 1322. The *average* price in the free market
in Central Russia of a kilogram of rye flour in that year could
hardly have been lower than 150 roubles, at a time when most
citizens earned less than this in a *week*. This is a measure of the
degree of suppressed inflation, of the shortage of goods in general,
and of the relatively restricted size of the free market.

Those peasants who were able to sell at these high prices
became rich. Many held on to their gains, waiting for the day
when there would be something to buy with their roubles. In
Chapter 11 it will be shown how they were prevented from
utilizing their gains.

As territories were reoccupied and agriculture and transport
gradually restored, conditions became somewhat easier. Reflecting this, free-market prices began to fall, after reaching a (season-
ally influenced) peak in April 1943, when they were 1602 (1940 =
100) for the country as a whole. In October 1943, after the harvest,
they fell to 1077, which was, however, higher than the figure
for October 1942. But in April 1944 prices rose only to 1488, and

there was a rapid fall to 758 in October 1944. In April 1945 the index fell to 737.[31] Evidently life was becoming bearable. The share of the free market in total turnover rose greatly; 46 per cent of the total *value* of all purchases in 1945 were in this market.[32]

Malafeyev's calculations led him to the conclusion that real wages in 1945 were roughly 40 per cent of their 1940 level,[33] after allowing for the low ration prices, high state 'commercial' off-ration prices and the free market. It is clear that, given the acute shortages, any such figures are necessarily extremely rough.

Costs of production in industry rose, particularly in labour-intensive industries, in which the impact of increased wages was combined with that of greater inefficiency, due to untrained labour, supply breakdowns and other difficulties. This contrasted with the large increase in productivity (and therefore reduced costs) in the armaments industries, which enabled substantial cuts to be made in prices of tanks, guns and aircraft. Prices of basic materials and fuels were unchanged, despite the fact that it was necessary to subsidize them out of the budget. Freight charges were also unaltered and also called for subsidies.

Agricultural procurement prices were unchanged throughout the war.

FINANCE

Wartime budgets were as follows:

	1941	1942	1943	1944	1945
	(Milliards of roubles)				
Revenue	177·0	165·0	204·4	268·7	302·0
Expenditure	191·4	182·8	210·0	264·0	298·6

(SOURCE: Malafeyev, op. cit., p. 234.)

It can be seen that there was a deficit in the first three years, which contributed to inflationary pressure. Revenues were adversely affected by the fall in turnover tax yield, owing to the drastic fall in consumers' goods output. It produced only 66 milliard roubles in 1942, against 104 milliard in 1940.[34] There was little chance to recoup this by price increases, since, with the important exception of vodka, the increases were small, until the introduction of 'commercial' shops in 1944, which contributed to an increase in revenues of 15 milliards. Altogether 70 per cent of all revenues during the war were raised from 'the national economy', in the form of turnover taxes and profits in the main. There were also large increases in personal taxation and in bond sales, and the share of these in the budget was far higher than in peacetime.

Thus a decree of 3 July 1941 added amounts ranging from 50 per cent to 200 per cent to existing income taxes and agricultural tax, with some exceptions for families of serving soldiers. This was replaced on 29 December 1941 by a special war tax; kolkhoz members paid a tax which varied from 150 to 600 roubles a year per member of household, and increased taxes on a sliding scale were imposed on workers and employees. Revised rates were introduced on 30 April 1943. On 21 November 1941, bachelors and childless persons were subjected to a special tax. Local taxes (on buildings, land, carts, livestock, markets) were introduced or consolidated on 10 April 1942. However, many of these local taxes were paid by institutions (e.g. buildings tax by trade enterprises).

The following table shows the pattern of revenue and the growth of taxes and bond sales during the war:

	1940	1942	1943	1944	1945
		(milliards of roubles)			
TOTAL REVENUE	180	165·0	204·4	268·7	302·0
of which: Turnover tax	104	66·4	71·0	94·9	123·1
Profits deductions	22	15·3	20·1	21·4	16·9
Taxes on citizens	9	21·6	28·6	37·0	39·8
Bond sales	11	15·3	25·5	32·6	29·0

(SOURCE: *Finansy SSSR*, (Moscow 1956), p. 123.)

Several issues of war loans were floated. Thus on 13 April 1942 a 10 milliard rouble local issue was quickly over-subscribed: a 12 milliard rouble issue in June 1943 in fact produced 20·3 milliards; in May 1944 bonds worth 28·1 milliards were sold in six days.[35] Presumably there was a mixture here of enthusiasm, persuasion and semi-compulsion. Agricultural taxes were also raised. Peasants were assessed in 1939 on a sliding scale related to estimated incomes based on the possession of a cow or other animals, and for every 1/100 hectare sown to a particular crop. Thus, for instance, a cow was deemed to be 'worth' an income of 3,500 roubles in 1943.[36] We will have occasion to examine this species of tax more closely in Chapter 11. There were lotteries, collections of valuables, mobilization of cash reserves belonging to state enterprises. Also the printing press was used, and the total currency circulation increased by 3·8.[37]

Social services were rapidly restored. Large sums had to be paid out to invalids and orphans, and in 1944, no doubt influenced by the vast losses of manpower, generous allowances (and medals) were introduced for mothers of many children. Those with ten children became known as 'mother heroines'.

LOSSES

It is hardly possible to compute the losses suffered by the economy and the population of the Soviet Union. One of the most important losses was in numbers of people, especially men. Exact figures have not been published, though numbers of about 20 million are mentioned at times. These include many civilians, in Leningrad, in other besieged and bombed cities, as well as in mass graves all over the western areas. They include many millions killed at the front, and certainly over 2 million soldiers who died in captivity. Although Soviet records show that no less than 5,457,856 ex-prisoners and civilian deportees returned or were returned to the Soviet Union, some, especially West Ukrainians and citizens of the Baltic states, did not return. There were many deaths among civilians who could not stand the hardships of war, and the birth rate fell drastically (this affected the labour-intake figures of the late fifties). True, the tribulations of war led to the training of millions formerly unskilled. But it is evident that the human loss was very heavy indeed.

There are statistics of material damage. Of the 11·6 million horses in occupied territory, 7 million were killed or taken away, as were 20 out of 23 million pigs. 137,000 tractors, 49,000 grain combines and large numbers of cowsheds and other farm buildings were destroyed. Transport was hit by the destruction of 65,000 kilometres of railway track, loss of or damage to

15,800 locomotives, 428,000 goods wagons, 4,280 river boats, and half of all the railway bridges in occupied territory. Almost 50 per cent of all urban living space in this territory, 1·2 million houses, was destroyed, as well as 3·5 million houses in rural areas.[38]

Many towns lay in ruins. Thousands of villages were smashed. People lived in holes in the ground. A great many factories, dams, bridges, which had been put up with so much sacrifice in the first five-year plan period, now had to be rebuilt. A daunting task awaited the survivors, once the victory celebrations were over. Indeed, the work of reconstruction was begun long before, though at first with few resources and fewer men.

The following is an example of wartime reconstruction, and of the means adopted to achieve it.

On the decision of the Leningrad city soviet and the *gorkom* [party] bureau it was stated that the entire able-bodied population of Leningrad, Kolpino, Petrodvorets, Pushkino and Kronstadt was to work on reconstruction, with the following work periods: for workers and employees with an 8-hour working day and those employed by military units: 30 hours a month; for workers and employees with a longer working day, and for students and schoolchildren: 10 hours a month; for citizens not working . . . 60 hours a month. Workers and employees are to work outside their usual working hours.[39]

No doubt many other local authorities took similar measures. Probably a great many citizens volunteered. Certainly this had been so when the survivors of Leningrad's dreadful winter of 1941–2 set about cleaning up their city in the spring. But there was also labour conscription and compulsory unpaid work.

In many cases, a remarkable job was done, and quickly. Priority was given to the reactivation of key industrial plant and mines, and press reports in 1944 are full of references to the restarting of production, especially in the industrial areas of the Ukraine.

On 9 May 1945 fighting with Germany was over, and the Red flag had been flying over the Reichstag in Berlin for a week. It is impossible to overstate the effect of this dearly bought victory on the morale and consciousness of the Russian people. Stalin was now the great war leader, who had led them to victory.

Were his policies therefore proved right by history? Some would argue so. A colleague once said: 'The result of the battle of Stalingrad showed that Stalin's basic line had been correct.' An unsympathetic critic retorted: 'Perhaps, if a different policy had been followed, the Germans would not have got as far as Stalingrad.'

There can be no way of judging between these two views. We know only that, in the end, Stalin's system stood up to the test of battle, though after heavy losses.

11. Recovery and Reaction

This chapter deals with the last years of Stalin's life. It is an oddly shapeless period. On the purely practical side, it was of course dominated by reconstruction and rebuilding, with the priorities affected increasingly by the impact of the cold war and the resultant arms race. However, economic policy, organization, ideas, rapidly became frozen into their pre-war mould. Stalin spoke seldom, party congresses were not called, even central committee meetings were rare (and virtually unreported) occasions. An oppressive censorship made public discussion of serious matters impossible. Numerous, usually unexplained, reorganizations of the ministerial structure made little difference to the actual functioning of the system. Central control was maintained, so was the policy of imposing disproportionate burdens on the peasants, but the issues involved in such policies were submerged beneath evasive formulae or self-congratulatory clichés. At no time, before or since, were Soviet publications more empty of real matter. This somewhat dreary intellectual scene contrasted with some remarkable achievements in rebuilding and re-equipping the economy.

THE ECONOMY IN 1945: THE FOURTH FIVE-YEAR PLAN

The Soviet state emerged in triumph from the trials and tribulations of war. The economy, though it had recovered from the low point of 1942, was seriously damaged by the war, and the people were exhausted. The task of reconversion and reconstruction lay ahead.

The western half of European Russia and virtually all the Ukraine and Belorussia were wrecked. 25 million people were homeless, 1,710 towns and 70,000 villages were classed as 'destroyed'.[1] Communications were disrupted. Ploughing had to be undertaken with cows or any other haulage-power that could be found or improvised. Millions of soldiers returned to the task of rebuilding their homes with their own hands. Millions, of course, never came back at all, and great numbers of widows and orphans, especially in villages, had to rebuild their lives as best they could alone. Many peasant soldiers had acquired new skills in the armed forces and went to work in industry. Shortage of men in the villages was henceforth a major social and economic problem.

The State Committee of Defence remained in full command as the all-powerful war cabinet, until it was abolished, and the regular governmental organs restored, on 4 September 1945. It issued orders for the rapid reconversion of war factories to civilian production; enterprises and the appropriate People's Commissariats were instructed to submit proposals as to what goods they should be producing. On 19 August 1945, while the Soviet army was completing its advance in Manchuria against the Japanese in that brief campaign, Gosplan was instructed to draft a five-year plan covering the period 1946–50. Its guiding light: to exceed pre-war output by 1950.

Various wartime measures were being relaxed. Thus on 30 June 1945 the granting of vacations was resumed, on 31 December the special war taxes were abandoned.

Reconstruction was hit by the precipitate cessation of lend-lease in August 1945. Not only was there much to repair, but the Soviet (like the British) balance of payments was in acute disarray as a result of the war. The importance of aid in 1945 can be measured by citing the foreign trade figures of that year:

	(millions of roubles)
Exports	1,433
Imports	14,805

(SOURCE: *Ekonomischeskaya zhizn SSSR*, p. 438.)
NOTE: Quite different figures are cited in foreign trade handbook; presumably they exclude lend-lease.

For most of 1945, therefore, American deliveries were still very significant. There was also aid for devastated regions under the auspices of UNRRA, and a senior American aid official, Marshal MacDuffie, has left us a vivid account of conditions in the Ukraine, where he had frequent dealings with the party's first secretary there, N. S. Khrushchev.[2]

Immediate action was taken to insist on reparations from ex-enemy countries, whether or not these now had pro-Soviet or even communist-led governments. Hungary, Bulgaria, Rumania, and especially Germany, were made to deliver all kinds of equipment and materials. Such German factories as the Zeiss works in Jena were dismantled and taken away to Russia, and some of the workers too, to train Russians in their highly specialized trades. Rails were taken up and used to rebuild Russian railways. At first these claims for reparations were recognized as applying to all of Germany, and some factories in the Ruhr, and other areas occupied by American and British forces, were dismantled also. However, with the growing tension of the cold war, direct reparations in Germany were largely concentrated, after 1945, in areas under Russian military control. Even in Manchuria, the incoming Chinese found that many factories had been taken away.

There have been controversies about the extent of reparation deliveries to the Soviet Union. Some Western analysts name huge sums, and claim that the success of Soviet reconstruction at this period owes much to this cause. There is no Soviet official total figure, and there seems little point in entering into arguments about just how much they did receive, especially as some of the gains took the form of half-shares in joint companies, the value of which (until the practice of such joint companies was abandoned ten years later) is difficult to compute. Similarly, we can only note, but cannot measure, the gain to the U.S.S.R. from trade treaties which worked out to the Soviet advantage to an unreasonable extent. But allowing for all this, surely the evidence is overwhelming that the achievement of reconstruction was due above all to the efforts of the Soviet people, though no one would deny that reparations deliveries helped. (More will be said in the next chapter about trade relations with other countries.)

On 9 February 1946 Stalin made his well-known election speech in Moscow. He extolled the achievements of the Soviet Union during the trials of war, cited many figures on the achievement of the armaments industry. He then looked forward to a long-term perspective in which industrial output would eventually be treble that of pre-war. He spoke of achieving, in the course of about three five-year periods, an annual production of 60 million tons of steel, 500 million tons of coal, 60 million tons of oil.[3] If these could be regarded as targets for the year 1960, then his prognostications were on the modest side, but at the time they seemed over-optimistic:

	1945	1960 (estimated)	1960 (actual)
Steel (million tons)	12·25	60	65
Coal (million tons)	149·3	500	513
Oil (million tons)	19·4	60	148

The immediate tasks were very largely reconstruction and reconversion. In 1945 the output of industries in the areas at one time occupied by the enemy – and this included most of the developed European territories of the Union – was only 30 per cent of pre-war. Stalin laid great stress, as he already had before the war, on the sinews of national power; to quote a Soviet textbook on the period, 'since the possibilities of financing and supplying capital construction at this period were limited, the major part of resources were concentrated on the most important sectors of the national economy – on the restoration and development of heavy industry and rail transport'. 87·9 per cent of industrial investments in 1945–50 were directed to the producers' goods sectors, only 12·1 per cent to the light and food industries.[4] Though much was, of course, done to rebuild the wrecked cities, on the whole factories took priority over dwelling-houses, and there was also a priority scheme in rebuilding, some historic cities and provincial capitals being given first attention. Thus when I was in the Ukraine in 1956 Kiev had been wholly, Poltava almost wholly, rebuilt. But Kremenchug, a sizeable town on the Dnieper, was still mostly a heap of ruins. Doubtless the builders got around to Kremenchug in 1957. These remarks are not in the least intended to decry the efforts made to make life bearable after so much suffering and destruction.

The fourth five-year plan provided for the following:

	1940	1945	1950 (plan)	1950 (actual)
National income (index)	100	83	138	164
Gross industrial production	100	92	148	173
Producers' goods	100	112	–	205
Consumers' goods	100	59	–	123
Gross agricultural production	100	60	127	99
Workers and employees (millions)	31·2	27·3	33·5	39·2
Average wages (per annum)	4,054	(5,000)	6,000	7,670
Railway goods traffic (milliard tons kms)	415	314	532	602·3
Coal (million tons)	165·9	149·3	250	261·1
Electricity (milliard Kwhs)	48·3	43·2	82	91·2
Oil (million tons)	31·1	19·4	35·4	37·9
Pig iron (million tons)	14·9	8·8	19·5	19·2
Steel (million tons)	18·3	12·3	25·4	27·3
Tractors (thousands)*	66·2	14·7	112	242·5
Cement (million tons)	5·7	1·8	10·5	10·2
Cotton fabrics (million metres)	3900	1617	4686	3899
Wool fabrics (million metres)	119·7	53·6	159	155·2
Leather footwear (million pairs)	211·0	63	240	203·4
Sugar (million tons)	2·2	0·46	2·4	2·5
Grain harvest (million tons) (barn)	95·6	47·3	–	81·2
('biological')	119	–	127	120

* 15 h.p. units.

(SOURCES: *Ek. Zh.*, pp. 437, 441, pp. 502–3; E. Lokshin, *Promyshlennost' SSSR, 1940–63* (Moscow, 1964), p. 150; *Pravda*, 16 March 1946; *Nar. khoz.*, *1965*, p. 567, p. 461, p. 311; *Nar. khoz., 1963*, p. 501.)

Yet again it is necessary to draw attention to the inflated nature of the total output indices for national income and gross industrial production. The 'inflation' is particularly great in the period 1946–50 because of the continued use of a completely obsolete set of price relationships, supposedly still based on 1926–7 prices, which greatly overweighted the rapidly-growing machinery and engineering sector. There were other reasons. This was the apogee of the Stalin despotism and of statistical suppression (very few of the figures cited in this book for this period were published until well after Stalin's death). One cannot imagine any Soviet statistician daring to challenge index numbers as being too high; by contrast, the physical output figures were actually used in planning and, with one conspicuous exception, have been regarded as reliable by virtually all scholars. The exception, of course, is grain, and a glance at the above table will show how very great was the disparity. (It is very much to the credit of the late Dr N. Jasny that, working alone in America, he not only correctly estimated the extent to which the 1937 crop was exaggerated, but also deduced the fact that the post-war harvest figures were even more exaggerated, though even he understated the extent of the 'inflation'.)

But while one has to take with a pinch of salt certain of the indices, a glance at the reliable output data shows that very rapid progress was made.

The first full peacetime year, 1946, proved a very difficult one. There was a severe drought, which (as will be shown) hindered recovery in a number of respects. The process of reconstruction encountered many problems. Productivity was adversely influenced by a general feeling of relaxation after the stern discipline and privation of wartime. It was affected too by the need to re-train the labour force, which had become accustomed to producing tanks and guns, to produce other goods. Many workers were on the move, back to their old homes, once the wartime conscription of labour was relaxed. Demobilized soldiers – 3 million or so returned to civilian employment in the first year of the plan[5] – required time to settle down and learn their new occupations. Some aged persons and others who had been mobilized or volunteered for wartime work left their employment. Conditions of work were often still very hard, and the organizational problems of reconversion taxed the ingenuity of the planners and managers. In the circumstances it is hardly surprising that the industrial output plan for 1946 was not fulfilled. Civilian output rose by 20 per cent (steel by 9 per cent, coal and electricity by 10 per cent, mineral fertilizer by no less than 50 per cent). However, this did not outweigh the fall in armaments production, so total industrial output fell, according to the official data, by almost 17 per cent compared with 1945.[6]

After 1946 industrial output increased by very high percentages. This was the consequence of successful reconversion, re-training, the bringing into operation of damaged mines and factories and very considerable new investments. Many metallurgical plants and heavy engineering works in particular were modernized in the course of their reconstruction.

The investment plan for 1946–50 was reported to have been surpassed by 22 per cent.[7] Investments were directed above all to the formerly occupied regions, and the tempos of reactivation of damaged mines and plant were most impressive. Thus despite the flooding and wrecking of mines by the Germans, the Donets basin managed by 1950 to exceed its 1940 output, as the following figures show:

1940	94·3 million tons
1945	38·4 million tons
1950	94·6 million tons

The Ukraine's metallurgical output also reached, or approached, its 1940 level by 1950. Since the capacity of the Urals and Siberia, expanded in wartime by new construction and the moving of plant from the west, continued to grow, the net result was that the five-year plan was over-fulfilled in these sectors and the 1940 level far surpassed. The great Dnieper dam was rebuilt and began to generate electricity as early as March 1947. In the Ukraine electricity generation in 1950 also exceeded that of 1940. All this required very hard work, under adverse circumstances.

The revival of consumers' goods industries from the exceedingly low levels of 1945 was rapid in all parts of the U.S.S.R. By 1948 the wool industry surpassed its 1940 levels; cotton fabrics and sugar achieved this in 1950, and footwear in 1951.[8] None the less, the textiles and footwear plan was not fulfilled, as the table on page 291 shows.

In 1950 planners had every ground for satisfaction. Errors and difficulties there were in plenty, but achievements could be said to be great. The U.S.S.R. could face the arms race, which in 1950 was again beginning, with a stronger industrial structure than before the war.

CHANGES IN ADMINISTRATION AND PLANNING

The end of the war saw the liquidation of some of the people's commissariats which had been responsible for armaments industries (e.g. for tank production), and the creation of new ones for civilian output (for transport equipment, agricultural machinery, construction and road-building machinery, etc.). There was again a great wave, in 1946 especially, of creating new people's commissariats by sub-division. On 15 March 1946 the designation 'ministry' was substituted for 'people's commissariat', but this had no significance beyond restoring a word formerly regarded as bourgeois. For a while sub-division continued. Some major economic ministries were divided geographically: there was one for the coal industry of the western regions of the U.S.S.R. and another for the coal industry of the eastern regions of the U.S.S.R. The ministries for the oil industry and for fisheries were also divided into two on geographical principles. The Ministry of Light Industry, on the other hand, was divided into Light Industry and Textile Industry. There were also created at this time a number of separate ministries for construction, to administer the growing number of specialized and permanent building enterprises. In 1946–7 the number of industrial and construction ministries alone reached thirty-three, against twenty-one in 1939. The number of other economic ministries increased markedly in these years. The Ministry of Agriculture, for instance, was divided; the department of labour reserves was turned into a ministry and so forth.

To some extent, as already indicated, it was a matter of changing labels, of promoting to ministerial status former deputy-ministers in charge of glavki, the glavk becoming a ministry. The process of increasing the number of ministries, which had already begun before the war, was associated with the appointment of party leaders of politbureau status as 'overlords', in charge of a sector within which there was a group of ministries. No longer, as in the days of Ordzhonikidze when he was Commissar of Heavy Industry, was the party leader directly (and colourfully) in charge of operations. Now the ministers were in reality non-political specialist heads of nationalized industries, with a party leader supervising them from the Kremlin. No doubt this change was due in part to the growing complexity and size of the economy, as well as to the fact that the days of trouble-shooting commissars, in industry at least, were largely over. Arbitrary intervention, especially at Stalin's whim, was common enough, but it seemed sensible to leave day-to-day administration to the specialist ministers.

The union republics never had so few powers over their own economies as at this period. Indeed, if oil in the south-east was administered by a separate ministry in Moscow, what function was left to the Azerbaidzhan authorities in Baku?

However, the process of sub-division evidently went too far, and in 1947–8 there were many reunifications of ministries, though sometimes with altered names and responsibilities. Thus on 28 December 1948 the two coal ministries became one, the two oil ministries plus the oil supplies department were united in one Ministry of Oil Industry, and so forth. Many ministers now became deputy-ministers. Changes occurred also in the other direction: thus the Ministry of Metallurgy was divided, in December 1950, into two departments, to deal with ferrous and non-ferrous metals.

The fourth plan and its implementation was undertaken under the authority of Voznesensky, whose tenure of office as chief of the planners was linked with membership of the politbureau. But odd and still unexplained developments affected both Gosplan and Voznesensky. In December 1947 Gosplan was renamed: instead of being the State Planning Commission it became a Committee and limited to planning. Its supply functions were transferred to a separate organization, Gossnab (State Supplies Committee), and its responsibilities on technical progress to *Gostekhnika* (State Committee on the introduction of new techniques into the national economy of the U.S.S.R.).[9] Then on 10 August 1948 the Central Statistical Office was taken out of Gosplan and put under the Council of Ministers. Voznesensky himself was dismissed in March 1949, and later shot. He was replaced by one of his deputies, Saburov. It is possible that there was a dispute involving him over economic policy or prices; we shall see that major price changes announced in 1948 were partially reversed in 1950. But it seems just as likely that the shooting was connected with an intrigue against some Leningrad-based party officials, and had nothing to do with economic affairs.

The situation of enterprises *vis-à-vis* the ever-changing designations of planning offices and ministries did not undergo any significant change in these years.

AGRICULTURAL POLICIES:
EVER-TIGHTER CONTROLS

1946 was a very difficult year for agriculture. Shortage of manpower, tractors, horses, fuel, seeds, transport (and, in areas affected by war, of houses) initially delayed recovery. In 1946 the total area of land sown was only 76 per cent of that of 1940 (75 per cent in 1945). As if all this was not enough, a severe drought hit many areas. According to recently published statistics the grain harvest of 1945 was 47·3 million tons. In 1946 it was only 39·6 million.[10] Many went very short of food; Khrushchev later claimed that Stalin ordered grain to be exported when people were starving.[11] Serious depletion of reserves of food caused delay in abolishing rationing.

During the war, supervision over the operation of kolkhozes had been somewhat relaxed. Members were allowed in some cases to sow crops on collective land, and kolkhoz autonomy increased: party and government officials were otherwise engaged, and it was obvious that kolkhozes had to be allowed to judge for themselves what was possible amid universal shortage of everything. The governmental and party agencies therefore confined their efforts to procurements, and winked at breaches of rules in other respects. It was in fact widely rumoured that peasant-soldiers were told that the kolkhoz system would be relaxed or even abolished as a reward for victory.

Alas, Stalin took an early opportunity to reassert control. On 19 September 1946 there was adopted the decree 'On measures to liquidate breaches of the [kolkhoz] statute'. All lands acquired by private persons or institutions had to be returned to the kolkhoz. Violations of internal democracy within a kolkhoz were not to be tolerated (they were, of course!). Further decrees reasserted the primary duty of compulsory deliveries to the state, and the power of the procurement organs. To enforce central control over agriculture there then took place a truly extraordinary administrative mish-mash. Following the decree of 1946, a Council of Kolkhoz Affairs was set up to prevent

breaches of the kolkhoz statute, and to exercise general supervision over kolkhozes and MTS. Andreyev, a member of the
politbureau, was its chairman, and at this period he was agricultural 'overlord' for the party. Earlier in 1946 the Ministry
of Agriculture had been divided into three parts: for Food
Crops (*zemledeliye*), Industrial Crops and Livestock. It may be
said that this was a mere extension into the Ministry of Agriculture of the disease of sub-division, rampant at the time and
already noted in industry. But there is a clear difference: in
industry any enterprise was subject to only one ministry at any
one time, whereas virtually all kolkhozes had some livestock,
some industrial crop (sugar-beet, cotton, flax, sunflower, etc.),
and a food crop, and so were under different ministries in
respect of each, as well as obeying the Council for Kolkhoz Affairs,
not to speak of the procurement agencies and the local party
committee. This bureaucratic tangle was ended in February 1947
with the re-creation of a single Ministry of Agriculture 'in order
to eliminate organizational imperfections and parallelism'.[12]
The Council for Kolkhoz Affairs withered away.

However, central interference was not reduced, but rather
strengthened. The February 1947 plenum of the central committee decided that the much-needed expansion of agricultural
output was to be very carefully regulated, and every kolkhoz
was to have sowing plans laid down not only for categories of
crops, such as grain, but even for each kind of grain. The same
plenum reasserted the duty of enforcing the kolkhoz statute and
the absolute priority of state procurements. Procurement quotas
could now be varied within each region in the light of circumstances, a provision which in effect legalized local arbitrariness in
making delivery demands on kolkhozes. The supervisory role
of the MTS was to be strengthened by the appointing of a
'deputy director, political' to each MTS.

At this period the travopolye (grass rotation) system was
thought to be a panacea, and farms were instructed to employ
crop rotation schemes incorporating grasses, regardless of local
conditions, resuming the policies interrupted by the war. Stalin
was convinced by some adviser in 1946–7 that spring wheat was
superior to winter wheat, and Khrushchev later on told how he
resisted by various expedients orders to expand spring wheat

sowings in the Ukraine, where winter-sown wheat gives a much higher yield. Then in 1948 Lysenko triumphed over all critics, with the help of the party machine, and his ideas were pressed upon the farms, while real geneticists were dismissed. (Lysenko was one of a species of pseudo-scientific charlatan, whose ideas had for many years a great appeal to party officials, seeking a cheap way out of agricultural difficulties.)

The financial condition of kolkhozes was deplorable; the amount available to pay members was so exceedingly low that to this day no information has been published about it. Yet not only did the government not increase agricultural procurement prices, but it put additional burdens on the kolkhozes: instead of being able to obtain seeds from the Ministry of Procurements, they had to maintain their own seed reserves (decree of 28 July 1947, repeated on 29 June 1950), taxes on kolkhozes were increased (11 August 1948), and they had to set aside a greater amount for capital investment (16 February 1952). Since taxes on private plots were also increased (see below), it is as if Stalin was determined to make the peasants pay for the necessary post-war reconstruction. A useful concession, beneficial to kolkhoz finance, was that retail cooperatives were allowed to sell at or near free-market prices and to set up stalls in towns for the purpose.[13] They bought produce from kolkhozes at higher prices. However, this was stopped in 1948.

The burdens on kolkhozes were increased in October 1948 by the adoption of a great 'Stalin plan for the transformation of nature', which laid upon the farms in the steppe areas the duty of planting vast forest shelter-belts – at their own expense, of course, as well as providing for canals and irrigation. It later became clear that the efforts were wasted; the trees refused to obey Stalin and did not grow; very few canals were built. Then in April 1949 there was adopted a decree on a three-year plan for livestock, demanding a great expansion of livestock holdings and a 50 per cent increase in the output of milk, dairy produce, eggs, etc. Procurement quotas were correspondingly raised. Yet prices remained so low that this further impoverished the kolkhozes and their peasants. The plan was, of course, not fulfilled, but procurements were considerably increased.

Top priority was given to industry, and the villages were left without building materials or electric power. It must have been a source of wry (but secret) humour to observe the innumerable portraits in oils of Stalin gazing upon electric tractors (in fact there *were* no electric tractors) at a time when it was forbidden for kolkhozes to obtain electricity from the state's power stations.

The one bright spot was the rapid recovery of tractor and combine production, which enabled the MTS to do their work better.

Instead of offering higher prices the authorities offered medals. There was a stream of orders concerning the award of honorific titles to heroes of labour in the villages.

On 21 May 1947 the decision was taken to collectivize agriculture in the Baltic states,[14] and similar measures were taken in other 'new' territories. This took three years. No Soviet analysis of this process has ever been published. Some unofficial eye-witness accounts speak of threats, coercion, and deportations. Past experience leads us to suppose it must have been so.

AGRICULTURAL PRICES AND PEASANT INCOMES

We have noted the unfavourable terms of trade imposed on the villages by the practice of compulsory procurements at low prices. If, as seems to be the case, there was no increase at all in procurement prices between 1940 and 1947, then at that date the disparity between retail and procurement prices reached its highest point. The retail price index, as we shall see, reached 2,045 (1928 = 100) in 1947. In 1952 the averages for grain, beef, and pigs were actually lower than in 1940.[15] The price paid for compulsorily-delivered potatoes was less than the cost of transporting them to the collecting point, and this cost had to be borne by the kolkhozes, who thus in effect got less than nothing! It is true that retail prices fell after 1947, and in the early fifties prices of industrial crops were increased. But against this must be set not only the extra burdens mentioned (page 298) but also the substantial rise in the prices of fuel and building materials.

To give some idea of the effect of these prices on net revenues of kolkhozes we can cite costs in state farms. The average procurement price for grain in 1940 was 8·63 roubles per quintal, state-farm costs were 29·70 roubles. In 1952 the procurement price paid to kolkhozes was 8·25 roubles, and state-farm costs were 62 roubles! Yet state farms were able to obtain their inputs at state wholesale prices, while kolkhozes had to pay the much higher retail prices. Consequently real costs in kolkhozes were higher, or rather would have been higher had state-farm wages been paid to the peasant members.

The latter bore the brunt of the loss. In 1952, a better year than 1950, their incomes from collective work were as follows:

Roubles per trudoden'	*Cash paid to peasants (millions of roubles)*	*Roubles per annum per household (cash)*
1·40	12·4	623

(SOURCE: N. Khrushchev, *Pravda*, 25 January 1958.)

At this period payments in kind were of much greater importance than cash payments, but even so these are exceedingly low figures, in real terms somewhat below a good pre-war year. It should be noted that gross incomes of kolkhozes increased greatly, from 16·8 milliard roubles in 1937 to 42·8 milliards in 1952, but whereas in 1937 almost half of this was paid out to the peasants, the proportion in 1952 was under 29 per cent, reflecting the much higher prices of inputs, higher investment expenditure, taxes, etc.

In 1948–50 cash incomes must have been much lower, perhaps one rouble per trudoden' (these, it must be emphasized, are *old* roubles, ten present-day kopecks). At that time, therefore, the *average* cash income from collective work was such that twenty-eight trudodni (say twenty days' work) was needed to buy a bottle of vodka; a kilogram of butter equalled sixty trudodni, a poor-quality suit required well over a year's average collective 'wage'.[16]

The system of assessing trudodni was altered, by increasing work norms and devising complex bonus systems for over-fulfilment of plans. However, this in no way affected the key problem, that of providing a sufficient amount with which to pay the peasants. Lack of effective incentive, due to underpayment, contributed significantly to the difficulties of agriculture in these years.

The peasants were able to survive because of their private plots and animals. But in the absence of sufficient incentives for collective work these were seen as undesirable distractions. The heavy taxes levied during the war on private cultivation and animals were retained. It will be recalled that the taxes were based on a nominal 'rateable valuation'. In 1943, for instance, a cow was deemed to bring in an income of 3,500 roubles per annum, and at this date the tax ranged from 8 per cent to 30 per cent. The valuations were somewhat reduced after the war. Thus a cow became supposedly worth only 2,540 roubles, a pig 800 roubles of income (against 1,500 in 1943), potatoes 180 roubles per hundredth of a hectare (against 350); orchards stayed at 160 roubles. But free-market prices fell to a much greater extent than this, and the tax rate charged on this nominal valuation was several times increased, eventually reaching 12 per cent to 48 per cent. Whereas in 1943 the tax on an income assessment of 5,000 roubles was 540 roubles, by 1951 it had risen to 820 roubles.[17] The high taxes led to peasants reducing their cultivation and livestock. When Khrushchev remonstrated with Stalin, and said that peasants were chopping down fruit trees to avoid the very heavy tax on them, Stalin replied (or so Khrushchev said in later years) that Khrushchev was guilty of a 'populist' (*narodnik*) deviation in his attitude to peasants.[18] The growth of livestock numbers slowed down, and numbers of cows remained static, or fell, after 1949, partly because of lack of fodder for collective livestock, but especially owing to the fall in private livestock ownership due to taxes and also to centrally imposed limitation on pasture facilities. Numbers of private livestock in 1952 were far below pre-war levels, as the following table shows:

Livestock owned by kolkhoz peasants
(per 100 households)

	1940	1952
Cattle	100	86
(cows)	(66)	(55)
Sheep and goats	164	88
Pigs	45	27

(SOURCE: *Kommunist*, No. 1, 1954.)

Between January 1950 and January 1952 the number of cattle in private ownership fell from 29·0 to 23·2 million (including those owned by persons other than kolkhoz peasants).

But peasants who sold their cows or pigs, or eliminated potato plantings, faced other severe troubles. For their duty to make compulsory deliveries to the state on account of their private holdings was not dependent on the possession of livestock or the cultivation of the crop in question, and these obligations were particularly onerous in these years. Thus every household had to deliver, on average, 210–250 litres of milk a year[19] (also meat, vegetables, eggs, wool, etc.). They had to beg, borrow, or buy this milk if they had no cow. Only just over half of kolkhoz households had cows.

Kolkhoz market sales were still at a high level, despite the fall in prices in the period 1945–50, as the following figures show (the 1946–9 figures do not seem to be available):

Kolkhoz market turnover

1940	29·1*
1950	49·2
1951	50·8
1952	53·7

*Earlier sources gave this figure as 41·2.
(SOURCE: *Sovetskaya torgovlya* (Moscow, 1956), p. 19.)

Since *total* cash income for collective work even in 1952 was only 12½ milliard roubles it is obvious that at this date the subsidiary private economies of the peasants supplied the larger part of their incomes.

However, it must never be forgotten that there were great regional variations. Some kolkhozes benefited from higher prices for industrial crops: for instance in the cotton-growing farms in Central Asia. Others happened to be located within easy reach of a big city and made most of their money (as did individual peasants too) out of sales in the free market. But kolkhozes located far from cities and compelled to specialize in food crops and livestock were in a state approaching beggary. Many peasants fled. The population of such rural areas declined rapidly. To cite Khrushchev again: 'One would go through a village and look around, and have the impression that Mamai and his [Tartar] hordes had passed that way. Not only was there no new construction, but old structures were not repaired.'[20]

SOME PRODUCTION DATA

It is hardly surprising that, after recovering from war damage, Soviet agriculture remained in a very weak state until a drastic change of policy occurred after Stalin's death. There is no escaping the conclusion that he delayed long-necessary changes of policy by his obstinately hostile attitude to the peasantry.

In October 1952, a year of fairly good weather, Malenkov (Stalin was sitting alongside him) announced that the grain problem had been solved, with a harvest of over 8 milliard *poods*, i.e. 130 million tons. Reality was very different. Here are the relevant statistics (omitting the bad years 1945–6, already referred to):

	1940	1947	1948	1949	1950	1951	1952
			(million tons)				
Grain harvest, claimed (biological)	119	–	115	124	124	121	130
Grain harvest, real	95·6	65·9	67·2	70·2	81·2	78·7	92·2
Potatoes	76·1	74·5	95·0	89·6	88·6	58·7	69·2
Cotton	2·2	1·7	2·2	2·5	3·5	3·7	3·8
			(million head)				
Cows	28·0	23·0	23·8	24·2	24·6	24·3	24·9

(SOURCE: *Nar. khoz., 1965*, p. 310; *Pravda*, 20 January 1949, 18 January 1950, 26 January 1951, 29 January 1952.)

The much better performance of cotton is, of course, directly to be linked with the higher prices and greater incentives.

INTERNAL REORGANIZATIONS

In February 1950 it was declared that the *zveno* ('link') system of organizing farm labour was being over-used, and that the 'brigade' should be the foundation of work in the fields. This seems to have been part of a move to push Andreyev, an old Stalinist stalwart, out of his position as the senior party official in charge of agriculture. The point of the controversy was apparently as follows ('apparently' because this was not a period in which real argument was heard or debates were possible). The zveno was a small group of peasants, probably between six and ten persons, who were given a particular area of land to cultivate or a particular job to do. Andreyev favoured this because it was a way of avoiding 'lack of personal responsibility' (*obezlichka*) and facilitated payment by results. A brigade was much larger, often up to 100 strong, under a 'brigadier'. The trouble seems to have been due to excessive sub-division of tasks and land, which may have impeded the operation of large-scale mechanization, in grain farming especially, and also to the fact that very small groups could be members of one family and so acquire a proprietary feeling to the bit of land allocated to them. Brigades were henceforth to be the basis of operations and of payment by results too, though the zveno as a unit within the brigade continued to exist. Andreyev apologized for his errors.

Shortly afterwards there was another and much larger shake-up. It was decided that the kolkhozes were too small. In January 1950 there were over 250,000 of them. Khrushchev, who had recently moved to Moscow from the Ukraine, appears to have begun a campaign of amalgamations. By the end of the year the number of kolkhozes was halved. The process continued in subsequent years. By 1952 numbers had fallen to 97,000.

Why amalgamate? Two reasons seem to have been predominant. One was the undoubted fact that in the northern half of Russia the average size of kolkhozes was small, much too small to permit the introduction of the then fashionable travopolye crop rotation scheme. The second was lack of control. It was not so difficult to find reliable kolkhoz chairmen if there were to be fewer of them, local party secretaries had their task simplified, the MTS had fewer kolkhoz administrations to deal with, a much higher proportion of kolkhozes could have party groups. A decision on amalgamations, taken by the central committee on 30 May 1950, emphasized, as is customary in such cases, that the process must be 'voluntary'. As usual, it was not. Indeed, on 31 July 1950 the central committee had to demand measures against the slaughter of livestock which accompanied 'the work of enlarging small kolkhozes'.[21]

These amalgamations were undoubtedly linked in Khrushchev's mind with the concept of 'agro-towns', or new large urban settlements to which peasants would move from their old-fashioned villages and hamlets. After some preparatory publicity for the idea Khrushchev wrote an article in *Pravda* praising agro-towns. The next day the article was disowned by the device of an editorial note to the effect that it was for discussion only.[22] Shortly afterwards minor officials criticized the proposals for their 'consumer-orientated' approach, at a time when more production was needed. Later the 'agro-town' was attacked by Malenkov, speaking to the nineteenth party congress in October 1952. There were some grounds for his criticisms. Conditions in 1950–52 were not such that one could contemplate uprooting the peasantry and building thousands of modern urban-type settlements for them to live in. In practice nothing happened, and peasants went on living in their somewhat primitive villages. But the effect of amalgamations was to make kolkhoz management more remote, both figuratively and physically, and to complicate transport and organization on the larger farms.

Thus the agricultural situation in Stalin's last years was exacerbated by ill-judged interventions of authority, excessive centralization of decisions, extremely low prices, insufficient investment and lack of adequate incentives. High taxation, levied on private cultivation, did further damage. These conclusions are now accepted by every Soviet scholar. A fog of inflated statistics and misinformation (or censorship) hid the true state of affairs at the time from all but the acutest observers.

PRICES AND WAGES

There is a remarkable parallel between the course of prices in 1932–6 and in 1945–9. In both cases incomes exceeded plan, costs rose, wholesale prices of basic industrial products were held down and large subsidies paid. Meanwhile rationing of consumers' goods was accompanied by the emergence of very high 'commercial' prices, and the abolition of rationing was preceded by large increases in the ration price, culminating in the fixing of unified prices, somewhat below the high 'commercial' levels. A year or two later an industrial price reform sought to eliminate subsidies by means of a very large increase in prices of basic industrial materials and fuels. Precisely this happened in the first post-war years, as in the aftermath of the great leap forward.

To deal first with industrial wholesale prices. Costs rose rapidly, and subsidies became intolerably high, since many prices (for instance, for coal, timber, metals) were half or even below half of the cost of production. The following figures speak for themselves:

Subsidies to industry
(milliards of roubles)

1945	13·9
1946	25·8
1947	34·1
1948	41·2
1949	2·9

(SOURCE: Malafeyev, *Istoriya tsenoobrazovaniya v SSSR* (Moscow, 1964), pp. 246, 252.)

From 1 January 1949 industrial wholesale prices were increased by an average of 60 per cent, but timber, coal, iron and steel prices had to be raised by 3 to 4 times. Freight charges were also raised. One effect, as we shall note when we discuss finance, was to increase profits as a source of revenue and diminish the share of turnover tax, which was also being cut back as a result of reductions in retail prices. Turnover tax on producers' goods was abolished, except on oil and electricity.

In 1950 two price reductions were decreed, and average industrial prices fell by 20·3 per cent.[23] Further cuts in 1952 brought wholesale prices 30 per cent below 1949 levels. These cuts could have been a reaction to the arrest of Voznesensky, who had been responsible for the increases. Malafeyev argues that the 1949 increases were 'clearly excessive', and another Soviet analyst, Kondrashev, attached some weight to the efforts of ministries to get higher prices accepted, to 'guarantee for themselves unplanned accumulations'.[24] None the less, some of the reductions were plainly unsound. Thus a 25 per cent cut in the price of sawn timber in 1950, when even in 1949 the timber industry still required a subsidy, can hardly be described as reasonable.

The picture is very different indeed regarding retail prices. We have seen that in 1944 the dual price system was introduced. As in 1932–4, there was a low 'ration' (or 'normal') price, and a very much higher 'commercial' price, which was close to the free-market price. No Soviet source has been found which cites the commercial prices at this period, and Malafeyev, on whom we drew copiously for evidence on the prices of the early thirties, devotes exactly one line to the fact that commercial prices were introduced in 1944, and does not quote a single instance. It is therefore necessary to refer to evidence collected by the US Embassy in Moscow, and published in the *Monthly Labor Review* in July 1947. In July 1944, according to this source, the ration price of beef was 14 roubles per kilo, the commercial price 320 roubles (three weeks' wages for an average worker!). The very scarce sugar was 5·50 roubles on the ration, 750 roubles in commercial stores.

The system continued through 1945, though the commercial prices were reduced. In 1946 it was intended to eliminate multiple pricing and to abolish rationing. However, this proved impossible because of the extremely poor harvest of that year and it was announced on 28 August 1946 that de-rationing would be postponed for a year. It was announced in September 1946 that as a first stage ration prices would be substantially increased, and commercial prices cut. This diminished, but did not come near abolishing, the difference between them. According to the above source, ration prices rose as follows:

	Ration prices up to September 1946	*New ration prices*
	(Roubles per kilogram)	
Rye bread	1·00	3·40
Beef	14·00	30·00
Sugar	5·50	15·00
Butter	28·00	66·00
Milk	2·50	8·00

Commercial prices were much lower compared with 1945: thus beef cost 90 roubles (against 140), sugar 70 (150), still well above the ration price but no longer astronomically so. There was a similar pattern for manufactured consumers' goods.

In compensation for the very large increase in the prices of rationed commodities, on which the lower-paid workers had almost totally to rely, there was the so-called 'bread supplement' wage increase. The maximum increase, 110 roubles a month, went to the lower-paid. The middle-grade worker gained 90 roubles. Those earning over 900 roubles per month received nothing extra. Perhaps it was surmised that the better-paid groups benefited from the fall in 'commercial' prices, which they alone could afford. Retail prices at this period, reflecting the acute shortages (of everything), were exceedingly hard on the lower-paid. Even including the increase, many of the less-skilled workers were earning under 300 roubles a month (as is evident from the number of those who benefited from minimum-wage legislation ten years later when the minimum was fixed at 300). For them life was harsh indeed.

High commercial prices survived, because there was an excess amount of money in circulation, much of it a product of wartime inflation, and the government decided that the abolition of rationing had to coincide with a currency reform. A currency reform was duly decreed on 14 December 1947. All cash in the possession of individuals was exchanged in the ratio of 1:10, so that cash hoards lost the bulk of their value, and many a peasant found his wartime savings wiped out. However, all holdings in savings banks below 3,000 roubles were exchanged at face value, 1:1, with lower ratios for larger sums. (The peasants seldom used savings banks.) All state bonds were converted at the ratio of 1:3, i.e. became worth a third of their value (and were converted into bonds carrying a lower rate of interest, 2 per cent). This reform must not be confused with the 'new rouble' (or 'new franc') reforms of subsequent years, which increased the *value* of the rouble ten-fold (or the franc 100-fold). Incomes remained unchanged: a salary of 1,000 roubles in November 1947 was still 1,000 roubles in January 1948. So the operation eliminated the bulk of cash holdings, and greatly reduced bond debts. Simultaneously rationing was ended. The state had accumulated enough material reserves to abolish 'commercial' prices and in some cases to fix the general level of unified retail prices a little below the greatly increased ration prices. Rye bread, so vital to the poorer strata, was reduced to 3·00 roubles per kilo, from 3·40 roubles. The net effect was an overall reduction of 17 per cent in state retail prices.[25] The new prices were realistic, on the whole, as is shown by the fact that free-market prices in 1948 were at or sometimes even below those ruling for foodstuffs in state shops.

Thus the Soviet leadership had the courage (or the nerve) to impose prices which fully reflected the great all-round shortages which characterized the first post-war years. The net effect on real incomes may be calculated by reference to Malafeyev: his price index for 1947 was 321 (1940 = 100).[26] Post-war wage rates remained a secret until recently, but the 1946 average is now known to have been 475 roubles per month. Allowing for the increase in the third quarter of 1946 this suggests a 1947 average of perhaps 550 per month, or 6,600 per annum. This gives a wage index of about 165, and a real wage index for the year of only 51. (It is true that prices of services increased by less than those of goods, but, if account is taken of the many shortages which are not reflected in price indices, conditions were fully as bad as is indicated by the index.) But from December 1947 conditions improved.

The practice of virtually compulsory bond subscriptions continued in these years, representing a further burden of three to four weeks' wages, when it was very hard to make ends meet.

However, as the flow of consumers' goods increased faster than the rise in wages, the Soviet authorities were able to reduce prices in subsequent years. Prices were cut every spring from 1948 to 1954. By March 1950 the average reduction in retail prices was 40 per cent, compared with the last quarter of 1947. The price of rye bread, which had been three roubles in December 1947, was reduced to 1·40 roubles by 1950. Wages rose, to an average of 7,668 roubles per annum in 1950, representing an increase of roughly 16 per cent over 1947,[27] and there was indeed a sharp and striking recovery in living standards. However, exaggerated claims have been made by Soviet statisticians. Thus incomes of 'workers and peasants' in 1950 were repeatedly alleged to have been 62 per cent above 1940. Yet, as we have seen, the peasants' incomes were adversely affected by government policies, while real wages had barely reached 1940 levels (the wage index stood at 191, the price index at 186, according to Malafeyev). One must allow for social and other services, but even so an increase of 62 per cent is plainly out of the question.

After 1950 the picture is complicated by a growing disparity between state and free-market prices, due to the fact that official price cuts were beyond the economically justified. Thus by 1953 free-market prices were roughly 30 per cent above official retail levels. Wages continued to increase, and must have reached 8,100 roubles per annum, while the official retail price index fell to 146. Taking into account the very high free-market prices of 1940 (they were then 75 per cent above official prices) and the relatively stable prices of services (rents, etc.), Soviet official sources claimed a cost-of-living index of 122 (1940 = 100) and that real wages in 1953 were 65 per cent above 1940. It is to the credit of a Soviet scholar, Figurnov, that he pointed out the inconsistency of such claims both with the volume of retail sales and with the output of consumers' goods. His own re-calculation, which allowed for higher taxes and the virtually compulsory bond purchases, led him to a real wage increase of 43 per cent.[28] But of course this too represented a notable improvement over 1940, and still more over the first post-war years. These gains were very unevenly distributed. While, as we have seen, the 1946 wage increase had the effect of reducing differentials between low-paid and high-paid, there was no systematic revision of wage rates, and drift and inter-ministerial competition for labour led to the emergence of very marked discrepancies and illogicalities, which, in the absence of a Ministry of Labour, it was no one's responsibility to correct. Not until 1956 was there a move to bring order into the wages structure.

None the less, real wages did show a remarkable and welcome improvement in 1947–52.

To some extent this was facilitated by the severe burdens placed on the peasants. One must also make allowance for the very great shortage of housing, which the modest building programmes of these years did nothing to remedy. Stalin was at this stage fond of encouraging the erection of new skyscrapers with decorated towers, while maintenance of existing houses was disgracefully neglected. It was quite normal for a four-room flat to be occupied by four families, one in each room, sharing kitchen and bathroom (if any). Rents were low, it is true, but the persistence of such conditions for a generation and more did much to cause unhappiness and to coarsen life.

The average 1950 wage, 7,668 roubles, was much higher than the 6,000 envisaged in the five-year plan. But the latter was drafted before the substantial increases in retail prices decreed in 1946, and the wage increases which accompanied them. In the same way, retail trade turnover for 1950 was planned on the assumption of much lower prices than in fact ruled. So the original plan was knocked sideways.

A source of pressure on goods and services was the larger number of 'workers and employees' in the state sector: 39·2 instead of the planned 33·5 million. This caused some unplanned increase in total disposable incomes, as well as an extra strain on the inadequate housing in towns.

No doubt this also contributed to a constant pressure on the poorly developed trade network. Queues were the rule rather than the exception. Stalin's rationalization of goods shortages was frequently quoted at this time: 'The increase of mass consumption [purchasing power] constantly outstrips the growth of production and pushes it forward.'[29] Queues therefore could be made to seem a progressive feature of a socialist economy.

FINANCE

During the war extra direct taxes replaced part of the turnover taxes which were lost through the reduction in turnover. We have seen that these additional taxes were abolished at the end of 1945. Thereafter, turnover tax became once again the dominant source of revenue, particularly after the very substantial retail price increases of 1946. Profits were low until after the price reform of 1949, which eliminated the bulk of the subsidies and greatly increased budgetary revenue from profits. This is reflected in the figures in the following table:

Revenue	1947	1948	1949	1950	1951	1952
	(milliard roubles)					
Total	386·2	410·5	437·0	422·8	470·3	499·9
of which: Turnover tax	239·9	247·3	245·5	236·1	247·8	246·9
Profits tax	22·6	27·2	42·2	40·4	48·0	58·5
Direct taxes	28·0	33·1	33·7	35·8	42·9	47·4
Loan revenues	25·7	23·9	27·6	31·0	34·5	35·7

(SOURCE: K. N. Plotnikov, *Ocherki istorii byudzheta Sovetskovo Gosudavstva* (Moscow, 1954), pp. 379, 466.)

It is noteworthy that turnover tax revenue was more or less stable after 1948, while sales of consumers' goods rose very substantially indeed (retail trade turnover in 1952 was 393·6 million roubles, 19 per cent above 1947 by value, but 135 per cent above it in volume).[30] The price reductions of those years were largely made possible by reductions in rates of tax while turnover tax revenues were maintained by the increased volume. Quite clearly, the tax burden on consumers' goods was much lower in 1952 than in 1947.

The increase in yield of direct taxes was due partly to the impact of unchanged income tax rates on increased incomes, partly to the upward revision in the 'agricultural tax' levied on peasant private plots (see p. 301). Bond sales were maintained at a high level.

This was a period of maximum centralization. The bulk of state investments were financed out of the budget. The republics' powers over the financing of enterprises created within their borders were minimal.

TRANSPORT

The reconstruction of the railways after war damage was a great achievement. As statistics cited (p. 291) have shown, the railways surpassed their freight plan. This required great efforts, and in September 1948 political departments were re-created on the railways.

It must be borne in mind that over-fulfilment of transport plans is not always a sign of health: it might mean unnecessary cross-hauls, or mistakes in the planning of industrial location. In more recent years the authorities have rightly realized this. However, in the first post-war years the volume of freight carried did represent some rough index of recovery of the system.

Freight was given priority, and passengers without official reasons for travelling often faced long delays and queues. (In fact as late as 1956, when travelling in the Ukraine, I met peasants who had waited forty-eight hours for a train, and then could travel only after paying a 'soft-class' supplement.)

The equipment of the railways had been vastly improved in the thirties. But the introduction of new ideas slowed down, possibly reflecting the increasing age of key party figures, notably Kaganovich, who retained a kind of overlordship with regard to transport. He apparently expressed a preference for steam traction, or so his enemies later alleged, and diesel and electrification developments were delayed.

Large-scale canal construction was undertaken in Stalin's last years. The Volga–Don canal was one of many such projects. It is doubtful whether their economic value repaid the heavy expenditure involved.

FOREIGN TRADE: COMECON

The Soviet Union emerged from the war no longer isolated, no longer the world's only communist-ruled state. We have seen how, in the very hard first years of peace, the policy was to dismantle and acquire by way of reparations anything that could be taken from an ex-enemy state, even if it had now become allied with the Soviet Union.

Meanwhile, the cold war was developing, and this affected the trade behaviour of both sides in the conflict. The Marshall Plan proposals were put to a conference in Paris on 27 June–2 July 1947. Molotov represented the Soviet Union, which turned down the proposals and exerted pressure to ensure that their allies did likewise. In retrospect, we could say that the Marshall Plan could scarcely have been accepted by Congress had the American government not presented it as a measure to combat communism. This suggests that Molotov could have caused tactical embarrassment to Washington by agreeing to the plan, but aid would not necessarily have been granted to the U.S.S.R. if he had done so.

Moscow reacted to this and other elements of the cold-war situation by tightening its grip on the political systems of the countries that were now fast becoming its satellites.

As these countries slid further into a condition of political subservience they became subject to Stalin's will in matters of trading relations with the Soviet Union. Since there was no precedent for a theory covering trade policy between socialist countries, and since the priority of the Soviet Union's interest had become an article of faith among communists at this time, there were unequal trade treaties from which the U.S.S.R.

benefited somewhat one-sidedly. The extent of such benefits has been exaggerated by propagandists, and cannot be precisely measured. Thus Poland supplied the U.S.S.R. with coal at extremely low prices, but against this it is necessary to set the fact that the Soviet Union also supplied Poland with materials below world market prices. None the less, the fact that two ministers of foreign trade (in Bulgaria and Czechoslovakia) were executed quite specifically for bargaining too hard with the Soviet Union suggests that bargaining was not really equal.[31] After Yugoslavia's defection (1948), Stalin became deeply suspicious of nationalist deviations, and a great many communists in all East European countries were shot or imprisoned for giving too great weight to their countries' national interests.

But some sort of answer was needed to the Marshall Plan. The Soviet Union made a series of gestures: credit agreements were negotiated with Yugoslavia (25 July 1947), Bulgaria (9 August 1948), Czechoslovakia (7 December 1948). Half of the Roumanian and Hungarian reparations debts were written off (9 June 1948). East German reparation debts were not halved until May 1950. And finally the Council of Mutual Economic Assistance (COMECON) was set up in Moscow in January 1949.[32] Yugoslavia was by then excluded – and subject to total trade embargo – following the Stalin–Tito break in 1948.

COMECON in fact led a sleepy and inactive existence until well after Stalin's death, and the U.S.S.R.'s relations with its satellites were conducted, at this period, almost exclusively on a bilateral basis.

Trade relations with the West were meanwhile becoming increasingly affected by the prevailing political tension. This culminated in the Korean war, and the imposition by the West of far-reaching restrictions on trade with communist countries. This compelled them all to trade to an increasing extent with each other. East Germany and Czechoslovakia in particular had highly developed industries, and they became major suppliers of machinery and equipment to the U.S.S.R.

The triumph of the communists in China in 1949 was followed by the granting of substantial credits by the Soviet Union; thus on 14 February 1950 the U.S.S.R. granted China a credit of $300 million at 1 per cent interest. These and other economic aid agreements were accompanied, during the Korean war, by military aid programmes designed to help China and North Korea.

No aid agreements involving countries outside the Soviet sphere of influence were negotiated until after Stalin's death. The dictator took the view that the process of decolonization was in some sense a fraud, that Nehru was probably a Western agent.

One odd feature of the year 1950 was the decision to increase the nominal gold and foreign exchange value of the already greatly overvalued rouble, from 5·30 to 4·00 roubles to the U.S. dollar. This made all Soviet prices much too high. However, trade with Western countries was conducted in Western currencies, and there was no connexion at all between internal prices, foreign trade decisions and the official exchange rate. Or rather the connexion was purely one of statistics and accountancy. The rate was used to convert foreign currency into roubles, and thus Soviet exporting corporations, under the Ministry of Foreign Trade, tended to make large losses, which had to be made good out of the budget, while importing corporations made large profits, which mostly were transferred to the budget. It was, of course, quite impossible to use the exchange rate as a basis for economic calculation.

THE ATMOSPHERE OF LATE-STALINISM: SCIENCE AND TECHNICAL PROGRESS

A Soviet commentator wrote:

The cult of Stalin's personality had a negative effect on the economic development of the country. The fact that Stalin decided all important questions himself led to errors in plans and the lessening of the creative activity of party, planning and managerial organs; many questions were decided without sufficiently wide discussion among the workers, engineer-technicians and scientists. In planning and direction of the economy over-centralization was dominant. Insufficient steps were taken to combat technological know-alls and conservatism, which infected part of the leading cadres of the party and higher management.[33]

What precisely did the Soviet critic have in mind? After all, statistics of growth were most satisfactory.

The quantitative gains were indeed impressive. But quality and technical progress both suffered. Everyone was rewarded above all for fulfilling output plans, and the planners, under pressure to expand production, proceeded on what has been called 'the ratchet principle': that more should be made of everything. This led to several defects. Firstly, the simplest way to produce more is to go on making the same designs. Therefore, unless the particular item was given detailed attention at the very top, there was a marked tendency to go on making obsolete equipment. Secondly, the *pattern* of production was to a great extent frozen: thus the output of coal, oil and electric current was increased in like proportions, whereas in America non-solid fuels were making spectacular relative gains. New products, such as plastics and synthetics, or a new and highly economical fuel, natural gas, were neglected. Such defects had other contributory causes. Thus planning by 'material balances', like the use of input-output tables, is of its nature based on past experience and is thus 'conservative'. Also the ministerial system of administration led to competition between ministries for investment resources, and in the absence of any usable or recognized economic criteria of choice, investments often went to industries with most political pull, or at best were simply distributed proportionately. A weak industry, such as chemicals, was probably because of its very weakness lacking in weight in the struggle for a share in investment allocations, and so, while in the West chemicals were bounding ahead, this sector remained underdeveloped in the U.S.S.R. Matters were not helped by the 'anti-cosmopolitan' campaign, which led to the claims that everything had been invented in Russia and that there was nothing to be learned in the decadent West.

A change of policy required decisions to be taken at the top. The matter could be quite small and technical. Thus, I accompanied a Soviet agricultural delegation in England in 1955, when Stalin's pathological anti-Westernism was over. They saw numerous small-wheeled tractors and their many attachments and uses. Yet, because the current party line favoured large caterpillar tractors, they simply did not dare to discuss or envisage their introduction in Russia, where they were in fact badly needed. (A few years later they were in mass production in the Soviet Union, but by then Khrushchev had brought his personal influence to bear.) Even ministers at this time could exercise few decision-making powers, judging from severe criticisms which accompanied the extension of these powers in 1953 (see next chapter). But the 'top' (Stalin and his immediate entourage) could take cognizance of only a fraction of problems that arose, and usually they could be prevailed upon to take a decision on a matter only if it had got badly out of hand. And the top men's behaviour was now very different from the leap-forward period. Stalin himself was more arbitrary and unpredictable than ever. His closest collaborators were no longer rabble-rousing innovators of the type of Ordzhonikidze, who could take and give responsibility. Stalin stayed in the Kremlin; his immediate subordinates became accustomed to a sedentary life. Even Kaganovich sat in his office, and we have already seen that he was blamed for technical conservatism in matters of rail transport. After the fall of Voznesensky, the top planners were mainly rather lustreless bureaucrats. Though the mass purges were not repeated, a great many people remained in camps, arrests were still a common occurrence, fear of responsibility was still a great cause of waste. So was a tendency to 'please the boss' by adopting spectacular rather than economically sensible methods. So initiative at lower levels was stultified, or distorted. Thus output rose, but the pattern and quality of products, and investment policy too, no longer measured up to the needs of a by now developed industrial economy. Whole areas of backwardness persisted.

In these years the political leaders sought to establish 'little Stalins' at the head of each branch of science and the arts. It is in this context that Lysenko was allowed or encouraged to destroy genetics. Contacts with world science were systematically broken off. Even Einstein's theories were attacked, but in most of the natural sciences and in mathematics the top scientists (who were of exceedingly high quality) succeeded in preserving their disciplines from serious damage.

It was not so with economics. The subject was altogether too close to politics, and any serious discussion of economic issues or of objective criteria was inconsistent with the political arbitrariness which reached its peak in the period 1947–53. It is true that there was a brief false spring in the first post-war year. Novozhilov published some original thoughts which foreshadowed his more recent doctrines (of which more in Chapter 12), and Varga, a leading international economist of Hungarian origin, produced an analysis of Western capitalism which challenged traditional dogmas. Both were severely castigated in 1947–8, and Varga's Institute of World Economics was closed down (but neither was arrested). Unofficial discussion of economic issues was further obstructed by an almost total closedown on publication of statistics. Incomes, outputs, labour, even the size of population, were hidden by a uniquely tight censorship.

In 1952 Stalin's last published work appeared. It was entitled 'Economic problems of socialism in the U.S.S.R.', and was a collection of his writings in connexion with the preparation of a textbook on political economy. (It may seem surprising that Stalin had time for such exercises, but it must be recalled that he aspired to the status of philosopher-king and therefore had to produce *obiter dicta* on a variety of issues. In 1950 it was on linguistics. No one else was allowed any intellectual innovation.)

In his last work Stalin expressed a number of thoughts which need not detain us in the present context. However, three points are directly relevant to our theme. One was his warning to officialdom that they must take economic laws into account – though he did not clarify how economic laws were to be identified. His assertion that transfer prices within the state sector were outside the ambit of the 'law of value' obscured the issue of rational prices. The second was his order to economists to keep out of practical affairs: 'The rational organization of the production forces, economic planning, etc., are not problems of political economy, but of the economic policy of the directing bodies. They are two provinces which must not be confused.'[34] Thirdly, he expressed his belief in the need 'gradually to raise kolkhoz property to the status of state property' and 'to replace commodity circulation [i.e. sales and purchases by kolkhozes] by a system of products exchange'.[35] This is an interesting throw-back to the ideas of 1920 and 1930, and may help to explain Stalin's reluctance to consider proposals to increase agricultural procurement prices (though he did nothing in his last years to convert kolkhozes into state farms). In the name of the same principle he rejected suggestions by the agricultural economists Venzher and Sanina to sell or transfer the machinery of the MTS to the kolkhozes: this would be disposing of state property to an inferior, merely cooperative group of enterprises, and furthermore it would result in increasing the role of money and trade, since the produce handed over in payment for MTS services would now be sold. He would have none of it. (The MTS were liquidated in 1958.)

STALIN'S LAST YEARS: THE NINETEENTH CONGRESS AND THE FIFTH FIVE-YEAR PLAN

Economic policy in 1951 and 1952 followed, with few exceptions, the lines already indicated. State retail prices were cut in March in both years, by a total of about 14 per cent. Yet already in 1950 free-market prices were 10 per cent above official prices. The disparity increased steadily: it was 17 per cent in 1951, 20 per cent in 1952.[36] Evidently the increase in real income in these years cannot be accurately based on the uncorrected use of the official index of retail prices. Price cuts had become a political habit.

The increased strain in retail trade, and in the supply of some materials, was no doubt contributed to by the effect of the cold war on arms expenditure and on manpower. The total strength of the armed forces increased from 2·874 million in 1948 to 5·763 million in 1955,[37] and presumably much of the increase had already occurred by 1952. Military budgets rose year by year:

	1950	1951	1952
TOTAL EXPENDITURE	427·9	451·5	476·9
of which: Military	79·4	96·4	113·8

(SOURCE: Annual budget reports in *Ek. Zh.*)

In real terms the increase was greater, since prices were falling.

No doubt for this reason the output of some products which used potentially military productive capacity fell in these years. Thus tractor production reached 116,700 in 1950, but fell to

93,100 and 98,900 in 1951 and 1952, and was only 111,300 in 1953. But the economy was now stronger than in the late thirties, and there was much less disruption of civilian production this time.

The new five-year plan was supposed to begin in 1951, but no announcement was made. Through most of 1952 not even a draft five-year plan was referred to. Finally, a plan covering the five years 1951–5 was presented to the nineteenth party congress in October 1952 and approved unanimously. We do not know whether an unpublished long-term plan existed in 1951–2, or whether the worsening of the international situation led to a postponement in the drafting of such a plan.

Industrial output was due to increase by 70 per cent in the five years, which represented a slowdown compared with the rate of growth claimed for the previous quinquennium, but this could be readily explained by the gradual end of post-war reactivation of damaged enterprises. (This was still a significant factor even as late as 1953, as is shown by the fact that five factories begun or completed before the war and important enough to be mentioned by name in a 'catalogue' of economic events covering the period 1917–1959 started production only in the first quarter of 1953.)[38] National income was to rise by 60 per cent, real wages by 35 per cent, peasant incomes by 40 per cent. Agricultural production was to increase by large percentages: the grain crop by 40–50 per cent, meat production by 80–90 per cent, milk by 45–50 per cent. (Details of plan targets and fulfilment will be given in Chapter 12.)

Yet, until after Stalin's death, nothing was done to increase the miserably inadequate income of the kolkhoz peasantry. Indeed, the tax screw was actually tightened, and the final twist was given in January 1953, with a demand that the 'agricultural tax' (on private plots) be paid earlier in the year than usual.[39] The livestock population began to show a downward trend. One is at a loss to explain such actions; did none of his comrades dare to tell Stalin what conditions were like in the villages?

But there is some evidence of the beginning of rethinking. Certain prices – for milk and flax, for instance – were increased in 1952, though with so little publicity that it was necessary to search for references to any such decisions;[40] and publication was allowed of very severe criticisms of the arbitrary and inefficient rural planning methods; this was ignored by the nineteenth party congress which met in the same month.[41]

There were relatively few organizational changes in Stalin's last two years. Major changes had to wait until after the death of the great dictator. On 5 March 1953 his henchmen found themselves successors to his heritage, a great country, the second military and industrial power, yet one with many weaknesses, unevenly developed. Great scientific achievements had been made but the housing situation was still appalling, consumers' goods of poor quality, the villages primitive. Even within a single sector, grain cultivation for instance, large modern combine-harvesters were used alongside totally unmechanized hand operations in the process of cleaning, drying, loading. What was to be done about over-centralization, lack of acceptable (or accepted) investment criteria, agricultural prices, the deficiencies of the trading network, the breakdowns in material supplies? How could it be tolerated that a country capable of making an A-bomb could not supply its citizens with eggs? How could necessary initiative be encouraged under conditions of terror?

On the morrow of Stalin's death his successors called for the avoidance of 'panic and disarray'. How they faced the problems which they inherited is the subject of the next chapter (while the whole question of the Stalin epoch and its significance is left to be discussed in a concluding assessment).

12. The Khrushchev Era

THE MALENKOV INTERREGNUM

Malenkov had presented the principal report at the nineteenth party congress, in October 1952. He, with Molotov and the security chief, Beria, formed a triumvirate when Stalin died. The means by which Malenkov, who seemed to be the appointed heir, was manoeuvred out of this position belong strictly to the realm of politics. Khrushchev had already become in March 1953 the senior of the party secretaries. This, plus the fact that Khrushchev's experience inclined him to work in and through the party machine, undoubtedly affected economic administration, and the relation between governmental and party organs at all levels in the years that followed.

Malenkov held on to the chairmanship of the Council of Ministers, and at first exercised a leading role in economic policy, except perhaps for agriculture, where his particular contribution to the reforms of 1953 is still obscure, for reasons to be noted.

Before discussing Malenkov's policies, we must note briefly a sudden but short-term change in the ministerial structure, affecting the economy and all other branches of administration. It seems that, in the aftermath of the dictator's death, the politbureau members decided to appoint key party leaders to head ministries, and for this purpose amalgamated the ministries. Thus the Ministries of Agriculture, State farms, Cotton-growing (such a ministry was set up in 1950), Procurements and Forestry were combined. So were the Ministries of Light, Fish, Food and Meat and Dairy Industries. So were those responsible for Ferrous and Non-ferrous Metals. The many ministries in the machinery and engineering sector (other than arms) were combined into only two, and so on.[1] This arrangement did not last long. The number of economic ministries began to grow again within six months, probably under the combined pressure of business and of Khrushchev's interest in weakening decision-making through ministries, though their exact number and coverage changed in the course of their disaggregation. One change proved more durable: the transfer back to Gosplan of the material-supply function (carried out since the end of 1947 by Gossnab for industrial materials, and since 1951 by Gosprodsnab in respect of consumers' goods). We shall see that Gosplan's key role involved so great a burden of responsibility that it was later again divided and re-divided, but not until 1965 was something very like Gossnab re-created, under very different conditions.

On 11 April 1953 a decree of the Council of Ministers increased the power of decision-making of ministries (and of glavki heads). They could, within stated limits, alter staff establishment of their own enterprises, redistribute equipment, materials and resources, approve plans for small- and medium-scale investment, and so on. Readers of the decree were perhaps surprised how few of such decisions were within the competence of ministries before this date. No doubt the decree was partly due to the fact that senior party leaders were now ministers, and many ex-ministers were now heads of glavki. However, these powers were somewhat enlarged after 1954, and it may be that this caused a weakness in coordination which led to the drastic reforms of 1957.

The collective leadership quickly came to the conclusion that the quality and quantity of consumers' goods, of housing, of services, the depressed state of the villages and of agriculture, insistently called for remedy. Furthermore, it was essential to show the people that this was indeed their intention. Political-struggle considerations played their role. Thus Malenkov came to be identified in the public mind with a consumer-orientated policy, no doubt deliberately. However, even Beria, the grim police chief, was said to have advocated a 'consumer' policy, in the few months before his comrades had him arrested and shot (for quite other reasons). The wind of change was blowing powerfully.

But material concessions require material means to implement them. One cannot increase consumption without providing more goods and services. The first steps of Malenkov suggested that he was willing to ignore this simple truth. Retail price cuts were announced on 1 April 1953, and they far exceeded the justifiable. On average, retail prices fell by 10 per cent, but many items were reduced by much more than this, without good reason. A 15 per cent cut in meat prices, for instance, made no sense when meat was short. Even more absurd was a 50 per cent reduction in prices of potatoes and vegetables. The result, as might have been foreseen, was queues for many goods, and an increasing gap between official and free-market prices, a disparity which reached 34 per cent on average in 1954,[2] and which was 100–150 per cent for potatoes and vegetables (in 1955 theii prices were increased to more reasonable dimensions). A retail price policy which, in the interests of publicity, ignored supply-and-demand considerations so completely was bound to increase the already great deficiencies of the trade network.

Another 'popular' act of Malenkov made matters worse. The 'compulsory-voluntary' bond sales were sharply reduced, from 35·7 milliards in 1952 to 15·3 milliards in 1953. This, together with a 3 per cent rise in average wages in that year, led to an increase in take-home pay of no less then 8 per cent and so contributed to the gap between purchasing power and goods available at established prices.[3]

The budget for 1953 was adopted unusually late, on 8 August. Its oddities must be briefly noted. It included an item of *expenditure*, 43·2 milliards, being the 'cost' of retail price cuts, and another 13·6 milliards representing the 'cost' of higher agricultural procurement prices and of agricultural tax cuts, about to be announced. Never before had decisions leading to loss of revenue – for all the above items represent revenue foregone – appeared in the budget as items of expenditure. Similar items appeared in the 1954 budget, but never again after Malenkov's fall.

Malenkov it was who announced, in his speech to the budget session of the Supreme Soviet in August, not only a reduction in agricultural tax but also the decision to raise procurement prices and to make sweeping changes in agricultural policy, details of which were to be announced by Khrushchev at the plenum of the central committee of the party in the following month (and which will be analysed at length when we discuss agriculture). When Malenkov resigned, it was stated that he was not the initiator of agricultural reforms, indeed that he had responsibility for the bad policies of the past. But by then he was in no position to defend himself, so we cannot know the truth. His views during Stalin's lifetime may simply have been a reflection of his superior's. What is one to make of Malenkov's announcement, in October 1952, that the grain problem was finally solved with a harvest of 130 million tons, when in August 1953 he stated that harvests had been greatly exaggerated by the 'biological yield' statistics? Did he not know in 1952 that his figures were quite misleading?

Malenkov propounded a new industrial policy. He held that, based on the successful creation of a powerful heavy industry, it now was both possible and desirable to speed the growth in the output of consumers' goods. It was to expand even faster than that of producers' goods. On 28 October 1953 *Pravda* announced new accelerated plans for expanding the output of the consumers' goods industries, over and above the provisions of the five-year plan, and ambitious targets were announced for 1954 and 1955, covering a wide range of goods. The following figures illustrate the scale of the proposed increases, and the extent to which they exceed reality:

	1952 (actual)	1955 (Malenkov)	1955 (actual)
Cotton textiles (million metres)	5044	6,267	5,905
Wool textiles (million metres)	190·5	271	252·3
Silk textiles (million metres)	224·6	573	525·8
Linen textiles (million metres)	256·5	406	305·5
Knitted underwear (million units)	234·9	382	346·5
Knitted outerwear (million units)	63·5	88	85·1
Hosiery (million pairs)	584·9	777	772·2
Leather footwear (million pairs)	237·7	318	274·3
Sewing machines (thousand units)	804·5	2,615	1,610·9
Bicycles (thousand units)	1,650·4	3,445	2,883·8
Motorcycles (thousand units)	104·4	225	244·5
Watches & clocks (thousand units)	10,486	22,000	19,705
Radios and TVs (thousand units)	1,331·9	4,527	4,024·6
Domestic refrigerators (thousand units)	–	330	151·4
Furniture (million roubles)	2,883*	6,958*	4,911*

* 1952 and 1955 (actual) are in 1 July 1955 enterprise wholesale prices. The 1955 planned figure was presumably in pre-1955 prices.

(SOURCES: *Pravda*, 28 October 1953; *Promyshlennost' SSSR* (1956), pp. 328, 343, 351, 362, 363; *Promyshlennost' SSSR* (1964), pp. 43, 411.)

Investments in industrial consumers' goods production in 1954 were planned at 5·85 milliard roubles, against only 3·14 milliard in 1953.

To meet the increased demand, every effort was made to achieve an immediate rise in consumable output. In 1953 consumers' goods production increased faster than that of producers' goods (13 per cent against 12 per cent, according to the official statistics). Total investment rose very little, by only 4 per cent, as the economy adjusted itself to new policies. In April 1954 there was a further reduction in retail prices. However, following the mediocre harvest of 1953, the reductions were almost wholly confined to manufactured goods (e.g. cotton fabrics by almost 15 per cent). The one significant cut in food-stuffs affected bread, which, at 1·12 roubles, was now a third of the price-ruling in 1947. This further over-stimulated demand for all goods.

The increased plans for consumers' goods were over-ambitious and were abandoned quickly after Malenkov's fall. Oddly enough, this episode is not mentioned in economic histories published in the U.S.S.R. covering this period. Malenkov's name is likewise systematically omitted, save in relation to earlier periods of his career (and then always in a critical context).

One simple but popular reform was announced on 29 August 1953. It laid down normal working hours in all offices, especially government offices. Stalin worked at night, and in consequence officials had to adjust their hours, so that many hardly ever saw their families, since they had to stay at their desks in case the boss or one of his underlings made an inquiry. At long last, this distortion was over.

Stalin's successors also declared an amnesty. The big releases of political prisoners did not occur until 1955, but the dismissal and death of Beria was followed by a drastic curtailment of police powers and the liquidation of the security authorities' economic empire.

Before attempting any assessment of the Malenkov period, we must look at the far-reaching agricultural reforms launched in 1953–4, the first instalment of a series of changes which are still continuing.

AGRICULTURE 1953-4: KHRUSHCHEV'S EMERGENCE

The changes foreshadowed by Malenkov's speech at the Supreme Soviet were announced by Khrushchev in a speech which showed him to be a key figure in the post-Stalin political arena. The central committee which heard him and approved his proposals also gave him the formal rank of first secretary (later First Secretary, the capital letter being, as they say, 'not accidental').

He began by making the first frank statement since collectivization concerning the state of Soviet farming. Productivity (per hectare, per cow, per peasant) was much too low. Livestock herds compared very unfavourably with 1928 and even 1916. The truth was hidden by statistical distortions, such as biological yield (though corrected figures were not published; in the period 1953 to 1959 grain statistics were kept secret). Peasants were paid too little, and investments by farms were inadequate. This was due to the fact that procurement prices were far too low. Taxes on private plots were having the effect of discouraging production and harming peasant interests. Agricultural planning was defective and bureaucratic, procurement quotas were arbitrarily altered by local officials. These and other defects were to be put right 'in the next 2–3 years' (Khrushchev was always to be in a hurry).

There followed a stream of measures designed to transform the situation. These can be classified under the following heads:

(a) Prices

There were very marked increases in compulsory-procurement prices and over-quota delivery prices for grain, potatoes and vegetables, meat, dairy produce, sunflower seed; smaller increases affected eggs and 'contract' prices for flax. At the same time the procurement quotas were reduced in a number of cases, so that a larger proportion of sales to the state took place at the higher over-quota delivery prices. In the full year 1954 the average prices paid by the state for produce were as follows:

(1952 = 100)	1954
All grains (average)	739
Flax fibre	166
Sunflower	626
Potatoes	369
Meat, all types (average)	579
Dairy produce (average, milk equivalent)	289
Raw cotton	102

(SOURCE: Malafeyev, *Istoriya tsenoobrazovaniya v SSSR* (Moscow, 1964), pp. 412–13.)

The effect of all this on the net revenue of kolkhozes requires careful assessment. Thus at least half of all grain deliveries were in payment for MTS services and so had no price at all. Increases in *total* state purchases, albeit at higher prices, sometimes cut down the amount available for sale in the free market at still higher prices. None the less, it is quite undeniable that the net effect was positive and appreciable.

It will be noted that commodities previously well treated, such as cotton, did not benefit from price increases. So heavy was their 'weight' in total procurements that the average price paid for *all* products in 1954 was only just over double that of 1952.

These prices were somewhat modified in 1955–6, the average being reduced somewhat for grain, greatly increased for potatoes, vegetables, sugar-beet. The average procurement prices for all products rose from 207 in 1954 (1952 = 100) to 251 in 1956.[4]

(b) Other material concessions to kolkhozes

The state was now to pay most or all of the transport costs involved in delivering produce to collecting points. Market charges were reduced. Past debts in cash or produce of kolkhozes to the state were written off. Payments for the work of the MTS were to be based on regionally differentiated fixed amounts, and no longer on percentages of the (usually exaggerated) harvest. Kolkhozes were, by the decree of 25 August 1953, to be increasingly linked with the public electricity generation system.

(c) New policy towards private plots

Tax was not only reduced substantially, so that its yield in 1954 was 60 per cent below the level of 1952, but its basis was altered. No longer did possession of an animal or the sowing of a particular crop attract tax on the theoretical income deemed to arise from them. The tax was levied henceforth on the area of the plot, differentiated by region. (It is noteworthy that the rate was lower, usually half of the usual rate, in the case of all territories annexed by the U.S.S.R. in 1939–40, in which collectivization was completed in 1950.) Peasants were encouraged to acquire livestock, promised better pasture facilities. Peasant families without livestock were freed from compulsory deliveries of meat, and all such deliveries by peasants were reduced. Of course, the peasants benefited along with kolkhozes from the higher prices paid by the state, but more important was their freedom to eat more of their own produce or to take it to market (in which market dues were cut).

(d) Increased inputs and investments

A large increase was announced in the planned output of tractors and other farm machinery, also in fertilizer. It would be necessary to provide more building materials to make a reality of an expanded investment programme on the farms.

(e) Discipline

So as to have the necessary labour available for higher output and investment, despite the counter-attraction of the private plot, the power of farms to compel their members to perform their duties was enhanced, until in March 1956 the principle of centrally determined compulsory minima of collective work was dropped altogether, and kolkhoz managements were empowered to fix the work minima required to fulfil plans for collective production. Penalties could include a reduction in the area of the private plot.

(f) Planning, personnel and the role of the MTS

There were injunctions to officials to take local conditions into account, to consult the kolkhoz management. Many qualified agronomists, administrators and other specialists were encouraged or directed into agriculture. The MTS permanent staff was greatly strengthened in quality and numbers, and many of the part-time employees hitherto borrowed for particular tasks from kolkhozes were to be employed permanently. The MTS director was to have important (but unfortunately ill-defined) supervisory powers over the kolkhozes served by him, and the party in rural areas was reorganized so that it exercised its great powers of control through 'secretaries for the MTS zone'; that is to say, a full-time party secretary, under the first secretary of the raion (district), was based on the MTS and was responsible for the party's control over kolkhozes within that MTS's service area, along with the kolkhoz party groups. The deputy-director, political, of the MTS was therefore superseded or replaced by this new system. At the same time the Ministry of Agriculture continued for several years to be in charge of agricultural planning, through republican, *oblast'* (provincial) and raion agricultural administrations. (This parallelism led, as we shall see, to several more reorganizations.)

The net effect of these measures on incentives and on investment was positive. We shall be assessing the first five years of Khrushchev's agricultural policies later in this chapter. The measures were accompanied by ambitious new production targets and plans for increasing livestock.[5]

However, these quickly ran into difficulties; there was too little fodder, and, particularly, far too little fodder grain. A plenum of the central committee was devoted to this subject at the end of February 1954. (Note how agricultural decisions tend to be made at party plenums.) Khrushchev made another big speech. It was decided to increase state procurements of grain through all channels), by 35–40 per cent. Everything was to be concentrated on getting quick results; the phrase 'in the next 2–3 years' occurs repeatedly in the resolution of the plenum. Measures to increase yields take time, the building of many new fertilizer factories is a slow business. The alternative: a spectacular expansion of sown area. This led Khrushchev to revive and then expand a plan which had already been adopted (but not put into effect) in 1940:[6] to bring under cultivation at least 13 million hectares of virgin and fallow land located on the southern fringes of the area of adequate rainfall, in northern Kazakhstan, the southern parts of Siberia, the south-east of European Russia. This decision was followed by another decree devoted solely to this theme.[7] It so happened that 1954 was a good year climatically in the regions concerned, the harvest was good on the relatively small area which could be sown that year, so the plan was more than doubled, to 28–30 million hectares.[8]

This became a great campaign. The young communist organization (*Komsomol*) was mobilized to encourage young persons to go east. Some volunteered, others went after it had been announced that this 'vital task set by the party and government' was 'an important patriotic duty'. Many went for two months to cope with the harvest peak: 200,000 were to go in 1956.[9] Much of the new cultivation was undertaken by state farms, since it was hardly realistic to expect people to go to work in a new and largely empty area, with a high drought risk, without an assured income and considerable state-financed investments. (Kolkhozes already existing in the area of the campaign greatly expanded sowing but few new ones were erected.) Between 1953 and 1956 the amount of cultivated land was increased by 35·9 million hectares, an area equivalent to the total cultivated land of Canada. World history knows nothing like it.

The number of migrants to the newly cultivated areas cannot be precisely determined (and some returned), but perhaps as many as 300,000 persons moved permanently. At first conditions were primitive, and for many years the press printed complaints about lack of housing, shops, amenities. The campaign also required the transfer of tractors, combine-harvesters and other machines, since the big new farms had to be highly mechanized if the job was to be done with the minimum of labour. It was also necessary to build new railway lines. All this naturally required resources of which the old-established regions were deprived.

The year 1955 was one of drought. I visited northern Kazakhstan in that year and saw some of the new farms under unfavourable conditions. However, 1956 produced good weather and an excellent harvest. The area is highly vulnerable to drought, and suffers from a short growing season, and the winter snow cover is insufficient to permit autumn sowing as frost would kill the plants. So there is a great rush to sow in the spring, soon enough to enable the grain to ripen. The aim was to grow grain, and almost all the land was sown with spring wheat. Because it was a party campaign, local officials vied with one another to plough up most land, and some quite unsuitable soils were brought under cultivation. We shall see that dangers of mono-culture and soil erosion were ignored for too long.

The addition of so much new land sown to wheat led Khrushchev to the second stage of his grain campaign: to provide fodder grain and concentrates in the principal livestock areas, i.e. in European territories of the U.S.S.R. The crop he preferred was maize. So the central committee plenum of January 1955 accepted Khrushchev's proposals to increase the area under maize 'in the next 2–3 years' from 3·5 to no less than 28 million hectares. More food-grains for human consumption from the virgin lands would leave scope for the use of more land in European territories of the U.S.S.R. for the fodder grain which promised the highest yields per hectare, maize.

The maize campaign, with which Khrushchev's name was particularly closely associated, will receive further analysis later in this chapter.

THE FALL OF MALENKOV

The policies outlined in the preceding pages were bound to lead to overstrain. Higher incomes in town and country, higher production and investment plans for consumers' goods, large agricultural programmes, an expansion of house-building and other consumer services, and simultaneously the continued growth of basic industry and the insistence of the military that the great U.S. superiority in aircraft and in bombs required rectifying, this was too vast a programme to sustain. Malenkov's regime, sapped politically by Khrushchev, fell because he was never strong enough to adjudicate between conflicting claimants on resources. Exactly what finally decided the majority of the leadership to get rid of Malenkov is still not clear, since no sort of open discussion took place.

We know, however, that the occasion of his disgrace was the allegedly incorrect views he held concerning the priority (or rather, lack of priority) of heavy industry, of producers' goods as against consumers' goods. This was a principal accusation against him on the occasion of his resignation in February 1955. Coincidentally, *Voprosy ekonomiki* published, in its January number, a slashing attack on these 'woe-begone economists' who had made the error of climbing on to the bandwagon and who had produced arguments in favour of the more rapid growth of consumers' goods output.

It would be quite misleading to conclude that Khrushchev was against consumers' goods, since his very ambitious and expensive agricultural programmes would, if successful, provide food, which is a consumers' product *par excellence*. Khrushchev allied himself with those planners and military men who considered that Malenkov was unbalancing the economy by an over-concentration on industrial consumers' goods, and thereafter he proceeded to emphasize the citizen's material interests; this was a political necessity after Stalin's death. It is also perhaps worth saying that no one, not even Stalin, was against a higher standard of living. All would welcome it, but for long periods other preoccupations took priority.

Malenkov was succeeded as prime minister by Bulganin. Khrushchev was dominant in his position as First Secretary, but, though until his fall he took practically every major policy initiative, he never achieved (if indeed he sought) the total dominance of a Stalin, and a number of steps taken in the period 1955–64 are explicable only by the fact that he could not always ride rough-shod over the opinions of his colleagues.

FURTHER AGRICULTURAL REFORMS AND SUCCESSES

Khrushchev used his enhanced political powers to press ahead with his agricultural policies, and it must be admitted that until 1958 output did rise substantially. Many analysts have expressed some suspicion about the statistics. The inflated figures of the past, especially the most recent past, may have been too drastically scaled down. The pressure on local party secretaries to report success may have led to exaggerations in reporting. Examples of such exaggerations have been reported from time to time, such as buying butter or meat in shops to deliver it to the state as new production, or including in grain

Agriculture 1953-8

	Production						State procurement	
	1953	1954	1955	1956	1957	1958	1953	1958
					(million tons)			
Total grain harvest	82.5	85.6	103.7	125.0	102.6	134.7	31.1	56.9
Virgin land areas	26.9	37.2	27.7	63.3	38.1	58.4	10.8	32.8
Potatoes	72.6	75.0	71.8	96.0	87.8	86.5	5.4	7.0
Sunflower	2.6	1.9	3.8	3.9	2.8	4.6	1.8	2.6
Flax	0.16	0.22	0.38	0.52	0.44	0.44	0.15	0.39
Cotton	3.9	4.2	3.9	4.3	4.2	4.4	3.9	4.4
Meat	5.8	6.3	6.3	6.6	7.4	7.7	3.6	5.7
Milk	36.5	38.1	43.1	49.1	54.8	58.9	10.6	22.1
Cows	24.3	25.2	26.4	27.7	29.0	31.4	–	–
Pigs	28.5	33.3	30.9	34.0	40.8	44.3	–	–
Sheep	94.3	99.8	99.0	103.3	108.2	120.2	–	–
Gross agricultural production (1953 = 100)	100	105	117	132	136	151	–	–

(SOURCES: *Sel'skoe khozyaistvo SSSR* (Moscow, 1960), pp. 90–91, 226–7, 228–9; *Nar. khoz.*, 1965, pp. 259, 325, 330, 332, 334, 367; *Nar. khoz.*, 1958, p. 107, 470.)

output what Khrushchev himself once described as 'mud, ice, snow and unthreshed stalks',[10] and other plan-fulfilment reporting devices of varying degrees of ingenuity. While on occasion there is clearly a case for adding a pinch of salt to official claims, no one doubts that both production and procurements did rise in the period, allowing for the inevitable year-by-year variations due to weather.

In March 1955 a decree ostensibly enlarged the decision-making powers of the kolkhozes. Plans henceforth would specify delivery obligations, not production; the extent of sown areas and livestock numbers were to be a matter for kolkhozes to decide. An end must be put to centralized decisions about what each kolkhoz should do. The job of assessing and reconciling draft plans received from below was given to Gosplan. This was the first of many blows to the powers of the Ministry of Agriculture. In practice, as later evidence shows, detailed regulation from above continued, but primarily via local party organs. As we shall see, most of the interference with kolkhozes was due to campaigns initiated by Khrushchev himself. Supervision within kolkhozes was facilitated by a campaign, begun in March 1958, to send to the villages, as kolkhoz chairmen, reliable party men from the towns. These were by no means necessarily specialists in agriculture, but they were given 'a three weeks' initial course and two months' training on the job (*stazhirovka*)'.[11]

The twentieth party congress approved, in February 1956, the sixth five-year plan, and one of its objectives for 1960 was a grain harvest of 180 million tons, and more maize. This was but one more of the over-ambitious objectives set before agriculture. In 1957 the campaign began to catch up the United States in the production of meat, milk and butter. In the case of meat this would mean at least trebling Soviet output. We shall be examining the harm done by endeavours to fulfil such impossible plans as these and to impose them on the farms. Khrushchev himself toured the country, talking to officials and peasants, cajoling, haranguing, sometimes dismissing. Thus on 8 March 1957 he was in Krasnodar, on 10 March at a meeting at Rostov, at the end of the month he was addressing officials of the central region in Moscow. On 2 April he was in Voronezh, supporting the promise of a kolkhoz chairman to increase spectacularly output of meat and milk per 100 hectares (with extraordinary indifference to considerations of cost, prices, labour inputs). Five days later Khrushchev was in Gorky on the meat and milk campaign. Yet at this very time, as we shall see, Khrushchev was engaged not only in pushing through an immense industrial-planning reform, but was also under heavy attack from his political enemies. This, with all its defects, was a form of leadership very unlike Stalin's.

An important development began to gather speed during 1957: this was the transformation ('voluntary', of course, except that it was party-directed) of kolkhozes into state farms. In some cases it was a consequence of efforts to create vegetable-and-dairy state farms around big cities. In others economically weak and depressed kolkhozes were made into state farms. In still others, local party secretaries pressed ahead with conversions, either because they believed it to be party policy, or because it was easier to obtain financial help from Moscow for state farms. The statistics of change are as follows (the increase between 1953–6 being principally in the virgin lands):

State farms*

	1953	1956	1965
Total area sown (million ha.)	18·2	35·3	97·43
Total labour employed (millions)	2·6	2·9	8·6
Loss to collective farms (million ha.) (1956–65)	–	–	(47·1)†
Loss to collective farms (million households) (1956–65)	–	–	(4·5)†

* Including other forms of state enterprise.

† These figures give the decline in the period, regardless of cause.

(SOURCES: *Nar. khoz.* 1965, pp. 288, 455; *Sel'skoe khozyaistvo* (1960), p. 46.)

Kolkhoz amalgamations were being pressed at the same time. After the first wave of amalgamations at the end of 1950 there were 125,000. In 1958 numbers of kolkhozes fell to 69,100 and continued to decline, and their size to increase, despite some criticism of unmanageability of big farms in areas of scattered small villages. There are now fewer than 36,000.

How did all these changes affect the incomes of kolkhoz

peasants? Cash distributions rose from a total 12·4 to 47·8 milliard roubles, or from 1·40 to 4·00 roubles per trudoden', from 1952 to 1957.[12] The practice of making 'advances', i.e. regular payments at frequent intervals (every month or every quarter) in advance and on account of the end-year distribution, became the practice in the majority of kolkhozes. Payments in kind, on the other hand, increased little if at all, because of the pressure to deliver produce to the state, a pressure which, despite repeated promises and injunctions to the contrary, was still arbitrarily increased by local officials whenever the procurement plan for the area was endangered. Indeed, all too frequently Party secretaries undertook to increase the planned deliveries, and all too often one read that 'overplan deliveries are continuing', as if this should have been a source of pride – which it could have been, had the deliveries been voluntary and had there been enough left for the needs of the village.

The combined income in cash and kind from collective work rose from 47·5 milliards in 1952 to 83·8 milliards in 1957,[13] or roughly from 5·40 to 7·55 roubles per trudoden (*old* roubles).

Peasant consumption of their own produce certainly rose. Private livestock holdings increased, and so private output of meat in 1958 was 35 per cent higher than in 1953, milk over 25 per cent higher. From 1 January 1958 private plots were freed of all compulsory delivery obligations. Sales in the free market fell from 53·7 to 39·6 milliard roubles between 1952 and 1957, perhaps because of lack of time to get to market as a result of tighter control over members' time. It is true that there developed a new facility: 'commission trading' by cooperatives, selling peasants' (and kolkhozes') surplus on their behalf in the market, usually at prices a little below those ruling in the free market. But this sensible idea made only slow progress, due partly to lack of enterprise by the cooperatives and partly to peasant distrust. It should be mentioned that for the peasant a journey to town meant an opportunity to buy manufactured goods which were not available in the rural trading network. So the very substantial increase in cash payments for collective work, and of consumption by peasants of their own produce, was to some extent balanced by insignificant increases in payments-in-kind and a decline in the value of market sales.

The effect on incomes of a transfer to state-farm status depended greatly on the circumstances. It will be recalled that kolkhoz incomes varied very widely, as did opportunities to earn extra by free-market sales. In general an average state-farm wage of 531 roubles per month in 1958,[14] plus social security benefits to which kolkhoz peasants were not entitled, was certainly a means of bettering one's lot. However, some relatively prosperous kolkhozes were forcibly converted, to their members' chagrin, and once they had become state-farm workers the peasants quickly found that their private plots and animals were subject to severe limitations.

In fact, already in 1956 there had appeared the first sign that the development of the private sector was worrying the authorities: on 27 August all townspeople (other than kolkhoz members) were subject to tax on any livestock they owned; this hit suburban dwellers. On the same day peasants were strictly forbidden to buy bread, potatoes and other foodstuffs in state stores to feed to their livestock.[15] They did this because of lack of other fodder, and also because the price of bread in particular was low in relation to the free-market price of meat.

Price relativities were indeed wrong. So, it became increasingly apparent, was the principle of purchasing products at two different prices. For instance, in 1957 the average quota price for grain was 25 roubles per quintal, the over-quota price 80 roubles. In practice, sales at both prices were compulsory, so it was not a matter of encouraging voluntary additional sales. The result of this was that poorer farms, with little to sell over the basic delivery quota, were in effect paid less per unit than the more successful farms. Over the country as a whole prices per ton in a good harvest year were higher than in time of scarcity, which did not make any kind of economic sense. Therefore the creation of a unified price (with zonal variations) was decided. (On a visit to the Ministry of Agriculture in September 1955, I was told that they were actively thinking of this even then.)

This decision was linked with another, which concerned the MTS. For a long time the dual position of the MTS, as hiring agency and a means of party-state control, gave rise to friction. The MTS were given plans, in terms of volume of work ('soft ploughing equivalents'), and they and the kolkhozes were often at odds; the kolkhozes wanted a good harvest with a minimum of payments in kind for the MTS, the latter tried to maximize the operations of the machinery, since they would not only fulfil the plan but obtain more payments in kind and so increase state procurements. There were 'two masters in the field', not infrequently at odds with one another. Khrushchev saw sense in the suggestion, rejected by Stalin in 1952 (see p. 319), that the MTS be abolished and their equipment sold to the kolkhozes (the state farms operated their own machinery from the beginning).

The price reform and the winding up of the MTS were decreed together, following the resolution of the February 1958 central committee plenum. Prices had to be adjusted to the fact that the kolkhozes were now to sell to the state the grain (and some other crops) formerly handed over in payment for MTS services; on the other hand the kolkhozes had to purchase machinery and then maintain it and buy replacements, spare parts, fuel, and to pay the employees of the MTS, who had received state wages in whole or in part, and who were to be given privileged status within the kolkhozes which they were now to join. Repair Technical Stations (RTS) were to undertake major technical tasks, on payment.

The plenum also decided to call a Kolkhoz Congress to revise the kolkhoz statutes, but it did not in fact meet.

The results of this reform were, in retrospect, extremely disappointing. But before examining the reason for this and the other unfavourable developments in agriculture after 1958, it is necessary to look at other branches of the economy.

INDUSTRY: PLAN FULFILMENT AND THE NEW (SIXTH) FIVE-YEAR PLAN

1955 was the last year of the fifth five-year plan, which had been adopted with Stalin very much in charge. We have seen that its industrial components were amended by Malenkov, but these amendments were forgotten after his fall. The key indicators of the plan were fulfilled as follows (with the next five-year plan targets given in the last column):

	1950	1955 plan	1955 actual	1960 plan
National income (1950 = 100)	100	160	171	160†
Gross industrial production	100	170	185	165†
Producers' goods	100	180	191	170†
Consumers' goods	100	165	176	160†
Coal (million tons)	261·1	373·4	389·9	592
Oil (million tons)	37·9	70·9	70·8	135
Electricity (milliard Kwhs)	91·2	164·2	170·2	320
Pig iron (million tons)	19·2	33·8	33·3	53
Steel (million tons)	27·3	44·2	45·3	68·3
Tractors (15 h.p. units)	246·1	292·9	314·0	–
Mineral fertilizer (million tons)	5·5	10·3	9·7	20
Cement (million tons)	10·2	22·4	22·5	55
Commercial timber (million cubic metres)	161·0	251·2	212·1	301
Cotton fabrics (million metres)	3899	6277	5905	–
Wool fabrics (million metres)	155·2	239·0	252·3	–
Leather footwear (million pairs)	203	315	271	–
Sugar (thousand tons)	2523	4491	3419	–
Fish (thousand tons)	1755	2773	2737	–
Total workers and employees (millions)	40·4	46·5	50·3	–
Housing (million square metres)	72·4*	105*	112·9*	205*
Retail trade turn-over (Index)	100	170	189	150†

* In the five years ending in 1950, 1955 and 1960 respectively.
† 1955 = 100.

(SOURCES: *Nar. khoz.*, 1965, pp. 130–39, 557; *Promyshlennost' SSSR* (1957), p. 43; *Direktivy XIX S'ezda Partii po pyatletnemu planu razvitiya SSSR na 1951–5 gody*, (1952), pp. 3–4, 25; *Dinektivy XX S'ezda KPSS po Shestomu pyatletnemu planu razvitiya narodnogo khozyaistva SSSR na 1956–60 gg*; *XX S'ezda KPSS stenotchet*, Vol. II (1956).)

NOTE: I am aware that, since 70 per cent of industrial output consisted of producers' goods, the 1960 plan indices for gross industrial production are inconsistent. But that is how they appeared in the plan.

It is unnecessary to repeat yet again the cautionary words with regard to aggregate indices of growth. But it is clear that this was a reasonably successful quinquennium, in quantitative terms. Growth was rapid, particularly in agricultural machinery, which continued (until 1958) to receive priority attention. Total investments exceeded the previous quinquennium by 93 per cent, and the plan by over 3 per cent.

The preparation of the sixth plan was particularly thorough. Attention was drawn to the erroneous practice of confining plan-drafting to a small group of senior officials. During 1955 the need was repeatedly stressed to incorporate the latest technical ideas, to study foreign achievements. Thus there was an all-union conference of industrial staffs in Moscow on 16–18 May, a decree on technical progress adopted on 28 May setting up the State Committee on New Technique (Gostekhnika); a central committee plenum of 4–12 July was devoted to this theme, and so on. In the same year the powers of union republics over the economic enterprises within their borders, over the allocation of materials and over a range of investment decisions were increased (decree of 4 May). A decree of 9 August enlarged the powers of directors of enterprises (in practice there was little change). Directors, local officials and even trade-union branches were to be brought into thorough discussion of draft proposals for the next five-year plan. After full consultation Gosplan was to submit to the central committee (*sic*) its draft of the five-year plan by 1 November 1955.[16]

Meanwhile, Gosplan was divided, in June 1955, between the State Committee on long-term (*perspektivnomu*) planning, still called Gosplan, and the State Economic Commission for current planning, known as *Gosekonomkomissiya*.[17] No doubt these tasks were distinct as well as vast, but this rearrangement in no way helped to ensure consistency between the long-term and current plans, which is essential if an investment programme is to be implemented. To carry out the growing construction programme more expeditiously there were major developments towards industrial methods and prefabrication in building. Specialized construction ministries were created, supervised by the State Committee on Construction, whose head, Kucherenko, was a deputy-premier.

The twentieth party congress duly adopted the sixth five-year plan; the key target figures appear in the above table. Among its most important elements was the decision to create what was called the third metallurgical base, in Kazakhstan and Siberia, which would produce 15–20 million tons of pig iron a year. The delegates also heard Khrushchev's famous 'secret speech' denouncing Stalin. Among his many objectives one was doubtless to break bureaucratic and petty-authoritarian habits of the Stalin era, habits which still affected the administration of the economy. Various decrees were issued at this period seeking to combat bureaucracy and overstaffing of government offices.

However, the sixth plan was the one Soviet long-term plan which was explicitly abandoned in peacetime. Within a year of its adoption it was decided to revise it, and then it never came to life again.

THE ECONOMIC–POLITICAL CRISIS (DECEMBER 1956–MAY 1957): SOVNARKHOZY

A plenum of the central committee met on 20 December 1956. It emerged, judging from its resolution, that the plan was out of balance, that there was 'excessive tautness', that there were defects in planning. A pamphlet published later[18] blamed the ministries for competing with each other for investment resources, and in effect trying to bite off more than they could chew. The investment programme required by the plan was impossible, in the sense that the material means of carrying it out in time were not present. This suggested lack of adequate coordinating powers. (It is only fair to add that a fairly senior Soviet economist once told me, unofficially, that the sixth plan was quite good, and the objections to it were political.) Anyhow, it was decided that the plan be revised and re-submitted.

To ensure that the ministries would pull together, it was proposed that Gosekonomkomissiya, headed by M. Pervukhin, should become a kind of super-ministry with powers to issue orders to all the economic ministries (in the first version of the proposals, even to the Ministry of Agriculture). Since this was a moment at which Khrushchev's political standing was weakened by the Hungarian and Polish events, for which his 'de-staliniza-tion' activities could be blamed, it is more than probable that this proposal, while a reaction to genuine economic-organiza-tional problems, was politically inspired. Khrushchev's powers of interference on economic affairs were to be drastically curtailed by the appointment of another man as overlord. Judging by subsequent events, this was an alliance against him of the old guard (Molotov, Malenkov, Kaganovich) and the professional planners (headed by Saburov, chairman of Gosplan, and Pervukhin).

By appropriate political manoeuvring Khrushchev hit back, successfully prevented these arrangements from materializing, and had already by February 1957 put up for 'discussion' totally different proposals. These, after several amendments, served as a basis for legislation adopted (with its usual unanimity) by the Supreme Soviet in May. The idea was to cope with the problem of ministerial empire-building and insufficient co-ordination by the drastic expedient of abolishing the industrial ministries altogether, and substituting a regional structure coordinated by Gosplan. Khrushchev was responding to a very real problem, but doing so in a manner calculated to harm his political opponents, and to weaken the state-ministerial hierarchy.

Several reasons were advanced for the need for such a change: there was duplication of supply arrangements and components manufacture between ministries, resulting in unnecessary cross-hauls. A steamer belonging to one ministry would proceed up the river Lena full and return empty, while another steamer, transporting goods for a different ministry, went down-river full and up-river empty. Local coordination and regional planning were impeded by the fact that all enterprises of all-union interest had lines of subordination to Moscow, and too many minor decisions required the assent of remote ministries. True, republican powers had been increased, but they were still insufficient, and in any case the Russian republic (RSFSR) was far too big, covering most of the Soviet Union. All these points did have some validity. Ministerial 'empire-building' in particular did lead to waste. This was due not so much to bureaucratic-personal ambitions as to uncertainty over supplies: one made or procured one's own components and materials in case they were otherwise unobtainable. Every enterprise tended to do this too.

The trouble, as we shall see, was that the cure was worse than the disease. In any case it is in the nature of all-round planning that any organizational solution carries with it certain disadvantages.

There are common problems of an industry, or a region, or

matters transcending both a given industry and a given region such as labour, investment policy, finance. Arrangements best suited for considering one of these matters seldom suit others. It is like writing an economic history: a strictly chronological account confuses the reader by fragmenting the discussion of, say, industrial reform or agricultural prices. But an arrangement based on subject may well confuse the chronology, and there is no perfect solution.

Under the reform, civilian industrial and building enterprises of other than purely local significance were placed under regional economic councils (*sovnarkhozy*, resurrecting the name devised in the war-communism period for regional sub-divisions of VSNKH). There were to be 105 of them, of which, initially, seventy would be in the RSFSR (there were minor changes in numbers in 1957–9). Of the other republics, only Kazakhstan (9 regions), the Ukraine (11) and Uzbekistan (4) were divided. The other eleven republics were each co-extensive with a sovnarkhoz.[19] At first the ministries concerned with armaments production, chemicals and electricity were left in being. Each sovnarkhoz was in general command of its enterprises. The sovnarkhozy were to be appointed by and responsible to the republican Councils of Ministers, with the republican Gosplans acting as coordinators in the four multi-sovnarkhoz republics. The all-union Gosplan was to be responsible for general planning, the coordination of plans, the allocation between republics of key commodities, though without executive authority, which resided formally in the all-union Council of Ministers. Gosekonomkomissiya was abolished.

Khrushchev's success in pushing through these proposals was followed in the next month by the effort of the so-called 'antiparty group' to unseat him. This ended the political careers of many old leaders. Bulganin, the premier, remained until the following year, though he too had opposed Khrushchev. On 28 March 1958 Khrushchev combined in himself the first secretaryship and the premiership. A minor party official, I. Kuzmin, was appointed to head Gosplan on 1 April 1958. So far, so good. Trouble was to come later.

LABOUR, WAGES, PRICES, SOCIAL SERVICES (1955-8)

Malenkov's fall had been due, *inter alia*, to alleged softness towards the consumer. The immediate response was to withdraw purchasing power by doubling the level of the virtually compulsory bond sales. The effect was, according to a well-informed Soviet analyst, that workers' take-home pay in 1955 was 1·8 per cent below 1954 levels.[20] But after this the advance was resumed. Average wages rose from 715 roubles a month in 1955 to 778 roubles in 1958.[21] As for compulsory bond sales, these were abandoned altogether from 1958, by decree of 19 April 1957, but simultaneously a moratorium ('for twenty years') was declared in respect of repayment of principal and of interest (in the form, mainly, of lottery winnings) on past bonds of this type. It was explained that the budget could not afford simply to abandon the bond sales while continuing repayments and interest.

It was clearly a matter of concern to the authorities that the retail price level should stay steady. Such increases as occurred (the doubling, for good reason, of potato and vegetable prices, a sharp increase in the price of vodka, etc.) were balanced by cuts, notably in prices of those manufactured goods which were relatively abundant (watches, sewing machines, certain textiles, etc.). The retail price index showed no appreciable change after 1954. Kolkhoz market prices remained, on average, 35 per cent to 45 per cent above official prices, indicating a tendency to under-price foodstuffs.

The wages structure was in a tangle. There had been no systematic overhaul of wages since long before the war. The prevalence of piece-rates and bonus schemes enabled some enterprises and some ministries to increase pay, while others, especially those on time-rates and fixed salaries, had been less lucky. Different ministries adopted different wage-zones. Pay of persons of similar skills and even identical occupations, in different ministries, varied without reason. Norms were over-fulfilled on average by 60 per cent to 70 per cent. The unskilled grades in many industries were virtually unstaffed; the least qualified workers were graded as semi-skilled, so as to pay them something nearer to a living wage. 'Progressive' piece-rates led to many abuses. Reform was essential, but was also exceedingly difficult.[22]

This situation led to the setting up, on 24 May 1955, of the State Committee on Labour and Wages. Its first head was L. Kaganovich. It proved to be his last job before his fall. A systematic rearrangement of the wages structure followed. Differentials were reduced, standard pay scales introduced for occupations regardless of departmental subordination, various anomalies ironed out. The obsession with individual piece-rates which characterized the Stalin period was modified. Extreme incentive schemes – such as payments at treble-rates for output over the norm in the coal industry – were eliminated. All this took many years. Indeed, the task could be said to have been completed only in 1965, when improved pay rates came into operation for most of the service sectors.

Reduction of differentials was speeded by the introduction of a minimum wage of 300 roubles per month in towns by a decree of 8 September 1956 (270 roubles in rural areas), a minimum subsequently raised. It is some reflection on the number of persons whose basic pay was lower than these exceedingly modest sums that, so we are told, the low-paid worker gained an average of 33 per cent.[23] At a reasonable conversion rate – remembering these are old roubles – 300 roubles equalled £9, or $25 per *month*. Khrushchev also took some steps against very high salaries: I was told that professors' basic pay went down from 6,000 to 5,000 roubles per month, for instance.

This was part of a wave of social legislation. Shorter hours for juveniles without loss of pay; a minimum of a month's holiday for those under 18;[24] the reduction of the working week by two hours, with more cuts to come,[25] a 7-hour day (6 hours on Saturdays) being introduced by stages; the lengthening of paid maternity-leave to 112 days,[26] the repeal of the criminal-law liability for leaving work without permission and for absenteeism,[27] the abolition of tuition fees in secondary and higher education,[28] and, most important, great improvements in pension and disability benefits, these being now calculated in relation to actual earnings, with a minimum old-age pension of 300 roubles per month (and an overall maximum of 1,200 roubles).[29] Until this date all but the specially favoured sectors were subject to a maximum reckonable earnings rule which reduced pensions to low and sometimes derisory levels. The effect of this law was to increase the average pension by 81 per cent, but the underprivileged gained much more than this. Finally, income tax exemptions on low incomes were extended.[30] A big increase in the rate of house construction began in 1956, and private house-building received rather more financial and material support:

Urban Housing
(million square metres total space, new construction)

	State and cooperative	*Private*
1955	25·0	8·4
1956	29·5	11·5
1957	38·5	13·5
1958	46·7	24·5
1959	53·5	27·2

(SOURCES: *Nar. khoz., 1963,* p. 514; *Nar. khoz.,* 1956, p. 176.)

It will be noted that the lower-paid were the principal beneficiaries of most of these reforms, which reversed much of Stalin's reactionary legislation of the 1938–40 period.

Workers were now free to leave their jobs, though subject to some limitations on movement owing to the passport system; thus it is still very hard to obtain permission to live in Moscow and some other big cities. None the less, greater unplanned labour mobility was a fact, and this, plus the abolition of most of the forced-labour camps, complicated the process of planning and made wage relativities of ever greater economic significance.

Trade-union activity was increased; their failure to represent and defend the workers was criticized, especially at the central committee plenum devoted to this subject held in December 1957. The powers of the factory trade-union committees were defined and extended by a new set of rules promulgated on 15 July 1958 by the Supreme Soviet presidium.

Wholesale prices of industry were lowered in July 1955, following some reduction in costs. Thus coal prices were cut by 5 per cent, oil and gas by 10 per cent, iron and steel by 10 per cent, electricity by 13 per cent, rail freight rates by 10·5 per cent, and so on.[31] The process of price fixing was not significantly altered. Though some anomalies were removed, there was still insufficient inducement to produce goods in demand or of high quality, or modern types of machinery. Timber prices were insufficient to cover costs, and they had to be increased several times in the following years. Coal was also produced at an increasing loss.

This was the last systematic price review implemented before 1967.

In 1961, possibly inspired by de Gaulle's 'heavy franc', the Soviet government decided to multiply the internal value of the rouble by ten. No purchasing power was withdrawn from circulation – this was not a repeat of 1947. New notes were exchanged on a 1:10 basis, but all prices and wages were altered proportionately: 1,000 old roubles became 100 new roubles. The opportunity was taken to devalue the external value of the rouble while seeming to increase it: the rate of 4 roubles = \$1 was altered to 0·90 rouble = \$1, i.e. by much less than internal incomes and prices.

The share of turnover tax and profits in the budget remained much as before, with a continued trend to a relative increase of profits as a source of revenue, despite the price reductions referred to above. Direct taxation became even less significant, following cuts in tax on low incomes, and also, in December 1957, in the 'bachelor and small family tax', introduced in wartime (single women and couples without children were now exempt).

The State Bank was separated from the Ministry of Finance (23 April 1954), and was then given wider powers to use credits to encourage financially efficient enterprises and to take measures, almost amounting to a form of 'socialist bankruptcy', to penalize those who incur debts (decree of 21 August 1954, and others).

FOREIGN TRADE: SOME SIGNIFICANT CHANGES

Relations with all categories of countries altered radically soon after Stalin's death.

There began a revision of the basis of exchanges with other communist-ruled states. Trade relations were resumed with

Yugoslavia in 1954. Reparations deliveries from East Germany were to cease in 1954 under an agreement negotiated in August 1953; this also provided for the handing over of Soviet-controlled enterprises on East German territory. In September 1953 the U.S.S.R. agreed to give more technical aid to China, and a similar agreement, in October 1954, also provided for the handing over of mixed Soviet-Chinese companies and granted a long-term credit of 520 million roubles. In March and September 1954 the U.S.S.R. sold and handed over the Soviet share in nearly all Soviet-Roumanian mixed companies. A credit at low interest rates was granted to Bulgaria (February 1956). This process of normalizing and regularizing trade relations received a powerful impetus from the troubles in Poland and Hungary. The Soviet leadership even felt it necessary, in its agreement with Poland in November 1956, to cancel Polish indebtedness on past credits as compensation for 'the full value of the coal supplies to the U.S.S.R. from Poland in the years 1946–53', a clear and public admission of past underpayments.[32] Hungary had to be given emergency aid.

Trade henceforth was based on prices ruling at various dates in world (i.e. capitalist) markets.

COMECON was revived as a functioning institution. It sprouted a number of working committees, and achieved some results in standardization of design, joint transport arrangements, a very few joint investment projects, pipelines, some specialization arrangements in the field of machinery. However, trade remained predominantly of a bilateral character, and there was no agreed criterion of specialization. The further evolution and prospects of COMECON belong to the present day rather than to history and so are not pursued here.

The U.S.S.R. began to perceive political and economic opportunities in its dealings with underdeveloped countries. Already on 5 August 1953 a trade agreement with Argentina had included some Soviet supplies on credit. But the U.S.S.R. seriously entered the aid business after the visit to India in 1955 of Khrushchev and Bulganin. A great increase in the volume of trade occurred, since Soviet aid took the form of supplying goods on credit, to be paid for in the goods normally exported by the recipient country. By 1958 the effect was sufficiently impressive to give rise to over-sanguine hopes in the mind of Khrushchev, and to exaggerated alarm in Western circles.

Trade with Western countries also increased, as relations unfroze. The combined effect of all these developments was to cause a sharp upward surge in the volume of Soviet foreign trade, as the following figures demonstrate:

Volume of Soviet Foreign Trade
(1955 = 100)

	Imports	*Exports*
1950	54·6	56·7
1958	148·4	130·0

(SOURCE: *Vneshnyaya torgovlya SSSR* (1961), p. 13.)

EDUCATION, TECHNICAL TRAINING, HEALTH

Among the most creditable achievements of the entire Soviet period has been the advance of education at all levels. Despite occasional setbacks, the ideological-political commitment to education, plus the need for more technically qualified staffs always led to yet another upward swing in educational spending. True, schools were too few, pupils attended in two or sometimes even three shifts. But teachers were produced at a most creditable rate, and the average number of pupils per teacher has for some time been much lower than in Britain or the United States. The following figures are impressive evidence (it should be noted that the total number of children educated was affected by the very large reduction in the birth-rate that occurred during and just after the war, and then the post-war increase):

School year	1940–41	1955–6	1958–9	1965–6
Total teachers (thousands)	1,237	1,733	1,900	2,497
Total pupils (thousands)	35,528	30,070	31,483	48,245
of which, ages 15–18	2,558	6,159	4,655	12,682

(SOURCES: *Nar.khoz.*, 1956, p.244; *SSSR v tsifrakh v 1965 godu*, p. 132.)

Particularly spectacular gains were recorded in the more backward national republics.

A large expansion of secondary education after the war culminated in the decision of the twentieth party congress (February 1956) to introduce full secondary education for all in towns, and gradually extend it to the country.

This led quickly to difficulties which remind one of the educational troubles of the late thirties. Soviet secondary education is academically orientated, and those who complete the full 10- or 11-year course (ages 7–17 or 18) and pass the examinations expect to enter higher education, and show little desire for ordinary technical training. But then, who is to work at the factory bench? There were declarations on 'polytechnization' of secondary education, i.e. the introduction of elementary work training in the upper forms, but lack of facilities and lack of interest among teachers made it virtually a dead letter. Khrushchev, who was instinctively suspicious of intellectuals and averse to privilege, also sought to widen the social intake into universities.

Stalin, it may be recalled, checked the spread of academic secondary education by introducing fees, and dealt with lack of trainees by compulsory call-up of juveniles. Khrushchev had just abolished fees, and compulsion was no longer politically practicable. So in December 1958, following a memorandum from Khrushchev, a new law was passed. This limited compulsory education to 8 years (7–15), after which the majority of pupils would go on to technical training of various kinds, either full-time or on the job. Higher education was to be part-time for most students, being combined with useful work, and was to be preceded by at least two years' employment.

This reform was not popular among either educationalists or students, and it is a fascinating study in social politics to observe how Khrushchev's intentions were evaded and then virtually nullified. Many thought up special reasons why an exception should be made for their subject (music, mathematics, physics, etc.) or their district, and so on. This was one of Khrushchev's first failures. But one essential element remained: the abandonment of the decision to extend full secondary education to all, the *de facto* acceptance of the principle of selection.

Total numbers of pupils receiving both higher and secondary-technical education advanced very rapidly, quadrupling between 1940 and 1964. However, it should be noted that in recent years roughly a third of both categories have been correspondence or external students, many of whom never complete their courses.

Health statistics are likewise very impressive:

	1940	*1958*	*1965*
Medical doctors (thousands)	134·9	347·6	484
Numbers of hospital beds (thousands)	791	1,533	2,224

(SOURCE: *SSR v tsifrakh v 1965 godu*, p. 154.)

NOTE: Figures exclude military.

Here, however, there was much to criticize in the quality of the service rendered, the availability and cost of drugs, the lack of hospital equipment. This was recognized when the price of drugs was halved as from August 1957, and a decree devoted to the medical industry output sought greatly to increase output of all kinds of items, from thermometers to penicillin.[33]

Nearly all teachers and medical practitioners were and are women. There are also many women engineers, technicians, judges. They predominate in retail trade.

During and just after the war the extreme shortage of young men caused the utilization of women in exceedingly heavy and disagreeable work, and while women are habitually engaged in manual labour in most peasant countries, this came increasingly to be regarded as undesirable; in July 1957 a decree forbade the employment of women in the mines, except in supervisory and service tasks.

Most wives continued to work and this had caused another social problem: small families became increasingly the rule in towns, especially as there was so little housing space. Increased créche facilities did little to counter this trend. The result was an immense gap between the birth-rates in Russia proper (15·8 per 1,000) or the Ukraine (15·3), and the far higher rates in Azerbaidzhan (36·4) or Uzbekistan (34·7).[34]

KHRUSHCHEV RIDES HIGH:
THE DRAFT SEVEN-YEAR PLAN

A revised version of the sixth five-year plan is now known to have been prepared by Gosplan and submitted for approval on 9 April 1957. Approval was not forthcoming. It was decided to prepare a new plan covering the seven-year period 1959–65.[35]

One reason given was the discovery of new mineral resources, and the new regional plans which would follow from the sovnarkhoz reform. Further, Khrushchev desired greatly to speed up the growth of the relatively backward chemical industry (this was decided by the plenum of May 1958), and also to transform the fuel balance of the U.S.S.R., which was too heavily orientated to coal and neglected oil (which was abundantly and cheaply available in the Volga-Urals fields) and above all natural gas, which was available in large quantities and very little used. The labour situation, while adversely affected by the consequences of low wartime births, was relieved by reductions in the armed forces.

This duly found its reflection in the new seven-year plan targets. Meanwhile the years 1956, 1957, and 1958 were, from the point of view of long-term plans, orphan years.

The seven-year plan envisaged the following:

	1958	1965 plan	1965 actual
National income	100	162–165	158
Gross industrial output (index)	100	180	184
Producers' goods	100	185–188	196
Consumers' goods	100	162–165	160
Iron ore (million tons)	88·8	150–160	153·4
Pig iron (million tons)	39·6	65–70	66·2
Steel (million tons)	54·9	86–91	91·0
Coal (million tons)	493	600–612	578
Oil (million tons)	113	230–240	242·9
Gas (milliard cubic metres)	29·9	150	129·3
Electricity (milliard Kwhs)	235	500–520	507
Mineral fertilizer (million tons)	12	35	31·6
Synthetic fibres (thousand tons)	166	666	407
Machine tools (thousands)	138	190–200	185
Tractors (thousands)	220	–	355
Commercial timber (million cubic metres)	251	275–280	273
Cement (million tons)	33·3	75–81	72·4
Cotton fabrics (million square metres)	5·79	7·7–8·0	7·08
Wool fabrics (million square metres)	303	1,485	365
Leather footwear (million pairs)	356·4	515	486
Grain harvest (million tons)	134·7	164–180	121·1
Meat (total) (million tons)	3·37	6·13	5·25
Workers and employees (millions)	56·0	66·5	76·9
Housing (million square metres)	71·2	650–660*	79·2

(SOURCES: *Nar. khoz.*, 1960, pp. 210–12; *Nar. khoz.*, 1965, pp. 136–9, 262, 557, 609; seven-year plan, *1959–65*.)

* Total for seven years 1959–65.

Apart from a particularly large increase in investment in chemicals and non-solid fuels, the plan envisaged an increased weight of investments in eastern areas, which were to receive over 40 per cent of the total funds. It was a repeated source of criticism that ministries had found it convenient to direct investments, whenever possible, to already developed regions, to save overheads. It was hoped that sovnarkhozy would see things differently.

Khrushchev, in his optimistic moods, soon found the new plan modest, and in October 1961, at the twenty-second congress, he announced some upward amendments (e.g. steel to 95–97 million tons). This was part of his mood of triumph. Success in making an impact on the imagination of the outside world, and his own self-confidence, was markedly helped by the launching, on 4 October 1957, of the first sputnik. Russia, the backward giant so often looked down upon by the developed West, not so long ago a country of peasant illiterates, had scooped the world. A fine advertisement for the Soviet system.

As the final column in the above table shows (and allowing yet again for exaggeration in indices), industrial progress was impressive during the period of the seven-year plan. There was also a major upward surge in output of many consumer durables. Yet there were some serious shortfalls, and critical deficiencies developed which, as we shall see, helped to topple Khrushchev. The precise fate of the various targets of the seven-year plan is too recent a question to be 'history'. However, the strains of recent years have been highly relevant to an understanding of the nature, functioning and problems of the Soviet planning system, as established under Stalin; a survey of the difficulties of Khrushchev's last years of power legitimately forms part of our study.

KHRUSHCHEV'S TROUBLES:
A CHRONIC CRISIS OF PLANNING

The sovnarkhoz (regional) structure contained within itself a very serious weakness. To say so is not to be wise after the event. It was no doubt clear to Khrushchev's opponents at the time – and to some outside observers too, including the present writer.[36]

The fundamental problem that faced the planning system under Stalin's successors was this: centralized decision-making could only encompass a portion of the multitude of decisions which, in an economy of the Stalin type, must be taken by the planner-administrators, and which, in the absence of any effective criteria other than plan-orders, logically cannot be taken elsewhere. Therefore many plans consisted of aggregated indicators (roubles of total output, or tons, or square metres), and many elements of the plan could be inconsistent with one another. For example, the supply plan frequently failed to match the production plan, the fulfilment of aggregated output targets was inconsistent with meeting user requirements; the labour, or wages, or financial plans were out of line with each other or with output plans, and so on. A large number of semi-anecdotal examples can readily be assembled to illustrate the resultant irrationalities. Steel sheet was made too heavy because the plan was in tons, and acceptance of orders from customers for thin sheet threatened plan fulfilment. Road transport vehicles made useless journeys to fulfil plans in ton-kilometres. Khrushchev himself quoted the examples of heavy chandeliers (plans in tons), and over-large sofas made by the furniture industry (the easiest way of fulfilling plans in roubles).[37] New designs or new methods were avoided, because the resultant temporary disruption of established practices would threaten the fulfilment of quantitative output targets. It would indeed be a miraculous coincidence if the product mix which accorded with the requirements of the user also happened to add up to the aggregate total required by the plan. Of course, ideally, the aggregate total is made up of the separate requirements of all the users. But there is never time or information available for such perfect planning. The planners in fact proceeded on the basis of statistics of past performance.

Yet, with prices not even in theory capable of fulfilling their role as economic 'signals', there was no other criterion than the plan. The central organs either themselves decided these plans, or laid down limits within which subordinate units could operate. Central decisions, though often incomplete and imperfect, were based on an assessment of what was needed. This in turn was based on political directives, past experience embodied in statistical returns, applications (indents) from below, and material balances designed to achieve an elementary input-output consistency. In the almost total absence of market forces this combination of information flows, requests and directives constituted the foundation upon which economic activity rested.

This activity, highly complex in its nature, became more so as the economy grew, and then in a sense outgrew the centralization upon which it was built. Tasks had to be divided, between ministries administering industries, Gosplan and its numerous sub-divisions (dealing with prices, investment policy, labour, supplies of key commodities), the Ministry of Finance, the State Committee on Construction, and so on. A further source of complication was the division of authority between government and party organs, at all levels. As already pointed out, all this led to acute strain in the process of seeking to ensure coordination, and sometimes plans were inconsistent, or impossible to carry out *in toto*. In the key area of investment this took the form of chronic over-spreading of resources on too many projects (*raspylenie sredstv*), causing serious delays in completion. A contributory cause was that capital was provided free out of the state budget, there was no capital charge, and so subordinate authorities over-applied for it and started all they could in the hope of getting more.

Under Stalin the top priority of heavy industry was ruthlessly enforced. Errors and omissions were borne by the less important sectors. Hence persistent neglect of agriculture, and the fact that even the modest housing plans were never fulfilled, despite the notorious degree of overcrowding.

But under his successors this was no longer so. Housing, agriculture, consumers' goods, trade, all became matters of importance, even of priority. So the task of planning became more complicated, because a system based on a few key priorities, resembling in this respect a Western war economy, could not work so effectively if priorities were diluted or multiplied.

Consumer demand, for so long ignored, became more important, as living standards improved and customers could exercise more choice. Some goods became unsaleable, either through poor quality or over-production. Yet the system was not designed to respond to demand, whether from consumers or indeed from enterprises (e.g. for some particular item of equipment, or metal of precisely the desired quality). It was built to respond to orders from above, and for the achievement of large-scale investment projects and expanding the volume of production. The financial and price systems were well able to extract the surpluses needed to sustain high growth-rates.

The planners tried various expedients. They issued instructions that user demand should be met. They modified the bonus systems so that the achievement of purely quantitative targets should not be sufficient, that the assortment plan had also to be fulfilled, that costs had to be reduced, the wages plan not exceeded, and so on. They experimented with a kind of value-added indicator known as 'normed value of processing'. Each of these 'success indicators' had its own defect, induced its own distortions. Thus, insistence on cost reduction often stood in the way of the making of a better-quality product. A book could be easily filled with a list of various expedients designed to encourage enterprises to act in the manner the planners wished, and the troubles to which each of them gave rise.[38] The greater the number of indicators, the more likely it was that they would be inconsistent. Similarly, the greater the number of items and sub-items planned and allocated by the centre, the greater the burden on the planners and the likelihood of error or delay. But, since the central plan was the basis of all activity, the absence of some item from the plan might have resulted in it not being provided. So efforts to reduce the number of centrally planned indicators tended to be futile. If, say, frying pans or electric irons were not in the plan, then they tended not to be produced, and productive capacity would be switched to make things in which the centre expressed an interest.

Errors in investment planning, shortfalls in technical progress, were to some extent offset, under Stalin, by unplanned movement of labour from the villages. In every plan period, except the second, more workers came to town than was envisaged. By the middle fifties there was much greater consciousness of the need for efficient utilization of resources, including the now scarcer labour resources, to meet the many competing needs. Matters were further complicated by labour's greater freedom of movement.

All these circumstances, and the liberating effect of Stalin's departure, led to a revival of economic thought. No longer barred from practical affairs, economists joined with the more intelligent planners in seeking new criteria – for investment decisions, for resource allocation by planners, and, last but not least, for decision-making at enterprise level. The role of prices, of what came to be called 'commodity–money relations', became a subject of lively discussion. Stalin's dictum that 'the law of value' (affecting exchange relations) does not operate in transactions between state enterprises was rejected.

In the light of this situation the 1957 sovnarkhoz reform was a step in the wrong direction, or at best a step sideways. The essentials of the system were unaltered, in the sense that the plan was the sole effective operational criterion, plan fulfilment the only important source of managerial bonuses. (Profits, it is true, formed the basis of a 'director's fund', renamed 'enterprise fund', but it was of minor importance as an incentive to act.) Yet the abolition of the ministries removed a vital element in the chain of command. Any one of the 105 sovnarkhozy was unable to assess the needs of the other 104, unless these were conveyed to them by the centre. For how could an official in, say, Kharkov or Omsk know the relative importance of requests received from all over the country? In any case, the central allocation of key commodities was preserved, and indeed had to be reinforced to avoid confusion. But one can only allocate what is produced, and so the centre quickly found itself immersed in what amounted to production decisions too, but without the ministerial mechanism. Gosplan, the one body with the necessary information, lacked power and was deluged with work.

Yet enterprises were subordinated to sovnarkhozy, and the latter had only one independent criterion: the needs of their own regions. In the absence of clear orders to the contrary they therefore allocated scarce resources to their own regions. Loud protests reached Moscow that established supply links were being broken. More and more items therefore had to be controlled by or through Gosplan. Similarly, investment funds under the control of sovnarkhozy were diverted to local needs, to the detriment of other regions. Evidently Khrushchev thought that the regional party secretaries would ensure that their sovnarkhozy enforced national priorities. This did not work out. The disease of 'localism' (*mestnichestvo*) was diagnosed, officials blamed, dismissed, threatened with punishment (e.g. by the decree of 4 August 1958). The same complaints were afterwards repeated, since the defect was built into the system.

As the sovnarkhozy gradually lost effective power – in 1962 the Estonian sovnarkhoz controlled only 0·2 per cent of the production of that republic[39] – the enterprises found themselves virtually without a master, or rather with many masters, since production and supply plans reached them from several production and allocation departments at all-union and republican levels.[40] It was fortunate that informal links of many kinds kept things more or less on an even keel most of the time. But some of these evidently worried the authorities, as they facilitated embezzlement, or just theft. The introduction on 7 May 1961 of the death sentence for a range of economic offences (as distinct from counter-revolutionary or treasonable activities) was a sign of alarm.

ORGANIZATION, REORGANIZATION, DISORGANIZATION

As the deficiencies of the 1957 reforms became apparent Khrushchev set about making administrative changes. To list and explain them all is impossible in the context of a general history. A bare list of main changes, affecting industry and construction, will be sufficient to demonstrate the confusion created in the process of trying to remedy confusion. They were part of a rather chaotic process of re-centralization.

Firstly, a long list of state committees was created which by 1962 had very much the same designations as the defunct ministries, but they had no executive authority. Enterprises were not subordinate to them. They considered. They advised, especially on investment and technical policy. They could not order.

Secondly, Gosplan was again split. At all-union level perspective planning and research were made the responsibility of an Economic science council (*Gosekonomsovet*) in 1960. A division on a different basis was tried out in the large republics from 1961: Gosplan there was given the task of planning, republican sovnarkhozy the task of implementation of plans, including material supplies. Then, in November 1962, Khrushchev announced a similar change at the centre: Gosplan would plan particularly in the longer term, while a new body, the all-union sovnarkhoz, would implement plans. This division was not a happy one, and there was much criticism of parallelism. So, after an obscure struggle, there emerged in February 1963 a super-coordinator of the coordinators which was given the historically memorable name of VSNKH, but which seems to have achieved little in its short life.

Thirdly, sovnarkhozy were merged, reduced in number from over 100 to 47 early in 1963. The four Central Asian republics were made into one sovnarkhoz, and a Russian was appointed to head it. There were also changes in sovnarkhoz powers: they lost construction, they gained local industry, previously under local soviets. Other changes included the creation of large planning regions (seventeen for the whole country), which achieved little in practice.

Fourthly, the party itself was split in 1962, into industrial and agricultural sections, which caused much perplexity. So did the fact that the enlarged sovnarkhozy seldom coincided with the geographical divisions of the party, so on the one hand each half of the divided party was enjoined to supervise the economic activities within its jurisdiction, but on the other it was made organizationally difficult to do this. Khrushchev seems to have feared party officials' collusion in 'localism'. Government organs were also, at provincial level at least, to be split into two. This was a shock to party and state bureaucracies alike, and made no administrative sense.

By 1963 no one knew quite where they were, or who was responsible for what. Pungent criticisms were appearing in the specialized press. Planning was being disrupted.

Khrushchev added to the troubles of the planners by loud accusations of conservatism, and by pressing on them a chemical industry investment programme which threatened the whole balance of the economy, and extended to industry a typical Khrushchevian campaign of a type which had done harm in agriculture. Output was to be trebled in seven years. The targets he forced through were absurd and were promptly abandoned by his successors, though no one denied the desirability of substantially expanding the chemical industry. To make room for them, a new plan for 1964–5 scaled down many other plan targets, from steel to housing. Khrushchev's campaigning zeal threatened serious shortages of steel, of coal, of bricks. It was typical that sensible changes – such as the substitution of non-solid fuels for coal, or prefabricated concrete for bricks – became much too drastic. Thus most brickworks were closed, 8,000 out of a total of 12,000, according to A. Birman (*Novy mir*, No. 1, 1967), causing grave shortages.

Further causes of trouble were the soaring expenses of the space and missile programme, and the sharp rise in military expenditure – by 30 per cent in 1961, which represented a heavy call on scarce skills and specialist equipment.

Other reasons too contributed to a slowdown in growth, which became quite noticeable especially after 1958. The rate of increase in investment greatly decreased, as the following figures amply demonstrate:

(per cent)

1958	+ 16
1959	+ 13
1960	+ 8
1961	+ 4
1962	+ 5
1963	+ 5

In 1963 and 1964, officially claimed industrial growth rates fell below 8 per cent, the lowest peacetime figures except 1933. Owing to poor agricultural performance, the national income in 1963 is said to have risen by only 4·2 per cent. This was when the C.I.A. made news by claiming that the real figure was now down to 2½ per cent, or well below even the United States. The view became widespread that Khrushchev's handling of affairs did not help.

As strains developed, so investment controls became more stringent, and the housing programme, to which much publicity had been devoted, was cut back. Particularly severe cuts were imposed on private building, as the following figures demonstrate:

Urban house-building
(million square metres of total space)

	State	Private
1960	55·8	27·0
1961	56·8	23·6
1962	59·8	20·7
1963	61·9	17·4
1964	58·9	16·2
1965	62·4	15·5

(SOURCE: *SSR v tsifrakh v 1965 godu*, p. 157.)

Khrushchev indeed mounted a campaign against ownership of private cottages, which resulted in some confiscations. He seemed in his later years to have been pursuing a campaign against 'bourgeois' property tendencies. He goaded on the planners and the management by loud-sounding promises about overtaking America within a few years, even by 1970, and promising to achieve some elements of full communism (free municipal housing, free canteen meals, free urban transport) by 1980. This was, no doubt, part of his effort to revive ideological fervour and inject dynamism into a system which, left to itself, was liable to fall victim to inertia. He made other promises too: a 35-hour week, the abolition of income tax, a minimum wage of 60 roubles, all by precise dates, at which they would not be implemented. (The 60-rouble minimum in fact came into force in 1968.)

KHRUSHCHEV'S AGRICULTURAL MISMANAGEMENT

Agricultural output during the seven-year plan should have increased by 70 per cent. This is what in fact happened:

	Total	Crops	Livestock
1958	100	100	100
1959	100·4	95	108
1960	103	99·4	107
1961	106	101	112
1962	107	101	115
1963	99	92	108
1964	113	119	106
1965	114	107	123
1965 plan	170	–	–

(SOURCE: *SSR v tsifrakh v 1965 godu*, p. 69.)

The 1963 and 1965 grain harvests were badly affected by adverse weather. But the picture as a whole was very disappointing, and was so treated by every Soviet analyst.

The following were the principal reasons for this – and speeches by Brezhnev and Matskevich made since Khrushchev's fall[41] show that the Soviet leaders would not disagree with the analysis made below.

Firstly, the 1958 reform imposed excessive burdens on kolkhozes, which the procurement prices did not fully cover. They had to pay too much, and too quickly, for machinery. The result was to compel a cutback in investments and decline in payments to peasants. That there was a decline in pay after 1957 is fully admitted; its precise extent is still covered by statistical silence.

Secondly, the abolition of the MTS adversely affected maintenance by dispersing sophisticated equipment around kolkhozes, few of which possessed either the workshops or the skilled manpower to maintain or repair it properly. Many mechanics and tractor-drivers had left the villages rather than become kolkhoz peasants. (The Repair Technical Stations provided for in the 1958 reform never got off, or rather on to, the ground.)

Thirdly, Khrushchev pressed campaigns upon the farms, using for this purpose the party machine, which interfered with little or no regard for local conditions, and did great damage. The list of campaigns and distortions is a long one. Maize was to be sown in areas in which the soil and climate were unsuitable, or where the necessary labour and machines were lacking, with miserable harvests as a result. Crop rotations were disrupted. Cattle was unnecessarily slaughtered to achieve spectacular short-term results in meat production, then permission to slaughter was withheld, to rebuild the herds, even if there was no point to it (other than statistical point). The campaign against Stalin's favourite travopolye (grass-rotation) reached such a height in 1961–3 that some farms were ordered to plough up fields of clover, for no better reason than to report to Moscow that the area under grasses had been reduced. Strong pressure from Moscow compelled monoculture (spring wheat) on the formerly virgin lands; permission to reduce sowings to preserve the soil from erosion and weed infestation was refused. Serious harm resulted. Khrushchev personally toured the country repeatedly to advocate these campaigns, and dismissed many local officials and experts who dared to disagree with him, or who failed to carry through impossibly ambitious plans. Procurement quotas tended to be based on unrealistic production targets, and this forced local party officials to act as they did in Stalin's time, taking whatever they could squeeze out, disregarding quota rules, ordering farms about. Centralization in procurements was first relaxed, then reimposed. Even methods of cultivation had their 'waves' of imposed fashions: two-stage harvesting (cutting first, threshing later), the use of peat compost for seedlings, square-cluster sowing of maize and potatoes, always regardless of the local situation.

Fourthly, as also in industrial planning, Khrushchev confused matters by repeated reorganizations. The 1958 reform had wiped out the MTS, and with it an important controlling mechanism. In the course of 1959–61 he had succeeded in turning the Ministry of Agriculture into a mere technical advisory service. At first he seemed to lay great stress on the role of the State Procurements Committee, as provider of guidance to farms by a contract system, while another state committee handled supplies of equipment and materials to agriculture (*Sel'khoztekhnika*). The rural party organs also interfered constantly. Impatient of defects and inefficiencies, Khrushchev then scrapped this inadequate administrative hotch-potch, decided on tighter controls, and in 1961 set up about a thousand Territorial Production Administrations (TPA), to control both kolkhozes and state farms within their areas. At first these TPA were outside the control of the raion (district) party secretaries, and in each there was to be a party plenipotentiary appointed by the republic. (It is as if in his old age Khrushchev remembered the political departments of the MTS, but only after having abolished the MTS.) Later he amalgamated the raiony to correspond to the TPA boundaries. Above the TPA were supposed to be provincial, republican, and all-union agricultural committees, and in each the senior party man was to be the boss. It seems that this was too much for some of his comrades, and though the TPA did actually exist, the all-union and republican agricultural committees were stillborn. But the powers of the TPA in relation to farm management and to the procurement and other administrative organs were far from clear. Confusion became worse when Khrushchev divided the party into two, so that the secretary in command of provincial or 'TPA-area' agriculture had no concern with local industry or services which supplied agricultural needs. In 1964 he threatened something quite intolerable, when he proposed a truly hare-brained scheme: to administer agriculture by all-union departments concerned with particular products. One such was set up: *Ptitseprom SSSR*, to deal with poultry and eggs. How this would fit in with the TPA, or indeed with the internal administration of the farm, was never explained.

Fifthly, after 1957, output of many kinds of farm machinery actually fell. Apparently it was instinctively felt that the kolkhozes, as non-state institutions, deserved a lower priority, and/or that the MTS had been over-supplied with equipment.

Sixthly, price relativities established in 1958 were in total conflict with the plan, especially the plan to expand the output of livestock products. Meat and milk prices were far too low, and Soviet economists had no difficulty in demonstrating that these items were produced at a loss. So the bigger the effort in this direction, the poorer the farm would become. This contributed to the reduction in peasant incomes as from 1958, already referred to. In 1962, procurement prices for these items were increased by upwards of 30 per cent, and this was accompanied by a similar increase in retail prices, which gave rise to loud protests in the towns. Yet even after this increase, and despite the fact that the gap between procurement and retail prices was now too small to cover expenses, necessitating a subsidy, livestock prices were still unattractive; they had to be further increased after Khrushchev's fall. Prices were also insufficiently differentiated between regions: the grain price was remunerative in the fertile North Caucasus, but too low in the centre, north and west; yet the authorities insisted on grain crops being sold also in these areas. In state farms there was a redistribution of net incomes by subsidy, and transfer of profits to the budget. In kolkhozes, in the absence of land rent, the effect was to cause excessive income differences between areas. These differences were enhanced by a further error of Khrushchev's which Brezhnev later pin-pointed: neglect of the 'non-black-earth' areas, which were left out in the cold in allocating equipment, investment funds, fertilizer. Yet despite their less fertile soil they did at least have reliable rainfall.

Finally, Khrushchev committed the grave error of attacking the private plots. We have seen earlier that suburban livestock owners had already been subjected to fiscal burdens in 1956 but the drive began seriously to get under way in 1958 when Khrushchev made his speech to the effect that peasants in his birthplace, Kalinovka, had voluntarily sold their cows to the kolkhoz. He emphasized the word 'voluntarily'; but soon enough thousands of party secretaries were exerting pressure on the peasants to do the same. Then there was the general fodder shortage, with priority enforced for collective and state animals. Pasture rights were limited. Taxes were imposed where numbers of livestock exceeded strict limits (especially on those owned by state-farm peasants). The net effect on labour inputs was negative. One analyst later showed that peasants worked *less* for the kolkhoz, because it took them longer to procure (or scrounge) fodder for their livestock.[42] An eye-witness told me of the following conversation overheard on a farm:

First peasant: 'That's smoke yonder; what's burning?'
Second peasant: 'Hay is burning.'
First peasant: 'Well, let it burn; if it all burns, maybe they will return us our cows.'

Statistics show that most peasants did retain their livestock, but the number of privately owned cows fell between 1959 and 1964 by 14 per cent while collective and state herds grew rapidly. All these irritations affected work morale, the more so as pay in the collectives declined at the same time.

The great pressure to increase livestock numbers while crop output was stagnant doubtless contributed to a run-down of reserves. The 1963 grain harvest was bad, and it proved necessary to import large amounts of wheat from capitalist countries. This was a heavy blow to prestige, and even though Khrushchev could blame the weather, his personal standing suffered a severe shock, since he had staked so much on his pet agricultural projects. Total livestock numbers fell very rapidly owing to fodder shortage (pigs from over 70 million to only 41 million). Despite imports there was some difficulty in supplying bread and, especially, flour. Many agricultural areas became short of food. The 1964 harvest proved much better, but it was too late to save Khrushchev. (His successors were so embarrassed that they for a time concealed the 1964 statistics, because they were too favourable.)

The basic problems of agriculture were not, of course, Khrushchev's creation. He interfered, reorganized and campaigned too much, but he had inherited a generation of neglect and impoverishment and a system in which change could come only by order from above, since it treated peasant or even farm-managerial initiative with instinctive suspicion. The word 'spontaneity' (*samotyok*) was never close to Bolshevik hearts, but it was treated as a particularly dirty word in agriculture. Dr Jasny, the outstanding émigré analyst, once said: 'More than any other single thing, Soviet agriculture needs samotyok.' In its absence, plans had to be imposed, and the methods of Soviet officialdom, created by the Stalin epoch, ensured that they be imposed on a set pattern, regardless of local circumstances, so as to report what Moscow wanted to hear. Hence campaigns always degenerated into excesses, even if they had within them some rational purpose – as they usually had.

Under Stalin too there had been campaigns. But in the last analysis he was really only interested in procurements, and interference of party officials was only spasmodic.

It may be asked: but if pay for collective work improved so much in comparison with the Stalin period (despite the downturn of 1958–60), and if prices paid by the state increased so very greatly, how is it that incentives remained inadequate? The answer lies in one seldom-understood fact. Under Stalin procurement of foodstuffs could be regarded as a heavy, burdensome tax in kind. Except in areas of industrial crops, such as cotton, where prices were much better, the peasants to all intents and purposes lived on what was left after the state had taken its share, and on the produce and cash derived from private plots. Their work for the kolkhoz, in the majority of cases, was just a part-time occupation. Therefore the fact that they could not exist on collective pay was not disastrous: it could be seen as almost unpaid service qualifying them to operate their little private enterprise, on which they lived. But under Khrushchev campaigns to increase output (especially in the almost un-mechanized and labour-intensive livestock sector), and to expand procurements led to a big rise in labour inputs on the farms, so for many peasants collective work became their primary occupation. For this purpose neither pay nor conditions were adequate. As late as 1967 there were still recommendations being made that shiftworking for milkmaids be adopted more widely, so as not to have work periods running from 3.45 a.m. to 9.45 p.m., which made all private life impossible.[43] The same point was still being made in 1975!

KHRUSHCHEV'S FALL

Khrushchev was dismissed in October 1964. The economic troubles just described were part-causes of his fall. His over-ambitious campaigning ('hare-brained schemes'), his exaggerated promises, his arbitrary methods, his disorganizing 'reorganizations', were too much. Yet he did achieve considerable successes, especially in his first five years, and his defects are explicable by his background and experience. He was politically 'educated' under Kaganovich, in the dramatic years of the early thirties. He inherited many perplexing problems, and his methods of tackling them belonged to a different epoch, and were now obsolete. He half understood the need and even the required direction of change, and often spoke of managerial autonomy in industry and agriculture, economic criteria, rational investment policy. He showed that he knew better than anyone how the bureaucratic apparatus of party and state could distort policy and paralyse desired initiative. But in the end he knew only the traditional methods.

13. Brezhnev: From Stability to Immobility

Brezhnev and Kosygin succeeded Krushchev. They undid what they considered (not without reason) to be his wrong-headed reforms: the agricultural TPAs were eliminated, the normal raion administrative organs were reconstituted, the Ministry of Agriculture reappeared with its former powers, the division of the party between industrial and agricultural parts was abandoned, its unity restored. In September 1965 the *sovnarkhozy* were finally abolished, and the industrial ministries reappeared; Gosplan again became the key planning organ, covering both long-term and short-term planning. However, regional supply depots remained, under the authority of the State Committee on Material Supplies (*Gossnab*). The country was promised an end to 'hare-brained schemes' and constant reorganizations.

AGRICULTURE A PRIORITY

The policies of the new leadership towards agriculture emerged in the March 1965 plenum of the central committee. There was severe criticism of past errors: excessive 'campaigning', neglect of the non-black-earth regions, restrictions on private livestock, inadequate supply of machinery, the unprofitability of the livestock sector. It was decided to increase procurement prices substantially, especially for livestock. In subsequent years there were further increases in prices, without any consequential rise in retail prices. As a result, there was a fundamental change in the role of agriculture in the economy. From being a source of accumulation of capital for investment in industry, agriculture became a net burden on the rest of the economy. The subsidy to cover losses in the livestock sector (i.e. the difference between the price paid to farms plus handling costs and receipts from retail sales) grew and grew, until by 1979 it exceeded 23 milliard roubles, i.e. more than the officially admitted level of military expenditure![1] In addition, a sum of 4 milliards annually is paid to compensate the suppliers of industrial inputs to agriculture for the low prices they were allowed to charge.[2] Investment in agriculture, both state and collective, rose substantially, both relatively and absolutely:

	1961–5	1966–70	1971–5
TOTAL INVESTMENTS (milliards of roubles)	247·6	353·8	501·6
of which: Agriculture*	48·6	82·2	131·5
per cent of total	19·6	23·2	26·2

(SOURCE: *Nar.khoz.* (1975), pp. 502, 510.)

* Productive and unproductive investments, including expenditure on repair workshops of *Sel'khoztekhnika*, research, establishments for processing farm products, irrigation workshops, etc.

Brezhnev affirmed and reaffirmed the high priority to be accorded to agriculture, and to industries serving the needs of agriculture. While scaling down Khrushchev's unrealistically overambitious targets for fertilizer production, he continued the policy of pressing ahead with the expansion of this and other sectors of the chemical industry.

With what result? One must try to eliminate the variations due to weather, which continue to have a large influence on the size of the harvest. Thus 1965, 1967, 1972, 1975 and 1979 were poor years, for reasons quite unconnected with plans and policies. The basic trend was upwards:

	Annual averages			
	1961–5	1966–70	1971–5	1976–80
Gross agricultural output (milliards of roubles)	66·3	80·5	91·0	99·9
Grain harvest (million tons)	130·3	167·6	181·6	205
Cotton (million tons)	4·9	6·1	7·7	8·9
Sugarbeet (million tons)	59·2	81·1	76·0	88·4
Potatoes (million tons)	81·6	94·8	89·6	(84)
Meat (million tons)	9·3	11·6	14·0	14·8
Milk (million tons)	64·7	80·6	87·4	92·6

(SOURCES: *Nar.khoz.*, 1975 and plan fulfilment reports.)

It may be said that, despite some recent slowdown, Soviet agricultural output has in fact risen faster than that of most other countries, especially Western countries. However, the situation is in fact deeply unsatisfactory, and this for the following reasons.

Firstly, the increases have been from very low levels, of yields and productivity, and measured by international standards they are still very low.

Secondly, the increases have been achieved at disproportionate cost. No major country devotes so high a proportion of its total investment to agriculture. Yet it still appears to be undercapitalized, due partly to unbalanced (*nekompleksnyi*) mechanization, partly to low quality of machines, and also to poor maintenance, so that equipment and vehicles wear out quickly.[3]

Thirdly, the reduction of the labour force was slow, while pay both in *kolkhozy* and *sovkhozy* increased substantially. This led to a sharp rise in costs of production, which added to the burdens on the state budget through the ever-increasing subsidies which had to be paid. There is still a shortage of labour at peak periods, and millions of urban citizens and soldiers are drafted in to help bring in the harvest.

Fourthly, agriculture proved incapable of supplying enough food to cover rapidly growing demands from the now better-paid citizens, especially as the peasants themselves were eating better. Demand was (and still is) artificially stimulated by the policy of low prices for food. Thus meat prices were last increased in 1962, and since then average wages have almost doubled. So even though production has gone up, shortages have become more acute.

Finally, the gap between supply and demand (especially of fodder grains, to sustain a larger livestock population) has led to increased dependence on imports from the capitalist world. Thus in 1973, following the poor harvest of the previous year, imports of grain reached 23 million tons.

It was often the case that successes in one sector were negatived by failure to provide complementary resources. For example fertilizer deliveries did rise rapidly, and this, together with increased expenditure on liming and land improvement, did contribute to the rise in the average harvests. However, as many sources show, there are shortages of bags, means of transport, storage facilities and spreading machines. Also, the type of fertilizer delivered often does not accord with local requirements, and under these conditions it piles up at railway yards and much of it is wasted.

Organizational changes gradually introduced include the conversion of more *kolkhozy* into *sovkhozy*; the latter now have more than half of all the arable land in the U.S.S.R. There has also been an increase in the number of enterprises run by several *kolkhozy* and/or *sovkhozy* jointly, e.g. to produce bricks, to process vegetables, to make implements and so on. The status of *kolkhozy* has been redefined following the long-postponed Kolkhoz Congress, which finally met in 1968, but this did not introduce any new principles – except for the setting up at local, republican and all-union levels of *kolkhoz* councils (*sovet kolkhozov*). This may have been originally intended as a representative body, but in practice its chairman is a state official (at all-union level it is the Minister of Agriculture), and there is little sign that the new body has made much impact, *except* that in the republic of Moldavia it carries out administrative and planning functions. It seems that official policy envisages the gradual merging of *kolkhozy* with *sovkhozy*, and this would explain the reluctance to give powers to bodies which consist only of *kolkhozy*. An impetus in the same direction is being given by the creation of 'agro-industrial complexes', which link the two categories of farms with state (and joint *kolkhoz–sovkhoz*) industrial processing establishments and retailing organs.

Meanwhile policy towards private plots has fluctuated. As we have seen, Brezhnev in 1965 took a more positive view of them than did Khrushchev in his last years. However, the press has frequently cited instances of restrictions, and the present (1980) policy is to eliminate these and to encourage private cultivation and especially livestock. But there are still obstacles to marketing; for example, Georgia now stops its peasants from taking fruit and vegetables across the borders of the republic and taking advantage of high free-market prices in Russian cities.

'REFORM' OF INDUSTRIAL PLANNING

A seemingly important reform was adopted in 1965 – a decree which had the declared intention of increasing managerial powers, reducing considerably the number of compulsory indicators 'passed down' from the centre. Prices were recalculated on the basis of cost plus a percentage of the value of capital assets, thus supposedly corresponding to Marx's 'price of production'. A major adjustment in prices was to eliminate losses in such sectors as coal and timber. A capital charge, averaging 6 per cent, was to be levied. This same decree finally eliminated the *sovnarkhozy* and restored the industrial ministries with almost the pre-*sovnarkhoz* powers, one important difference being the attempt to concentrate in *Gossnab* (the state material-technical supply committee) the function of disposal which departments of ministries formerly exercised. Most of the ministries were all-union rather than union-republican; in other words, although there was a move towards greater devolution to management, the clear intention was to centralize planning and ministerial powers in Moscow. Republican and other regional planning organs lost powers.

It soon became evident that the reforms of 1965 were internally inconsistent. Ministerial powers over 'their' enterprises were those of superiors vis-a-vis subordinates, and orders could be, and were, issued on a wide variety of topics supposedly within the competence of management. Elaborate schemes intended to relate managerial bonuses, and payments into various incentive funds, to profitability and sales (disposals) were ineffective, for a variety of reasons. Thus plans were anything but stable, repeatedly altered within the period of their currency. The rules governing bonuses were changed arbitrarily and frequently. While management was supposed to be encouraged to aim high, success and bonuses still depended primarily on plan *fulfilment*, which meant that it always 'paid' to try to have a modest, 'fulfillable' plan. Persistent shortages of many inputs led to a number of negative phenomena: hoarding, over-application for material allocations, production and construction delays. It also meant that ministries and enterprises made their own supply and procurement arrangements, and unofficial 'expeditors' (*tolkachi*) supplemented the supply system by semi-legal deals. The stress on the profit motive, which seemed to be a feature of the 1965 reform, was also negatived by several factors: the fact that any additional profits were likely to be transferred to the budget (the 'free remainder', *svobodnyi ostatok pribyli*), the much greater stress on the fulfilment of quantitative targets, and, finally, the inherent illogicality of using profits as a guideline when prices were uninfluenced by demand or need – a point to which we will return when we consider the latest attempt to introduce reforms.

By 1970 not much was left of the additional managerial powers ostensibly granted by the 1965 reform. So the plans and policies of the decade of the seventies were applied within the traditional system of centralized planning, with multiple obligatory targets imposed on management, and administrative allocation of inputs; a task divided three ways: between Gosplan, Gossnab and ministries. As will be argued later in detail, the inherent impossibility of efficient and effective centralization underlay many if not most of the problems faced by the economy.

INDUSTRIAL GROWTH SLOWS DOWN

The targets of the eighth five-year plan (1966–70) may be seen in the table on page 377, together with those of the next two quinquennia. While ambitious, it avoided extremes. The outcome seemed satisfactory in aggregate. Aggregate growth rates were similar to those achieved in the previous quinquennium (50 per cent in 1966–70, 51 per cent in 1961–5), with a relative improvement in the growth of the output of consumers' goods (by 49 per cent, against 36 per cent in 1961–5).[4] Some industries did notably well. Oil output rose by 45 per cent, with a large contribution from the newly-operational wells in north-west Siberia, fertilizer by over 100 per cent (though the plan target was even higher). Of course, this was due in large part to the investments made in earlier periods, but industry as well as agriculture benefited from greater stability, the restoration of habitual lines of command, the end of extremes of 'campaigning' so characteristic of the Khrushchev period.

On closer examination, the table gives grounds for some perplexity. How could the overall plan for national income and for industry have been over-fulfilled, when there were serious shortfalls in energy (except oil), steel, and indeed virtually every other product except footwear (agriculture was also short of its plan targets)? Of course, the usual warning is necessary: aggregate indices tend to exaggerate growth rates. But why should this

The eighth, ninth and tenth five-year plans (*1965–70, 1971–5, 1976–80*)

A. *Index numbers*

	(1965 = 100)			(1970 = 100)		(1975 = 100)	
	1965	*1970 plan*	*1970 actual*	*1975 plan*	*1975 actual*	*1980 plan*	*1980 actual*
National income (utilized)*	100	139·5	141	138·6	128	126	120
Industrial production	100	148·5	150	147·0	143	137	124
Producers' goods	100	150·5	151	146·3	146	140	126
Consumers' goods	100	144·5	149	148·6	137	131	121

B. *Quantities*

	1965	*1970 plan*	*1970 actual*	*1975 plan*	*1975 actual*	*1980 plan*	*1980 actual*
Electricity (milliard kwhs)	507	840	740	1065	1039	1380	1290
Oil (million tons)†	243	350	353	505	491	640	604
Gas (milliard cub. metres)	129	233	200	320	289	435	435
Coal (million tons)	578	670	624	694·9	701	800	719
Steel (million tons)	91	126	116	146·4	141	168	155
Fertilizer (million tons)‡	31	63·5	55	90	90·2	143	104
Motor vehicles (million tons)‡	616	1385	916	2100	1964	2296	2199
Tractors (million tons)‡	355	612	458	575	550	590	(562)
Cement (million tons)‡	72·4	102·5	95	125	122	144·5	125
Fabrics (million sq. metres)	7500	9650	8852	11100	9956	12800	10700
Leather footwear (million pairs)	486	620	676	830	698	(n.a.)	—

(SOURCES: *Pravda*, 14 December 1975, 7 March 1976, and statistical returns for the relevant years; *UPSS v rezolyutsiyakh*, Vol. 9, 1972, pp. 42, 49, 50.)

NOTES: Some plan figures are midpoints of ranges.

* Material product (utilized).
† Including gas-condensate.
‡ Gross weight.

(N.B.: 1980 'actual' are estimates to be replaced by the 'real' figures.)

factor be especially important in *this* quinquennium? One reason suggests itself: this was a period of a particularly rapid rise in military hardware production, which is included in the aggregate index of industrial production, but not in the physical-output statistics, which are confined to civilian production.

The same table shows a significant degree of under-fulfilment, even in terms of aggregate indices, as well as in key industrial sectors, in the period after 1970. The ninth five-year plan (1971–5) envisaged an industrial growth rate almost as high as that previously claimed (47 per cent in the five years), with consumers' goods output slightly outpacing that of producers' goods. But this was not to be; the consumers' goods target proved far beyond reach, due partly to the shortfalls in agriculture (and so in the food industry), and partly to increasing delays in construction; the latter factor has become of great importance, with a persistent rise in the volume of uncompleted investments and in the time taken to bring new productive capacity into operation. The official index for national income showed an unusually large gap between plan (+ 38·6 per cent) and actual (+ 28 per cent). Since personal incomes rose faster than planned, shortages of consumers' goods grew gradually more acute.

These tendencies were even more noticeable in the tenth quinquennium (1976–80): a bigger gap still between the (already more modest) plan and its fulfilment, in aggregate and for almost every product of factory and mine. In 1980 only gas and passenger cars reached their five-year targets. Other sources of energy, and also steel, fertilizer, tractors, many other items of civilian machinery, locomotives, building materials, and virtually all consumers' goods, were far behind expectations.

Why? Was this evidence of a developing crisis in the Soviet planning system? The following appear to be key elements of any explanation.

One element is demographic. Because of low birth rates, the working population has virtually stopped increasing. Shortages of labour are serious, even though there is overmanning and waste in some sectors. The high birth-rate in Central Asia is of little help, since the native peoples tend to be immobile and to remain in their villages. Increased output now depends almost wholly on increased productivity per head.

This has been characterized by many in East and West as the need to shift from extensive to intensive growth. Technical progress, more and better mechanization and greater efficiency in the use of resources, become more and more vital.

The efforts to achieve parity in the military field have added to the strain on resources. The military sector absorbs a large segment of the output of the Soviet machinery, engineering and electronic industries; just how large we cannot precisely know owing to lack of published figures.

A further drain on resources has been the big investment effort in agriculture, already referred to. Of perhaps even greater importance are the magnitude and cost of investments in Siberia. These investments are undoubtedly necessary: vast resources of energy and many valuable minerals lie in the largely empty and climatically unpleasant areas north of the Trans-Siberian railway. It has been repeatedly pointed out that while 90 per cent of energy utilization is west of the Urals, 90 per cent of energy resources are east of the Urals. With the exhaustion of oil wells in European areas, it is essential to rely increasingly on Siberian sources, and the richest sources of natural gas lie in the frozen north-west of Siberia. But these investments are very costly, even while in the longer run their importance will be very great.

All this presents a challenge to the 'traditional' centralized planning system, much more attuned to quantity than to quality, poor at relating production to user demand, and, above all, increasingly overwhelmed by its own complexity. Inefficiency and waste, the slow diffusion of technical progress, imbalances and bottlenecks, continue to plague the economy, and may even have grown worse. The party leadership continues to adopt resolutions demanding greater attention to the requirements of the customer, whether the customer be industry itself or the citizen. However, production continues to adapt itself not to user demand but to the 'success indicators', i.e. to plan fulfilment statistics in roubles, tons, square metres. It has proved extremely difficult to incorporate desirable forms of technical progress into 'directive' plan-orders. New products or new machines require to be approved by numerous bureaucratic instances, and then fitted into plan target figures, and the allocated material inputs frequently fail to match requirements. Information flows remain distorted by the interest of information-providers in being given a plan that is easy to fulfil. Unreliability of supplies encourages hoarding and over-application for input allocations; frequent changes of plan encourage the hoarding of labour too. Brezhnev complained of the strength of what he called *vedomstvennost'* and *mestnichestvo*, 'departmentalism' and 'localism': ministries, and regional state and party organs, divert resources in the light of their sectional interests, make their own components (often in ill-equipped backyard workshops), and prove too strong for the overworked coordinating organs. Far too many investment projects are begun, despite repeated orders to stop this practice. Plans are too often unbalanced, in that the necessary inputs cannot be provided or do not arrive, and in the most recent years a serious bottleneck has been rail transport. The many calls on resources have greatly reduced the rate of growth of investment; in 1979, a particularly difficult year, it was only 1 per cent. It is possible that in fact the volume of investment declined, owing to increased costs and prices.

For all these reasons, growth has shrunk to modest levels, and the system is creaking at the joints. Some fundamental change seems inevitable, but the present leadership is not prepared to contemplate a departure from the essentials of the centralized model. The changes so far made are variations on the same basic theme. Mergers of enterprises into a smaller number of so-called *obyedineniya* ('associations') may relieve the centre by reducing the number of productive units to be planned, and much is hoped of computerization. Talented mathematical economists, including Nobel prize winner Kantorovich, have made important contributions to the theory of optimal planning and programming, but practical applications of their methods have so far been limited. In July 1979 a number of reforms were adopted: a greater emphasis on ('normed') net output as a success indicator, greater stress on balanced and stable plans, more emphasis on carrying out (planned) delivery obligations, rewards for improved quality and technical progress, and a review of prices. However, the effect is, if anything, to increase the number of indicators laid down by the plan agencies, and hence the burden of their work, with the quasi-certainty of contradictions and inconsistencies between plan indicators. The price reform now in progress (1981) is still based on cost-plus, and prices are to remain unchanged for five years at least; i.e. will not be flexible and will not reflect scarcity or supply-and-demand imbalances. Any move towards decentralization, with greater reliance on the market mechanism, seems to be blocked, despite the fact that a number of the most intelligent Soviet economists (and this author too) consider that this must be an essential feature of any effective reform of what is now a clumsy bureaucratic system of 'directive planning'. Unfortunately the recent worsening of the international situation, and the consequent high priority of the 'military-industrial complex', is a further obstacle to any decentralization: the present system is well able to ensure that top-priority sectors receive the resources they need. Brezhnev and official economists describe the present system as *mature socialism*, a complacent formulation which is contradicted by a rising tide of unresolved problems.

LIVING STANDARDS AND INFLATION

There has developed in the last ten years a growing disparity between personal incomes and the volume of goods and services at official prices. There are several statistical indications of this. One is to compare the growth of personal incomes with the level of consumers' goods and services. Another is to observe the extremely rapid rise in savings bank deposits. Still another is to note the gap between official and free-market prices of food: the latter exceeded the former by 37 per cent in 1965, 54·5 per cent in 1970, by 77 per cent in 1975 and by 100 per cent in 1978.[5] Indeed a *Pravda* editorial (27 July 1979) stated that 'the output of goods rose much more slowly than incomes' in the previous eight years. All this provides powerful confirmation of the 'anecdotal' evidence about growing shortages of many goods. These shortages seem to be due to a combination of two causes: failure to adjust prices and output to demand for particular commodities, and to macro-imbalance due to a global excess of purchasing power. A particularly obvious example of the first of these causes relates to meat. Prices have, as a matter of policy, remained unchanged since 1962. Average wages have almost doubled since that date. Supplies of meat have risen, but by a much smaller percentage. No wonder there is a shortage, at the low (heavily subsidized) price. Perhaps mindful of disturbances in Poland which were caused by attempts to fix a more realistic price for meat, the Soviet government has continued its policy of low prices for basic foodstuffs, thereby contributing to a chronic disequilibrium between demand and supply.

Shortages give rise to many undesirable phenomena: bribery, corruption, under-the-counter sales, disregard for the customer, hoarding. This is causing increasing concern, and could become a threat to order and stability. The urgent need for change has, so far, met a conservative reluctance to engage in any fundamental reforms.

FOREIGN TRADE, COMECON AND DÉTENTE

A feature of the Brezhnev period was a substantial expansion of Soviet foreign trade, particularly with the capitalist world. This was facilitated by an improved political atmosphere ('détente'), which encouraged Western firms and banks to grant long-term credits on a large scale, and, from 1973, also by the rise in prices of oil (by far the largest Soviet export to the West) and some other commodities, as well as of gold. Soviet purchases of Western (and Japanese) technology increased very sharply, and there were also large purchases of grain. Soviet aid programmes for developing countries did not expand, but trade grew, especially Soviet sales of arms to third-world countries.

Efforts were also made to enlarge the functions of Comecon. In 1971 an ambitious programme of integration was formally adopted, but, while the volume of trade expanded, it remained predominantly bilateral, with only slow progress in joint planning of investments. Though there is formal provision for convertibility, and Comecon banks are active, in practice the currencies remain non-convertible, and the so-called transferable rouble is a unit of account and not in fact transferable. In the most recent years the U.S.S.R., as the major supplier of fuel and material to the industrialized members of Comecon, has faced a dilemma: with the slow-down of growth of output, it has not been possible to supply the needs of the Comecon partners – for instance for oil – without adversely affecting earnings from sales to Western markets and/or failing to cope with rising domestic demand.

Comecon countries then have to buy more oil and other commodities (for instance, grain) for hard currency in the capitalist world, and therefore to export more to the West, which acts against closer integration with the U.S.S.R. and the other Comecon partners.

The following table shows the scale on which trade has increased, with all categories of countries:

	1965		1970		1975		1978	
	Exp.	Imp.	Exp.	Imp.	Exp.	Imp.	Exp.	Imp.
Volume index								
ALL TRADE	100	100	156	139	220	242	271	300
Socialist countries	100	100	162	128	215	193	253	250
Capitalist & developing countries	100	100	157	160	243	345	327	411
Value (milliards of roubles)								
ALL TRADE	7357	7252	11520	10558	24030	26669	33668	34557
Socialist countries	5001	5049	7530	6873	14584	13968	21254	20744
Capitalist countries	1346	1469	2154	2540	6140	9703	8699	10981
Developing countries	1010	734	1836	1146	3306	2999	5715	2831

(SOURCE: *Vneshnaya torgovlya SSSR*, of the above dates.)

NOTE: Indices based on 1965 and 1970 have been 'chained'; the total volume indices are slightly different with 1970 weights.

The policy of greater reliance on supplies from the West may have received a check as a result of the events which followed Soviet involvement in Afghanistan.

CONCLUSION

Let us now look back at what has been described, as far as this was possible in the space of 380 pages. What actually happened, and why?

A great country, already in the process of industrialization, collapsed under the combined stress of social-economic change and a great war. Whether or not the outbreak of war at that time caused the downfall of Tsarism is not a matter into which

it is fruitful to inquire. It collapsed. War and economic disorganization rendered impossible the task of the Provisional government,
which was unable to make peace, to give free rein to peasant land-
grabbing anarchy, or to control the urban mob. It lacked not
only unity and political determination, but also legitimacy in the
minds of the bulk of the citizens, whether army officers or
simple peasants. Moderates were swept away; Lenin rode the
storm, boldly using the peasant *jacquerie* to achieve power.
There followed civil war, ruin, starvation.

Then the reconstruction and economic development of
Russia proceeded, under Bolshevik leadership. Industrialization,
therefore, was to be achieved, without capitalists, in a peasant
country, under a government and party ruling dictatorially in
the name of a small and, by 1921, largely exhausted and dispersed proletariat. The country would be economically and
socially transformed *from above*, by the political leadership,
by an active minority of party zealots mobilizing the still inert
and ignorant mass, for its own good. Opposition to the leadership seemed to divert the people from the true path, known to
those who understood historical processes, who had the sure
compass of Marxism-Leninism to show the way to socialism
and communism. The party and government thus became the
instrument of industrialization, of modernization. Dissent was
not tolerated, the need for discipline stressed.

But the party and government were composed mostly of
partially educated men, rapidly promoted, who learned the
methods of governing the hard way in the ruthless school of
civil war. The relatively thin veneer of cosmopolitan intellectuals
could provide only part of the necessary ruling stratum, and in
any case they were by temperament better at discussion and clever
polemics than they were at organizing the state or the economy.
In due course both they and the more idealistic (or genuinely
ideologically-orientated) of the old party members of working-
class origin were pushed away from the seats of power and
ultimately liquidated almost to a man. The completion of this
process of change is well described in Svetlana Alliluyeva's
book.

It is interesting to note that for many of the old party intellectuals (for instance, Preobrazhensky, Larin), as well as for the

civil-war commissar type, NEP was a distasteful though necessary compromise; the market and economic laws were, if not
dirty words, then part of the definition of what they were
fighting against.

When the offensive was in fact resumed at the end of the
twenties, Stalin had already achieved personal dominance and
was placing in positions of leadership men more distinguished
by ruthlessness in action than by any sort of cultural background,
men whose Marxism-Leninism was a catechism, a card-index
of quotations, used either with conscious cynicism or with
blinkered sincerity, according to temperament. No doubt they
all claimed to believe in communism, and some did.

In emphasizing in this way the personal factor, it is not in the
least intended to adopt the 'struggle for power' interpretation
of history, let alone of economic and social history. The types of
people who provide leadership – not just the politbureau or the
government,[6] but also officialdom as a whole – do deeply affect
the way in which public affairs are run, and when most of the
economy and social organization are within the public sector,
this deeply affects them too. This is where both Russia's backwardness and her historical tradition are highly relevant: the
degree of crude ignorance among a great many of the people
is surpassed only by the quality and devotion of the best of the
intelligentsia and of the small idealistic group of skilled
workers.

Bumptious, boorish and crude followers of Stalin actually
carried out the leap forward, and possibly the ghost of Stalin
would claim in extenuation that no other instrument was to
hand.

The key problem, as Lenin had well understood, was that of
the peasants. We have seen how Stalin and his henchmen dealt
with it. Is this what Lenin had in mind when he spoke of the
ultimate resumption of the advance? Did Lenin also envisage
the ultra-centralized command economy which emerged in
the early thirties?

Lenin, as has been pointed out by many, was a unique mixture
of the intellectual and the ruthless organizer. Where Martov,
the Menshevik, had feared the cruel and ignorant peasant
masses (denouncing *Pugachevshchina*, i.e. elemental risings like

that of Pugachev against Catherine II) and hated terror, Lenin was willing to use both. But in respect of the peasants he seemed, at least in his last years, to harbour illusions. Even if world revolution did not rescue the Russian comrades from their dilemma, machinery would be the *deus ex machina*, so to speak. 'If we had 100,000 tractors . . . then the peasants would say: we are for communism.' This proved not to be the case. So, while Lenin would doubtless have sought to avoid typically Stalinist lies and brutalities, who knows what he, or Trotsky for that matter, would have done, faced with the realities of 1928? True, Lenin in his last years, and Trotsky in opposition, inveighed against bureaucratic deformations. But how could change from above be achieved other than through a party-supervised or party-staffed bureaucracy? It is like blaming the typical defects of military organization on the existence of a general staff or an officers' corps. (And are there not lessons to be learned from the chaos which has followed Mao Tse-tung's attempt to use the masses against party bureaucracy?)

Change from above, and in the direction of modernization, is well established in Russian historical tradition. It is not for nothing that Stalin ordered the glorification of Ivan the Terrible and Peter the Great. Many historians have made the point that it was lack of spontaneously functioning social forces, and of effective mercantile and (later) bourgeois groups, which compelled the state, in its own and the national interest, to substitute action from above for the weak initiatives of private persons. In doing so, of course, it snuffed out what initiative there was. In this respect Lenin was no exception. By the time he gained power private enterprise was developing rapidly, but it was still economically, socially, politically incapable of resisting either Tsarist despotism or its more determined and more ruthless successors.[7]

Would Lenin have organized a 'command economy'? In the midst of civil war he certainly helped in the attempt to do so. Did he learn by experience that this was a disastrous error? A school of Soviet publicists now claims that he did. They hold that, while expecting and hoping for a renewed offensive against Nepmen, i.e. against private enterprise, Lenin never thought of subordinating *state* enterprises and trusts to the centre in the

Stalinist manner, or of wiping out *socialist* market forces.[8] This is still a minority view. It cannot be either proved or disproved by the minutest study of Lenin's works. Clearly the 'leap forward' approach required stern centralization. But was the 'leap forward' required?

The period 1929–34 marked a great cataclysm, which shook the entire society of Russia to its foundations. It was then, facing a dramatic lowering of living standards in towns and coercion in the villages, that the security police secured the dominant place in society which they retained until Stalin's death. No doubt someone might object that the police were a key factor also in the twenties. The answer is: not nearly to the same extent. A Soviet novelist recalls how, in a rural area, 'the Eye whose task it was to be vigilant' used to sit quietly at the back of party bureau meetings. Then, gradually, he moved to the front row, and by the middle thirties he had his elbow on the table and glared at the frightened bureau members.

The atmosphere of terror affected the economy deeply. It was not only a matter of arbitrariness, it was the peculiar form that arbitrariness took: the fear of truthful reporting, the concentration of all effort on blind fulfilment of orders, the snuffing out of inconvenient ideas and of those who might be suspected of advocating them (plus their friends and relations for good measure). This meant not only that Stalin insisted on the incorrect siting of some factory, but also that a petty party secretary in the depths of the Tambov province appointed plenipotentiaries to see that next year's selected seed grain was handed over to the state, so that no one could accuse him of softness in pursuit of the party's plan for *zagotovki* (procurements). It put heavy pressure on those whose job it was to provide statistics: if they spoke the truth they might be dismissed or even arrested, but if they were discovered to be lying they might also be dismissed or even arrested. . . . Life was far from easy for managers, planners and officials generally. Stalin's capricious and hierarchically unpredictable methods involved everyone in risk and strain.

Yet in the midst of all this there were enthusiastic, young, talented organizers, who, along with scientists of the older generation, did do great things. Ordzhonikidze, until his suicide in 1937, was a hard-swearing, hard-drinking and inspiring leader of such men. Even Kaganovich in his trouble-shooting days was an efficient organizer. There were many others. Vast new industries were not created merely by threats. It must again be emphasized that we have been describing a system in which, despite the terror, many devoted people worked hard for the Cause, or for Russia, or even for their own advancement, but worked with a will.

One feature of the system as it emerged seldom appears in history books, but is surely of great social importance: the new men were remarkably indifferent to the welfare of the masses. Again, it is not just that Stalin imposed high investment targets or diverted brick supplies from house to factory building. Petty officials, managers, army officers, disregarded the most elementary needs of their subordinates to a degree almost past belief, in a country supposedly under a working-class dictatorship. The obverse of this is, of course, petty (and not so petty) privileges for the *nachal'stvo* (bosses). One cannot document this, but no one who knows Russia, or reads the more realistic novels, can fail to know it is so. Dudintsev (in *Not by Bread Alone*) described how dozens of women were cleared from a hospital ward when the wife of some local party secretary was brought in to give birth to a baby. A disobedient peasant family was punished by a local official by being forbidden to buy food at the local shop.[9] I myself have seen drivers of official cars left at the wheel all day without anyone thinking of providing them with an opportunity to eat. Canteens were often unspeakably bad. The most elementary services and amenities were lacking – except for those entitled to use official facilities. Far less pleasant examples could be multiplied over and over again.

Why did this happen? Well, it is an illusion that men promoted from the ranks care for those left in the ranks. Sergeants are not noted for softness towards erstwhile associates. The fact that the revolution led to such social effects does not prove it was not a real people's revolution; on the contrary. Many of the more ambitious and energetic workers became *nachal'stvo*. They tended to look down upon their more passive and unsuccessful fellows. Then, in the early thirties, came the great migration of unskilled peasants from the villages. They needed discipline.

Acute observations about this were put by Koestler (in *Darkness at Noon*) into the mouth of his policeman, Gletkin. They had no notion of time-keeping, they were inefficient. In everyone's interest, they had to be compelled. And this was a time of great shortage. The men who ran things, who worked exceedingly hard, were entitled not to have to stand in queues. Therefore they had special privileges. Then it was all institutionalized, became a habit, a vested interest. And no protest from below was possible, because all organized opposition of any kind had become, by definition, treasonable.

That, of course, was part of the logic of a one-party state, and such a state was part of the logic of a communist regime ruling an overwhelmingly peasant country.

Economic historians should note the high cost of elimination of all opposition. By this is meant not only the human cost, in the diversion to camps of unknown millions, of whom a high proportion were above average in intelligence, energy and technical knowledge. Heavy losses were imposed by the mere fact that exaggerated and impossible plans could not be criticized, even by experts, lest they be suspected of deviation. This certainly contributed a great deal to the excesses of the 'leap forward' period, in industry and agriculture alike, and later also in China.

Yet the success of the Soviet Union, albeit by totalitarian and economically inefficient methods, in making of itself the world's second industrial and military power is indisputable. Therein lies its attraction to the 'third world'. Therein, too, may be found many lessons for economist and historian. What if, given Russia's whole historical experience and the irreversible fact of revolution, there was in fact no alternative path for her? By this is not meant that any one act of cruelty or oppression was in some sense predestined, but rather that modernization from above, by crude and sometimes barbarous methods, was rendered highly probably by the circumstances of the time. Might not some, or many, of the excesses, stupidities, errors, be part of the *cost* of industrializing in this manner? Most of the deformations of Soviet planning could be observed in the war economies of Western states. It did not follow that the wars should have been run on free-market principles. In this case at least, it must be conceded that the errors and omissions inherent in bureaucratic centralization were an integral part of the cost of running a war economy, and in this case at least most people accept the cost as justified. Oskar Lange did describe the 'Stalin' economic system as 'a war economy *sui generis*'.

For what is a rational way of organizing the rapid development of a backward country? What content can be given to the word 'rational'? Surely not the achievement of a purely economic optimum which, in any country, is rendered impossible by considerations of political feasibility and social circumstances. One cannot leave out of account the existence of classes and groups, the nature and qualities of the administrative machine, the ideology in the name of which the political leadership seeks to mobilize itself and the masses for the difficult task of changing society. Nor should the more purely military aspect be forgotten: to some extent Stalin really was engaged in building up the industrial base of a war economy in peacetime.

No doubt development economists will study Russian experience for many years. It abounds in lessons (and warnings) of the very greatest interest. They may well conclude that the political terror, the pace of economic development, the problem of capital accumulation and of the peasants were very intimately interconnected. A 'softer' economic policy would have given more scope for the consideration of specifically economic efficacy, required fewer sacrifices, and so weakened the arguments for full-scale police terror. But a sense of danger contributed to the decision to go all out, to industrialize very quickly, to concentrate on heavy industry at any cost. Nor can one overlook the fact that there was no precedent, that bitter lessons had to be learnt from experience, and that the Western economy was in sad disarray at the time when the 'leap forward' was being effected. There was no easy path. The one chosen was not the result of accident or personal whim.

Historians may also conclude that the system, whatever its original logic or rationale, has for some time (literally) outgrown itself. If they are Marxists, they may speak of productive forces coming into contradition with productive relations, necessitating change in the direction of a market economy. The Soviet development model will exercise, should exercise, considerable fascination. But many, including communists, who study its evolution, especially the key period which began in 1928, might well feel that somewhere in those years there was a wrong turning. And that no one should follow the trail blazed by Stalin, with its terrible sacrifices, unless some overriding set of circumstances makes other paths impracticable. It is said that one cannot make omelettes without breaking eggs. In that case, perhaps one should not make omelettes, if the menu happens to provide other choices. Perhaps it is Russia's tragedy that these choices were absent, and a measure of her achievement that, despite all that happened, so much has been built, and not a few cultural values preserved and handed on to a vastly more literate population. It will also be Russia's tragedy if, in the name of the preservation of 'mature socialism', the ageing and conservative leadership continue to resist necessary change, since the resultant paralysis would have highly unpleasant consequences. But, in the absence of a crystal ball, let us not speculate.

Appendix: A Note on Growth Rates

How can it be – someone is bound to say – that an economic history of Russia can be written without any answer to the question which everyone asks: how fast did production grow? Soviet indices have been quoted, it is true, but always with qualifying phrases which are clearly intended to warn the reader against taking them literally. But if the official claims are misleading, what can take their place?

The problem might become clearer if one were to go a little more carefully into the reasons for the inadequacy of the official indices, and then consider the acceptability of recomputations undertaken by various Western scholars. It will be assumed that, except perhaps in agriculture, reporting from below is reasonably accurate. This does not exclude the possibility of cheating by officials anxious to claim plan fulfilment, but the telling of lies about output of commodities would disrupt plans and is therefore only possible to a limited extent, since a lack of available goods would speedily be noticed (e.g. in supplying users) and steps taken against the authors of false reports. Besides, it sometimes pays to conceal output (e.g. to build up reserve stocks), as well as to overstate it. Furthermore, false reporting at two different dates does not affect growth rates, unless there is more false reporting at one date than at another. This is what could be called 'the law of equal cheating': unless there is any reason to assume it is unequal, the mere presence of cheating should not affect statistics of growth.

Any doubts are not based on false reports; in fact all the recomputations made by Western scholars use Soviet physical output data as their raw materials. The problems begin when one looks at how the data are aggregated, and also at what the figures reported from below actually mean.

Every economist is, or should be, acquainted with the index number problem. Very briefly it can be illustrated as follows. Suppose that the industrial output of Scotland consists of whisky and bagpipes and has altered as follows:

	1930	1960
Whisky (thousand gallons)	500	1500
Bagpipes (thousands)	130	130

Obviously, the increase in total Scottish output will depend on the relative prices of whisky and bagpipes. Suppose in 1930 bagpipes cost £10 and whisky £10, but in 1960 bagpipes cost £100 and whisky £20. Then in 1930 prices the respective values were as follows:

(£)

	1930	1960	
Whisky	5,000	15,000	
Bagpipes	1,300	1,300	
Total	6,300	16,300	Index = 257

But in 1960 prices the answer would be very different.

(£)

	1930	1960	
Whisky	10,000	30,000	
Bagpipes	13,000	13,000	
Total	23,000	43,000	Index = 187

Neither answer is 'true', or rather both are. But most statisticians would change the weights as they become obsolete.

In the Soviet Union, there was a very sharp change in relative prices after 1928. The fastest-growing sectors were highly priced in the twenties, and so the preservation of 'obsolete' weights ensured high overall growth indices, which is no doubt why they were preserved until 1950. Much of the change in relative prices was concentrated into the drama-filled period 1928–37, and Bergson, in his very thorough calculations, gives some very different answers for these years.

At '1937 rouble factor cost', growth rates in 1928–37 work out at 5·5 per cent per annum (or 4·8 per cent on an alternative assumption). Using 1928 prices, the growth rate more than doubled, to 11·9 per cent per annum. Since the growth in numbers employed is a constant in both calculations, Bergson's labour productivity figures differ even more widely; output per worker increases either by 1·7 per cent or by 7·9 per cent per annum, depending on the prices used. It is worth adding that, if the share of investment in the national income in 1937 were computed in 1928 prices, it would exceed 44 per cent, whereas Bergson makes it 25·9 per cent in terms of 1937 roubles.[1] Needless to say, Soviet statisticians have never cited this high investment percentage, preferring in *this* instance to use current prices.

So far we have been discussing only the distortion which might be due to using obsolete weights, a practice which no longer applied to figures for periods since 1950. But this is far from being the end of the story.

All output indices are conventional aggregations of constantly shifting product mixes not strictly comparable with one another. Thus, in any country, statisticians have to compare clothing, cars, aircraft, turret lathes, etc., which have altered in important particulars. What is the comparison, for volume index purposes, between a DC3 and DC8 aircraft, or a 1950 and a 1967 Jaguar car? Obviously not 1:1. If prices had been unaltered then we could deduce a ratio from the relative prices at the two dates. But plainly we cannot do that. Then we deflate by a price index? But what price index? Clearly we cannot apply some general index of wholesale prices, since aircraft and cars could well have risen (or fallen) in price by more (or less) than the general average. So we need to devise an index for the items being compared. But we are then back where we started, since no DC3 was made in 1967 and no DC8 in 1950. Can we then deduce the price deflation from changes in costs of production of analogous objects, in this case other forms of transport equipment? But this is dangerous, since (especially in machinery and engineering) many of the items made in both 1950 and 1967 were only just being developed in 1950, and so had a high initial cost, later reduced as they fully expanded into mass production. The use of such a price deflator would overstate the rate of growth.

At this point one must introduce deliberate 'statistical' bias, and this both at the centre and at the reporting agencies (ministries, glavki, enterprises). First of all, whereas Western statistical offices face the same methodological problems, they try to allow for them and are not under political pressure to adopt, among a number of possibly valid ways of comparing the non-comparable, the ones which give high growth rates.

Secondly, the methods of aggregation are known, and Soviet management tries to choose a pattern of output which results in a high growth rate. This is not the same as cheating, for it is playing according to the rules. A given design, product mix, material, was often chosen with one eye on its effect on the output index. Only where the commodity is totally homogeneous (for instance, kilowatt-hours of electricity) would this not matter. If particular kinds of clothing, tractors, goods wagons, aircraft or prefabricated cement blocks were chosen *because of the effect of the choice on the statistical measurement of the volume of output*, then that volume would show a tendency to exaggeration. In the West the enterprises are uninterested in the statisticians' measurements; the pattern which yields a high profit may or may not be that which fits the conventions of measuring output statistics.

The fact that the official price indices understate price rises, or even show that prices have fallen when they have risen, has been pointed out by Soviet critics – the latest and most striking article on the subject being that by V. Krasovsky (*Voprosy ekonomiki*, No. 1, 1980). But if the price index fails to reflect the real increase in prices, then the volume index upon which it is based must to that extent overstate the rate of growth.

Some of the other indicators used in Soviet planning can also cause major distortions. Let us take the system under which large bonuses are paid for reducing costs. This encourages the following practice. Let us say an item has been for some time in mass production, and is popular with the consumers, but all means of reducing production costs have been already exhausted. It is then rational to stop producing this item, and to switch to another, dearer model, due only to one reason; for the next few years, as it passes from prototype to regular-flow production, its costs will fall, which would earn the managers a bonus for cost-reduction. Let us ignore for the moment the wasteful nature of the procedure (after all, we have wasteful 'novelties' in the West). The statistical point is that there would be apparent reductions in costs and prices, and so affect the price-deflator used in calculating aggregate indices, whereas in fact a change has been made to a dearer model, and purely for statistical reasons.

Another example concerns industrial production. The index used is 'gross', in the sense of being arrived at by adding together the total value of the output of any enterprise, and then deflating by the appropriate price index. This means that the index would be affected by the division of any process between enterprises. To take an example: if a complete sewing machine were made in one enterprise, and its price was 100 roubles, this equals 100 roubles of gross output (although the fuel and materials used had been counted as part of the output of other enterprises). But suppose enterprise A makes the actual machine, enterprise B makes the stand and pedals, and enterprise C assembles the sewing machine. Then the gross output would be something like this:

Enterprise A (machine parts)	60
Enterprise B (stand and pedals)	25
Enterprise C (assembly)	100
	185

Of course this would not affect the rate of growth, if the degree to which the process was divided (i.e. of vertical integration or disintegration) remained the same. But the system tends to encourage this kind of arrangement, while discouraging integration. Thus all three enterprises, in the example given above, are under the same ministry; the ministry's plan is expressed in value of gross output. Thus the above arrangement 'pays' the ministry.

Some recent developments could have had the opposite effect. Thus, large numbers of industrial associations (*obyedineniya*) have been created, reducing the number of enterprises by mergers and amalgamations, so that in some instances the gross output measurement can be adversely affected. One reads of the resistance of ministries to such changes, and one reason for such resistance might well be precisely their statistical effect,[2] and of course if market-type reforms were in fact adopted, this would have the effect of removing many of the objections to the index: profit-seeking managers would produce for the customer and not for the statistics.

This leads me to another point, which might appear 'philosophical' but is in fact fundamental, as every Soviet reformer-economist well knows. What is the value of output? Can it be based merely on cost? Why do we say that goods A, B and C are worth £50 million? Surely because two factors are present in the process of price formation: costs and a market valuation of the product. This leads one straight into the heart of recent Soviet discussions on value and price. Thus an admirable article by two leading mathematical economists quotes Marx to the effect that under communism 'time spent on production would be determined by the degree of social utility of this or that product', and that it is obvious anyhow that two differing product mixes that happen to cost the same to produce cannot, in any meaningful sense, be held to be of equal value or worth merely for that reason.[3]

In other words, Soviet industrial or farm prices, which do not reflect (save by chance) supply and demand conditions, are in principle rather inadequate 'weights', a poor measuring-rod for measuring output and its growth. Of course, existing Western prices suffer from distortions, and it would be absurd to deny that numerous imperfections affect the picture here too. But some sort of valuation through a market does more or less find reflection in most of our prices, certainly in theory, partially in practice.

Then why not use a Western recomputation? To this easy way out there are the following objections:

Firstly, many of them (e.g. Bergson's) use Soviet prices, and the adjustment for subsidies and turnover tax does not eliminate the 'philosophical' doubts.

Secondly, even the most careful and painstaking use of Soviet price lists – and Bergson, Moorsteen, Nutter and other American analysts are most thorough and use every scrap of evidence available – does not free them from the necessity of devising price indices to act as 'deflators', and this is a task fraught with great difficulty for reasons already given.

Thirdly, parts of the statistical data are simply missing. Some are covered by a security curtain (those for arms, ships, aircraft, electronics, even some non-ferrous metals). But others are available in too aggregated a form for effective use in an index: even such simple things as bread, furniture, clothing, wine, or more complex items such as instruments and implements, lathes, refrigerators come in very many different shapes, sizes and qualities. A change in product mix within each of these aggregated categories ought to find its reflection in any index, and the necessary information is not available. It is one of the faults in Nutter's very interesting recomputation of industrial output that many end-products are simply omitted altogether (there are no figures for bread, books or newspapers, clothing, furniture, etc.), while the materials of which they are made are included, though it is well known that there has been a relative increase precisely in the final stage of manufacture. Nutter did not, of course, omit these items through evil intent or inadvertence, but because figures were not to hand which he could use for this purpose. In the case of national income a big difference to the result is made by the weight assigned to agriculture, which pulls down the general growth index. Owing to the peculiarities of agricultural prices, official indices tend to underweight this sector. But what is the correct weight?

Finally, there is some disagreement between experts, estimates of industrial growth in the period 1928–50 ranging from an index of 725 by Seton to 376 by Nutter (the incredible official index figure is 1123).

The following table may be a useful summary of industrial progress:

	1928	1940	1950	1960	1970
Electricity (milliard Kwhs)	5·0	48·3	91·2	292·3	740
Steel (million tons)	4·3	18·3	27·3	65·3	116
Oil (million tons)	11·6	31·1	37·9	147·9	353
Gas (milliard cubic metres)	0·3	3·4	6·2	47·2	200
Coal (million tons)	35·5	166·0	261·1	509·6	624
Cement (million tons)	1·8	5·7	10·2	45·5	95·2
Machine-tools (thousands)	2·0	58·4	70·6	155·9	(240)
Motor vehicles (thousands)	0·8	145·4	362·9	523·6	916
Tractors (thousands)	1·3	31·6	116·7	238·5	459
Mineral fertilizer (million gross tons)	0·1	3·2	5·5	13·9	55·4
All fabrics (milliard metres)	3·0	4·5	4·5	8·2	10·2
Knitwear (million units)	8·3	183	197	584	1134
Leather footwear (million pairs)	58	211	203	419	676
Beet sugar (million tons)	1·3	2·2	2·5	5·3	9·4
Radios and radiograms (thousands)	—	160	1072	4165	7800
Television sets (thousands)	—	0·3	11·9	1726	6700
Domestic refrigerators (thousands)	—	3·5	1·5	529	4100

The list, up to 1960, is taken from a fiftieth-anniversary article in *Kommunist* (No. 11, 1967); the figures for 1970 are from *Pravda*, 4 February 1971. It leaves out some slower-growing items, such as timber and railway equipment, but some rapidly expanding items too. Of course, it should not be forgotten that many other countries expanded their industries rapidly in the same period, nor that in 1913, with all its weaknesses, the Russian Empire was the world's fifth industrial power (in total, not, of course, *per capita*).

We could perhaps agree that the U.S.S.R. did industrialize rapidly after 1928, that in doing so it had to overcome grave difficulties of a kind which the United States did not have to face (social, political, geographic, very different historical traditions, etc.) and that the word 'rapidly' cannot, from our present information, be given precision. A few other countries appear to have grown as fast, and to have recovered from disasters too: Japan, for instance. But it makes little sense to derive from this a moral concerning the efficacy of systems and government, as if Japanese methods could have been applied in Russia (or Pakistan, or Mexico). Each country's growth path and development potential are a function of many factors unique to them, and copying is frequently quite impracticable, or would lead to very different results in an uncongenial environment. Nor should resource endowment be overlooked as a factor: the highest rates of all were achieved in Kuwait.

If we really wished to measure the specifically Soviet contribution to Russia's growth, we would need to know what would have happened if some other regime had presided over the process of development. But since, for all we know or can know, the alternative to Lenin in the given circumstances was disintegration, the exercise could scarcely be meaningfully attempted.

Bibliography

The following is a short, selected bibliography of works in English:

GENERAL BOOKS ON SOVIET HISTORY

E. H. CARR: *A History of Soviet Russia* (Macmillan). Of the many volumes already published or prepared, of particular importance to economic historians are the following:
—*The Bolshevik Revolution, 1917–23*, Vol. 2 (1952, also Pelican, 1966)
—*Socialism in One Country*, Vol. 1 (1958)
—*Foundations of a Planned Economy, 1926–9*, Vol. 1 (with R. W. Davies) (1969, also Pelican, 1974)
L. KOCHAN: *The Making of Modern Russia* (Jonathan Cape, 1962, and Pelican, 1963)
P. LYASHCHENKO: *The History of the National Economy of Russia to the 1917 Revolution* (Macmillan, N.Y., 1949)
R. W. PETHYBRIDGE: *A History of Postwar Russia* (Allen & Unwin, 1966)

THE ECONOMY, GENERAL

R. W. CAMPBELL: *Soviet-type Economies* (Macmillan, 1974)
M. DOBB: *Soviet Economic Development since 1917*, 3rd edition (Routledge, 1966)
—*Russian Economic Development since the Revolution* (Routledge, 1928)
M. ELLMAN: *Socialist Planning* (Cambridge, 1978)
A. GERSCHENKRON: *Economic Backwardness in Historical Perspective* (Harvard University Press, 1962; Praeger, 1965)
P. R. GREGORY and R. C. STUART: *Soviet Economic Structure and Performance* (Harper & Row, 1974)
M. LAVIGNE: *The Socialist Economies* (Martin Robertson, 1974)
A. NOVE: *The Soviet Economic System* (Allen & Unwin, 1977)
—*Political Economy and Soviet Socialism* (Allen & Unwin, 1979)
H. SCHWARTZ: *Russia's Soviet Economy* (Prentice Hall, 1954)

PARTY AND GOVERNMENT

M. FAINSOD: *How Russia is Ruled* (Oxford University Press, 1963)
M. MCAULEY: *Politics and the Soviet State* (Pelican, 1977)
L. B. SCHAPIRO: *Government and Politics in the Soviet Union* (Hutchinson, 1965)
—*The Communist Party of the Soviet Union* (Eyre & Spottiswoode, 1960)

THE ECONOMY – PARTICULAR PERIODS

A. ERLICH: *The Soviet Industrialization Debate* (Harvard University Press, 1960)

C. BETTELHEIM: *Class Struggles in the U.S.S.R. (1917–23 and 1924–30)* (Harvester Press, 1977 and 1979)

J. R. MILLAR (ed.): *The Soviet Rural Community* (University of Illinois Press, 1971)

N. SPULBER: *Soviet Strategy for Economic Growth* (Indiana University Press, 1964)

H. SCHWARTZ: *The Soviet Economy since Stalin* (Gollancz, 1965)

L. SZAMUELY: *First Models of Socialist Economic Systems* (Budapest, 1974)

T. VON LAUE: *Sergei Witte and the Industrialization of Russia* (Columbia University Press, 1963)

IDEOLOGY AND PERSONALITIES

S. COHEN: *Bukharin* (Wildwood House, 1973)

R. DAY: *Leon Trotsky and the Politics of Economic Isolation* (Cambridge, 1973)

I. DEUTSCHER: *Stalin* (Oxford University Press, 1949)

I. DEUTSCHER: *Trotsky* (Oxford University Press, 1954, 1959, 1963)

H. MARCUSE: *Soviet Marxism* (Routledge, 1958)

A. MEYER: *Leninism* (Harvard University Press, 1957)

A. NOVE: *Stalinism and After* (Allen & Unwin, 1975)

L. B. SCHAPIRO and P. REDDAWAY (eds.): *Lenin* (Pall Mall Press, 1967)

A. ULAM: *The Bolsheviks* (Macmillan, 1965)

—*Stalin* (Allen Lane The Penguin Press, 1974)

G. WETTER: *Dialectical Materialism* (Routledge, 1958)

LABOUR

E. C. BROWN: *Soviet Trade Unions and Labour Relations* (Oxford University Press, 1960)

I. DEUTSCHER: *Soviet Trade Unions* (Royal Institute of International Affairs, 1950)

M. DEWAR: *Labour Policy in the USSR* (Royal Institute of International Affairs, 1956)

S. SCHWARZ: *Labour in the Soviet Union* (Cresset Press, 1953)

ECONOMIC SECTORS, REGIONS, FINANCE

R. W. DAVIES: *The Development of the Soviet Budgetary System* (Cambridge University Press, 1958)

—*The Industrialization of Soviet Russia*, Vol. 1: *The Socialist Offensive* (Macmillan, 1980)

H. HUNTER: *Soviet Transportation Policy* (Harvard University Press, 1957)

N. JASNY: *The Socialized Agriculture of the USSR* (Stanford University Press, 1949)

M. LEWIN: *Russian Peasants and Soviet Power* (Allen & Unwin, 1968)

A. NOVE and J. A. NEWTH: *The Soviet Middle East* (Allen & Unwin, 1966)

L. VOLIN: *A Century of Russian Agriculture* (Harvard, 1970)

GEOGRAPHY

J. P. COLE: *Geography of the USSR* (Penguin, 1967)

J. C. DEWDNEY: *Studies in Industrial Geography: the USSR* (Wm. Dawson & Sons, 1976)

STATISTICS AND GROWTH

A. BERGSON: *The Real National Income of Soviet Russia since 1928* (Harvard University Press, 1961)

J. CHAPMAN: *Real Wages in the Soviet Union* (Harvard University Press, 1963)

N. JASNY: *Soviet Industrialization 1928–52* (Chicago University Press, 1961)

R. MOORSTEEN: *Prices and Production of Machinery in the Soviet Union* (Harvard University Press, 1962)

G. W. NUTTER: *The Growth of Industrial Production in the Soviet Union* (NBER and Princeton University Press, 1962)

References

Chapter 1

1 Those who want to study the methods and to note the existence of alternative figures are referred to Goldsmith's article in *Economic development and cultural change*, Vol. IX, No. 3 (April 1961).
2 *Istoriya narodnovo khozyaistva*, edited by F. Polyansky (Moscow, 1960), p. 367.
3 *Annales* (Paris), November–December 1965.
4 *Opyt ischisleniya narodnovo dokhoda v Evropeiskoi Rossii* (Moscow, 1918).
5 *Poslevoennye perspektivy Russkoi promyshlennosti* (Kharkov, 1919), p. 100.
6 Cited from Rybnikov by D. Shapiro, *Problemi ekonomiki*, No. 7–8 (1929), p. 126.
7 *Materialy po istorii SSSR*, Vol. VI.
8 A. Gerschenkron: *Economic Backwardness in Historical Perspective* (Harvard University Press, 1962), p. 125.
9 N. Timasheff: *The Great Retreat* (New York, 1946), p. 29.
10 Cited from B. Kerblay's article on Chayanov, *Cahiers du monde russe et soviétique* (October–December 1964), p. 414.
11 A. Tyumenev: *Vestnik Kommunisticheskoi Akademii* (hereafter VKA), No. 8 (1924), p. 209 ff.
12 Gerschenkron: p. 130.

Chapter 2

1 Grinevetsky: *Poslevoennye perspektivy Russkoi promyshlennosti* (Kharkov, 1919), p. 33.
2 ibid., p. 49.
3 ibid.
4 Marx, Vol. 35 (Russian edition), pp. 136–7 (original in French).
5 Fully discussed in I. Getzler: *Martov* (Cambridge University Press, 1967), p. 102–3.
6 G. Zinoviev: *V. I. Lenin* (Kharkov, 1924), pp. 11, 13.
7 Lenin, Vol. 25, pp. 277–80, 11 September 1917. (Unless otherwise stated, Lenin will be quoted from the English translation of the Russian 4th edition; if marked as Russian, the 5th edition is cited.)
8 ibid. Vol. 25, p. 123.
9 ibid., Vol. 25, p. 21.
10 *Pravda*, 17 June 1917.
11 Lenin, Vol. 25, p. 44.
12 ibid., p. 69.
13 ibid., p. 336.

14 ibid., pp. 426, 427.
15 Lenin, Vol. 26, pp. 105–7.
16 ibid., pp. 170, 172, 173.
17 *Russian Economic Development since the Revolution* (Routledge, 1928), p. 28.
18 Lenin, Vol. 26, p. 365.

Chapter 3

1 Lenin, Vol. 32, p. 22.
2 The text of the decree is unclear about whether peasant holdings were nationalized.
3 *Istoriya narodnovo khozyaistva SSSR, 1917–59*, edited by A. Pogrebinsky (Moscow, 1960), p. 13.
4 E. H. Carr: *The Bolshevik Revolution*, Vol. 2 (Pelican, 1966), p. 27.
5 F. Samokhvalov: *Sovety narodnovo khozyaistva v 1917–32 gg.* (Moscow, 1964), p. 27.
6 ibid., pp. 34, 35.
7 A. Venediktov: *Organizatsiya gosudarstvennoi promyshlennosti v SSSR* (Leningrad, 1957), Vol. 1, pp. 180–1.
8 VKA, No. 19 (1924), p. 26.
9 Venediktov, pp. 186–7; Lenin, Vol. 26, p. 352.
10 *Etapy ekonomicheskoi politiki SSSR*, edited by P. Vaisberg (Moscow, 1934), p. 49.
11 Lenin, Vol. 26, p. 519.
12 ibid., pp. 501–2.
13 Pogrebinsky, p. 30.
14 E. Plimak: *Novy mir*, No. 6, 1966, p. 183. The quotation within the quotation is from a book by A. Matyes on 1793–4.
15 Lenin, Vol. 27, pp. 212–14.
16 M. Dobb: *Russian Economic Development since the Revolution* (Routledge, 1928), p. 49.
17 English translation (MacGibbon & Kee), p. 258.
18 Vaisberg, p. 50.
19 H. G. Wells, *Russia in the Shadows* (London, 1921), p. 137.
20 Venediktov, p. 445, quoting the congress report.
21 Venediktov, p. 446.
22 R. W. Davies: *Development of the Soviet Budgetary System* (Cambridge University Press, 1958).
23 *Novy mir*, No. 12 (1966), p. 259.
24 N. Bukharin: *Ekonomika perekhodnovo perioda* (1920).
25 *Sovetskoe narodnoe khozyaistvo, 1921–5*, edited by I. Gladkov (Moscow, 1960), p. 531.
26 Bukharin and Preobrazhensky: *Azbuka komunizma* (1919), pp. 195–6.
27 Editorial note to Lenin (Russian), Vol. 42, p. 397.
28 Vaisberg, p. 97.
29 ibid., p. 102.
30 Lenin (Russian), Vol. 40, p. 395, editorial footnote.

31 Lenin, Vol. 30, p. 314 (1 February 1920).
32 ibid., Vol. 32, pp. 44–5.
33 ibid., Vol. 31, p. 121.
34 Lenin (Russian), Vol. 40, p. 329.
35 ibid., Vol. 42, p. 193.
36 ibid., p. 185.
37 ibid., Vol. 43, p. 433, editorial notes.
38 ibid., Vol. 42, p. 349.
39 Lenin, Vol. 30, p. 332.
40 ibid., Vol. 32, p. 156.
41 *Vestnik Sotsialisticheskoi Akademii* (VSA), No. 1 (1922), p. 138.
42 ibid., p. 146.
43 ibid., pp. 148–9.
44 ibid., No. 2 (1923), p. 195.
45 New Economic Policy. See next chapter.
46 VSA, No. 2 (1923).
47 VKA, No. 9 (1924), pp. 105, 107.
48 VSA, Nos. 6, 7 (1923–4), p. 332.
49 *The defence of Terrorism* (Allen & Unwin, 1935), pp. 94–5.
50 *Political Economy of Communism* (Blackwell, 1963).
51 Lenin, Vol. 31, p. 339.
52 Lenin (Russian), Vol. 40, pp. 337–9.
53 ibid., Vol. 43, p. 385 and again p. 387.

Chapter 4

1 Lenin, Vol. 32, p. 156.
2 Gladkov (1921–5), p. 239, also G. Konyukhov: *KPSS v borbe s khlebnymi zatrudneniyami v strane* (Moscow, 1960), p. 38.
3 Lenin, Vol. 32, p. 185.
4 ibid., Vol. 33, p. 104.
5 VSA, No. 2 (1923), p. 173.
6 Samokhvalov: p. 121.
7 Gladkov (1921–5), pp. 193, 194.
8 ibid., pp. 316–17.
9 Polyansky, p. 484; also editorial notes to Lenin (Russian), Vol. 44 p. 544.
10 Vaisberg, p. 63.
11 Polyansky, pp. 485–6.
12 Gladkov (1921–5), p. 208.
13 ibid., p. 522.
14 ibid., pp. 374, 377, 380, 387, 389.
15 Lenin, Vol. 54 (Russian), p. 139.
16 VSA, No. 4 (1923), p. 29.
17 Polyansky, p. 496.
18 E.g. see VSA (1923); O. Yu. Shmidt: 'Mathematical laws of currency issue'; E. Preobrazhensky: 'The theoretical basis of arguments on the

gold and commodity rouble'; V. Bazarov: 'On the methodology of currency issue,' etc.

[18a] L. N. Yurovsky: *Denezhnaya Politika Sovetskoi Vlasti* (Moscow, 1928), p. 297. Some other sources quote somewhat different figures.

[19] These and other details have been taken from Polyansky, p. 498.

[20] Vaisberg, p. 57.

[21] VSA, Nos. 6–7 (1923–4), pp. 316–17.

[22] L. Gatovsky in Vaisberg, p. 60.

[23] Quoted from *Sovety narodnovo khozyaistva i planovye organy v tsentre i na mestakh, 1917–32* (Moscow, 1957), pp. 131–2.

[24] ibid., p. 154 and Samokhvalov, p. 153.

[25] E. H. Carr and R. W. Davies: *Foundations of a Planned Economy, 1926–9* (Macmillan, 1969).

[26] Samokhvalov, p. 97.

[27] ibid., p. 159.

[28] See in particular the valuable work of Malafeyev, *Istoriya tsenoobrazovaniya v SSSR* (Moscow, 1964), Chapter 1.

[29] L. Kritsman: *Problemy ekonomiki*, No. 4–5 (1930), p. 17.

[30] Gladkov (1921–5), p. 197.

[31] Kritsman, pp. 7, 9, 17.

[32] Vaisberg, p. 63.

[33] Cited from M. Lewin: *La paysannerie et la pouvoir soviétique* (Mouton, Paris, 1966) and *Russian Peasants and Soviet Power* (its English translation), (Allen & Unwin, 1968).

[34] *A History of Soviet Russia: Socialism in one Country*, Vol. 1, p. 99.

[35] VKA, No. 8 (1924), p. 40.

[36] Yu. Moshkov: *Zernovaya problema v gody sploshnoi kollektivizatsii* (Moscow University, 1966), p. 23; J. Karcz: *Soviet Studies* (April, 1967); and R. W. Davies, in *Soviet Studies* (January, 1970), pp. 314–29.

[37] Vaisberg, p. 308.

[38] Polyansky, p. 478.

[39] Cited from S. Schwarz: *Labour in the Soviet Union* (Cresset, 1953), p. 6.

[40] Vaisberg, p. 273.

[41] ibid., p. 274.

[42] Data are reprinted in M. Dewar: *Labour Policy in the USSR* (R.I.I.A. 1956).

[43] S. Gimmelfarb: *Problemy ekonomiki*, No. 45 (1931), p. 31.

[44] Details in Vaisberg, p. 275.

[45] ibid., p. 276.

[46] ibid., p. 198.

Chapter 5

[1] Lenin, Vol. 32, pp. 429–30.

[2] English translation by Brian Pearce, *The New Economics* (Oxford University Press, 1964).

[3] 'Economic equilibrium in the system of the USSR', published in

VKA. No 22 (1927). Translated in N. Spulber: *Foundations of Soviet Strategy for Economic Growth* (Indiana University Press, 1964).

[4] For a more systematic account, see A. Nove: 'New light on Trotsky's economic view', *American Slavic Review* (December 1980). Readers should also be aware of R. Day: *Leon Trotsky and the Politics of Economic Isolation* (Cambridge, 1973).

[5] *Politique des investissements et calcul économique* (Cujas, Paris, 1964).

[6] ibid., p 33.

[7] E. Domar: *Essays in the Theory of Economic Growth* (New York, 1957), pp. 223–61.

[8] F. Holzman: 'The Soviet Ural-Kuznetsk combine', Quarterly Journal of Economics (August 1957); and *Le développement du bassin du Kouznetsk*, edited by H. Chambre (Cahiers de l'I.S.E.A., Paris), No. 100.

[9] Z. K. Zvezdin, in *Voprosy istorii KPSS*, No. 3 (1967), p. 55.

Chapter 6

[1] (Five-year plan document) *Pyatiletnii plan nar.-khoz. stroitelstva SSSR* (Moscow, 1930), Vol. 2, p. 36.

[2] I can find no *law* earlier than 1932.

[3] Malafeyev, p. 133.

[4] G. Konyukhov: *KPSS v borbe s khlebnymi zatrudneniyami v strane* (Moscow, 1960).

[5] V. Novozhilov: *Vestnik finansov*, No. 2, 1926.

[6] Yu. Novakovsky: *Potrebitel'skaya kooperatsiya na 13 godu oktyabr'skoi revolutsii* (Tsentrosoyuz, Moscow, 1929), pp. 5–12.

[7] *Ekonomicheskaya zhizn' SSSR*, p. 188 (hereinafter *Ek.zh.*).

[8] *Planovoye khozyaistvo*, No. 2, p. 10 ff.

[9] Yet Maurice Dobb, in his 1966 edition of *Soviet Economic Development since 1917*, continues to argue that the plan might have been feasible but for subsequent changes in both the plan and the situation.

[10] J. Stalin, *Works* (in English), Vol. 10, p. 312.

[11] Detailed figures by regions may be found in Konyukhov, pp. 64–5.

[12] Cited by Konyukhov, p. 66. The Smolensk archives also provide evidence.

[13] ibid., p. 68.

[14] Stalin, Vol. 11, pp. 5, 6.

[15] Konyukhov, p. 72.

[16] ibid., p. 78.

[17] ibid., p. 128.

[18] ibid., pp. 146–7.

[19] ibid., p. 119.

[20] Stalin, Vol. 11, p. 7. Note that this was not *published* until 1949.

[21] ibid., Vol. 11, p. 166.

[22] The lower figure is given in the five-year plan document, Vol. 2, p. 30.

[23] Malafeyev, p. 119.

[24] *Pravda*, 25 May 1929.

[25] Stalin, Vol. 12, p. 43.

[26] ibid., p. 152.

[27] N. Valentinov: *Sotsialisticheskii Vestnik* (New York, April 1961).

Chapter 7

1 N. Ivnitsky: *Voprosy istorii KPSS*, No. 4 (1962), p. 56.
2 For example, see M. Vyltsan et al.: *Voprosy istorii*, No. 3 (1965), p. 5.
3 Moshkov, pp. 56, 59, 63–6.
4 ibid., p. 65.
5 ibid., p. 69.
6 M. Bogdenko, in *Istoricheskie zapiski*, No. 76, p. 20, and S. Trapeznikov: *Istoricheskii opyt KPSS* . . . (Moscow, 1959), p. 175.
7 All this from Ivnitsky, pp. 61–5. (Not published until over thirty years after the event.)
8 Vyltsan et al., p. 6.
9 ibid., p. 7.
10 Ivnitsky, p. 64.
11 Bogdenko, pp. 21, 26.
12 Numerous citations may be found in Lewin, chapter 17.
13 Stalin, Vol. 12, pp. 176, 177.
14 N. Ivnitsky: *Klassovaya bor'ba v derevnye i likvidatsiya kulachestva kak klassa* (Moscow, 1972), pp. 298–9.
15 Vyltsan et al., p. 18.
16 Stalin, Vol. 12, p. 194.
17 I. Trifonov: *Ocherki istorii klassovoi bor'by v SSSR v gody Nepa* (Moscow, 1960), p. 237.
18 See Chapter 12 in M. Fainsod: *Smolensk under Soviet Rule* (Macmillan, 1958). (The Smolensk archives were captured by the Germans and then by the Americans.)
19 *Soviet Studies* (July 1966), pp. 20–37.
20 Bogdenko, pp. 22, 24.
21 ibid., p. 29.
22 Moshkov, pp. 82–3.
23 Bogdenko, p. 20.
24 Stalin, Vol. 12, p. 199, published in *Pravda*, 2 March 1930.
25 Vyltsan et al., p. 7.
26 Bogdenko, p. 28.
27 ibid., p. 30.
28 ibid., p. 36.
29 Vyltsan et al., p. 4.
30 This was criticized in a central committee decision of 26 March 1932 but of course was due to centrally imposed policies.
31 *The Soil Upturned* (English translation, Moscow, 1934), p. 152.
32 Vyltsan et al., p. 10.
33 Moshkov, p. 112.
34 ibid., pp. 155–6.
35 ibid., pp. 101, 156, 176.
36 ibid., p. 176.
37 Quoted from *RSFSR, Ugolovnyi Kodeks* (1936), pp. 120–21.
38 *Pravda*, 10 March 1963.
39 Moshkov, p. 227.

[40] ibid., p. 137.
[41] ibid., pp. 190–91.
[42] ibid., p. 201.
[43] ibid., p. 215, quoting archive materials.
[44] ibid., p. 215.
[45] Zelenin: *Istoricheskie Zapiski*, No. 76 (1965), p. 53.
[46] Moshkov, p. 217.
[47] Zelenin, p. 44.
[48] *Fulfilment of second five-year plan*, p. 269.
[49] Zelenin, p. 47. See also Dana Dalrymple: *Soviet Studies* (January 1964), pp. 250–84.
[50] Zelenin: *Istoriya SSSR*, No. 5 (1964), p. 24.
[51] Decree of 20 December 1929, *Ek.zh.*, pp. 223–4.
[52] Zelenin: *Istoricheskie Zapiski*, No. 76, p. 52.
[53] ibid., p. 58.
[54] In his speech of 23 February 1933.
[55] *Ek.zh.*,'p. 276.
[56] ibid., p. 284.
[57] For instance Millar's articles in *Soviet Studies* (July 1970) and *Slavic Review* (December 1974), and Ellman's *Socialist Planning* (Cambridge, 1978), Chapter 4, and *Economic Journal* (December 1975).
[58] Reports by Cairns, Public Record Office, F.O. 371/16329 (1932), gives a devastating picture of the conditions of the time.

Chapter 8

[1] Stalin, Vol. 12, p. 31.
[2] ibid., p. 40.
[3] I. Babel: *Izbrannoe* (Moscow, 1966), p. 281.
[4] P. Sinitsyn: *Novy mir*, No. 8 (1967), p. 4.
[5] Pogrebinsky, p. 112.
[6] I. Gladkov: *Postroenie fundamenta sotsialisticheskoi ekonomiki v SSSR, 1926–32*.
[7] Malafeyev, p. 139.
[8] ibid., p. 148.
[9] Decree of 22 November 1930.
[10] Malafeyev, p. 167.
[11] ibid., p. 167.
[12] G. Neiman: *Planovoe khozyaistvo*, No. 6, 7 (1932), pp. 118–19.
[13] Malafeyev, p. 166. (Archive materials not previously published.)
[14] ibid., pp. 172–3.
[15] ibid., pp. 165, 189.
[16] ibid., p. 193.
[17] ibid., p. 402.
[18] ibid., p. 193 ('compulsory' products were those which customers had to buy as a condition for buying what they really wanted).
[19] ibid., p. 174.
[20] ibid.

[21] ibid.

[22] Stalin, Vol. 13, p. 204.

[23] Decrees of 20 October 1930, 15 November 1932, and others. See Schwarz, p. 98 ff.

[24] *Ek.zh.*, p. 234.

[25] Malafeyev, pp. 181–2.

[26] ibid., p. 182.

[27] Stalin, Vol. 13, p. 283.

[28] Samokhvalov, p. 283.

[29] All these particulars are well set out in Samokhvalov, p. 238 and pp. 280–92.

[30] E.g., decrees of 14 January, 20 March, and 23 July 1931.

[31] Seventeenth party congress, *stenotchet*, p. 67.

[32] ibid., p. 227.

[33] ibid., p. 62. Breakdowns therefore occurred elsewhere!

[34] ibid., p. 227.

[35] Y. Nikulikhin: VKA No. 14 (1933), pp. 57–8.

[36] See G. Neiman, op. cit., No. 1–2 and 5–6 (1933), which devotes several articles to the railway crisis.

[37] ibid., Nos. 5–6 (1933), p. 15.

[38] ibid., Nos. 5–6, pp. 273–7.

[39] Stalin: Letter to *Proletarskaya revolyutsiya*, reprinted in his *Works*, Vol. 13, pp. 86 ff.

[40] Kaverin's memoirs, *Novy mir*, No. 11 (1966), p. 138.

[41] 'Conferences' used to meet between 'congresses', and are of lower status.

[42] Seventeenth congress, *stenotchet*, p. 238.

[43] ibid., p. 250.

[44] ibid., p. 258.

Chapter 9

[1] *Sots. nar. khoz. v 1933–40*, pp. 204–5.

[2] ibid., p. 624.

[3] ibid., p. 221.

[4] ibid., p. 225.

[5] ibid., p. 127.

[6] ibid., p. 125.

[7] *Nar. khoz. SSSR* (1956), p. 251.

[8] *Sots. nar. knoz. v 1933–40*, p. 107.

[9] ibid., pp. 108, 109.

[10] Regulation cited in A. Baykov: *The Development of the Soviet Economic System* (Cambridge University Press, 1946), p. 231.

[11] *Nar. khoz.* (1956), pp. 172, 204.

[12] *Sots. nar. khoz. v 1933–40*, p. 506. These 'ton-kilometres' combine freight and passenger traffic by a simple formula.

[13] ibid., pp. 516, 518.

[14] ibid., pp. 204–5.

[15] ibid., p. 206.
[16] ibid., p. 87.
[17] ibid., p. 375.
[18] Malafeyev, pp. 41, 179.
[19] *Sots. nar. knoz. v 1933–40*, p. 459.
[20] ibid., p. 374.
[21] I. Zelenin: *Zernovye sovkhozy SSSR* (Moscow, 1966), pp. 18–19.
[22] ibid., p. 126.
[23] Malafeyev, p. 403, citing archives.
[24] *Sots. nar. khoz. v 1933–40*, pp. 388–9.
[25] ibid., p. 460.
[26] Malafeyev, p. 192.
[27] ibid., pp. 203–5.
[28] ibid., p. 205.
[29] Janet Chapman: *Real Wages in Soviet Russia* (Harvard University Press, 1963), p. 153.
[30] *Sots. nar. khoz. v 1933–40*, p. 565.
[31] ibid., p. 558.
[32] ibid., p. 560.
[33] V. Petrov et al., *Spravochnik po stavkam naloga s oborota i byudzhetnykh natsenok* (Moscow, 1955).
[34] G. Maryakhin: *Nalogovaya sistema SSSR* (Moscow, 1952).
[35] ibid., p. 592.
[36] *Sots. nar. khoz. v 1933–40*, p. 20.
[37] ibid., p. 517 (quoting archives).
[38] ibid., p. 213.
[39] ibid., p. 380.
[40] See, for instance, *Sotsialistıcheskaya zakonnost'*, No. 11 (1940), pp. 40–42.
[41] *Sots. nar. khoz. v 1933–40*, p. 388.
[42] ibid., p. 477.
[43] Schwarz, pp. 104–15.
[44] *Sotsialisticheskaya zakonnost'*, No. 9, pp. 9–14, 23.
[45] ibid., pp. 11, 17, 66.
[46] ibid., pp. 17, 62.
[47] ibid., No. 10, p. 51.
[48] ibid., No. 9, p. 13.
[49] ibid., No. 10, p. 52.
[50] ibid., No. 11, pp. 66–7.
[51] ibid.
[52] ibid., No. 2, p. 81.
[53] At some dates a Chief Supplies Department (Glavsnab) had a separate existence.

Chapter 10

[1] Ya. Chadayev: *Ekonomika SSSR, 1941–45* (Moscow, 1965), pp. 59–60.
[2] All this is fully documented in the official *Istoriya velikoi otechestvennoi voiny, 1941–5*, Vol. I, pp. 412–14.

3 There is a fascinating description of how the defence industries operated in the memoirs of V. Emelyanov in *Novy mir*, No. 1 (1967), pp. 5–82.

4 *Istoriya . . ., 1941–5*, Vol. II, p. 148.

5 Chadayev, p. 75.

6 E. Lokshin: *Promyshlennost' SSSR, 1940–63* (Moscow, 1964), pp. 40, 43.

7 Chadayev, p. 65.

8 Lokshin, p. 50.

9 ibid., p. 53.

10 *Istoriya . . ., 1941–5*, Vol. VI, p. 46.

11 Chadayev, p. 93.

12 ibid., p. 143.

13 Malafeyev, p. 218.

14 *Istoriya . . ., 1941–5*, Vol. VI, p. 48.

15 ibid., p. 72.

16 ibid., p. 62.

17 *Sovetskoye krestyanstvo v gody Velikoi otechestvennoi voiny* (Moscow, 1970).

18 Arutunyan, pp. 203, 204.

19 ibid., pp. 339, 340.

20 *Istoriya . . ., 1941–5*, Vol. VI, p. 98.

21 For a full account, see A. Dallin: *German Rule in Russia, 1941–45*, a study in occupation policies (Macmillan, 1957).

22 N. Voznesensky: *Voennaya ekonomika SSSR v period otechestvennoi voiny* (Moscow, 1948), p. 110.

23 Malafeyev, p. 225.

24 *Istoriya . . ., 1941–5*, Vol. VI, p. 75.

25 Voznesensky, p. 117–18.

26 *Trud v SSSR* (Moscow, 1968), 1938.

27 Voznesensky, p. 113.

28 *Istoriya . . ., 1941–5*, Vol. VI, p. 75.

29 Malefeyev, pp. 228–9.

30 Chadayev, p. 108.

31 Malafeyev, p. 235.

32 ibid., p. 234.

33 ibid., p. 235.

34 *Finansy i kredit SSSR* (Moscow, 1956), p. 123.

35 *Ek.zh.*, pp. 397, 409, 421.

36 Decree of 13 June 1943.

37 *Istoriya . . ., 1941–5*, Vol. VI, p. 79.

38 Voznesensky, p. 160–61.

39 *Ek.zh.*, p. 421 (quoting a decision of 27 May 1944).

Chapter 11

1 *Ek.zh.*, p. 434.

2 M. MacDuffie: *Red Carpet: 10,000 miles through Russia on a visa from Khrushchev* (London, 1955).

3 *Pravda*, 10 February 1946.

4 *Ekonomika SSSR v poslevoennyi period*, edited by A. Efimov (Moscow, 1962), pp. 19, 125.

5 Lokshin, p. 111.

6 ibid., pp. 121–2.

7 ibid., p. 124.

8 ibid., p. 148.

9 *Ek.zh.*, p. 462.

10 *Nar. khoz.* (1965), p. 311.

11 *Pravda*, 10 December 1963.

12 *Ek.zh.*, p. 453.

13 Decree of 9 November 1946, *Ek.zh.*, p. 443.

14 ibid., p. 457.

15 Malafeyev, pp. 266–7.

16 See pages 307, 309–10, for more details of retail prices.

17 Full details of sources in A. Nove: 'Rural taxation in the USSR', *Soviet Studies* (October 1953).

18 Khrushchev: *Stroitel'stvo kommunizma v SSSR* (Moscow, 1964), Vol. 2, p. 463.

19 This is taken from an obscure legal textbook, *Spravochnik po zakonodatelstvu dlya sudebno-prokurorskikh rabotnikov* (Moscow, 1949).

20 Khrushchev, Vol. 13, p. 529.

21 *Ek.zh.*, p. 496.

22 *Pravda*, 4 and 5 March 1950.

23 Malafeyev, p. 255.

24 D. Kondrashev: *Tsenoobrazovanie v promyshlennosti SSSR* (Moscow, 1956).

25 Malafeyev, p. 244.

26 ibid., p. 258.

27 The 1950 figure was published in *Nar. khoz.* (1965), p. 567. It was secret at the time. The 1947 figure still needs to be roughly estimated.

28 S. Figurnov: *Real'naya zarabotnaya plata* (Moscow, 1960), p. 198.

29 Stalin, Vol. 12, p. 332.

30 Malafeyev, p. 407.

31 The system is well described by F. Pryor: *The Communist foreign trade system* (Allen & Unwin, 1963).

32 *Izvestiya*, 25 January 1949.

33 Lokshin, p. 157.

34 J. Stalin: *Economic Problems of Socialism in the USSR* (Moscow, 1952), p. 81, in English.

35 ibid., p. 75.

36 Malafeyev, p. 265.

37 Khrushchev, *Pravda*, 15 January 1960.

38 *Ek.zh.*, pp. 523, 524, 533.

39 A. Nove: 'Rural taxation in the USSR', *Soviet Studies* (October 1953), p. 164.

40 Such increases are implied in the table cited by Malafeyev, p. 267.

41 V. Ovechkin: 'Rayonnye budni', *Novy mir*, No. 9 (1952), pp. 204–21.

Chapter 12

1 Full list in *Pravda*, 16 March 1953.
2 Malafeyev, p. 261.
3 Figurnov, p. 192.
4 Malafeyev, p. 412.
5 For full details of all this, see Khrushchev's speech to the September 1953 plenum, *Pravda*, 3 September 1953; for texts of decrees and resolutions see *Direktivy KPSS i sovetskovo pravitel'stva po khozyaistvennym voprosam* (Moscow, 1958), Vol. 4, pp. 16–18, 62–162.
6 *Ek.zh.*, p. 369.
7 *Direktivy . . .*, Vol. 4, pp. 193–205.
8 Decree of 25 December 1954, text in *Direktivy . . .*, Vol. 5, pp. 319–28.
9 *Komsomol'skaya Pravda*, 22 June 1956.
10 Speech at Novosibirsk. See *Stroitel'stvo Kommunizma v SSSR i razvitie sel'skovo khozyaistva* (speeches by N. Khrushchev), Vol. 6, p. 205.
11 *Ek.zh.*, p. 561.
12 Sources set out in A. Nove: 'The income of Soviet peasants', *Slavonic Review* (June 1960).
13 N. Khrushchev, speaking to the December 1958 plenum. He did not make it clear at what price the in-kind income was valued.
14 *SSSR v tsifrakh v 1965 godu*, p. 126.
15 *Ek.zh.*, p. 593.
16 *Direktivy . . .*, Vol. 14, p. 451 ff.
17 ibid., p. 421.
18 I. Kulev: *O dal'neishem sovershenstvovanii planirovaniya i rukovodstva narodnym khozyaistvom* (Moscow, 1957).
19 Full details in the *Stenografichesky otchet* (report) of the Supreme Soviet session of 7–10 May 1957.
20 Figurnov, p. 192.
21 *SSSR v tsifrakh v 1965 godu*, p. 125.
22 See particularly N. Bulganin's speech, *Pravda*, 17 July 1955, and *Sotsialistichesky trud*, No. 1 (1956), pp. 24 ff.; N. Maslova in *Voprosy ekonomiki*, No. 8 (1955).
23 V. Mayer: *Zarabotnaya plata v periode perekhoda k Kommunizmu* (Moscow, 1963), p. 61.
24 Decree of 15 August 1955.
25 Decree of 8 March 1956.
26 Decree of 26 March 1956.
27 25 April 1956.
28 6 June 1956.
29 Adopted by Supreme Soviet, 14 July 1956.
30 Decree of 23 March 1957.
31 Malafeyev, p. 313.
32 *Ek.zh.*, p. 596.
33 *Ek.zh.*, p. 613, and *Pravda*, 20 January 1960.
34 *Nar. khoz.*, 1965, p. 47.
35 *Ek.zh.*, p. 621.

[36] See A. Nove: 'Soviet industrial reorganization', *Problems of Communism* (November–December 1957).

[37] *Pravda*, 2 July 1959.

[38] Particularly useful is O. Antonov: *Dlya vsekh i dlya sebya* (Moscow, 1965).

[39] *Ekonomicheskaya gazeta*, 10 November 1962, p. 8.

[40] For a fully documented account of the resulting muddle, see A. Nove: 'The industrial planning system: reforms in prospect', *Soviet Studies* (July 1962).

[41] L. Brezhnev's speech to the March 1965 plenum, and V. Matskevich (Minister of Agriculture) in *Voprosy ekonomiki*, No. 6 (1965). To prove that the author did not have to wait for such evidence, see A. Nove: 'Soviet agriculture marks time', *Foreign Affairs* (July 1962).

[42] G. Shmelev: *Voprosy ekonomiki*, No. 4 (1965).

[43] *Ekonomicheskaya gazeta*, No. 15 (April 1967), p. 32.

Chapter 13

[1] According to the Minister of Finance, Garbuzov, this subsidy was to total 100 milliard roubles in the five years 1976–80 (*Plan. khoz.*, No. 7, 1977, p. 17). A further increase in prices in 1978 was to 'cost' a further 3 milliards a year, according to *Plan. khoz.*, No. 10, 1978.

[2] N. Glushkov: *Plan. khoz.*, No. 1, 1980, p. 10.

[3] For evidence see for instance V. Dobrynin, *Vop. ekon.*, No. 11, 1974, and S. Khoicman in E.K.O., Nos. 5 and 6, 1980.

[4] *Nar. khoz.*, 1970, p. 131.

[5] Calculated from data in *Nar.khoz.*, 1978, p. 433.

[6] For excellent biographies of some key planners of the twenties, see R. W. Davies: 'Some Soviet economic controllers', *Soviet Studies* (January, April and July 1960).

[7] For a very interesting historical survey of the role of authority in Russia, see T. H. Rigby: 'Security and modernization', *Survey* (July 1967).

[8] G. Lisichkin: *Plan i rynok* (Moscow, 1966).

[9] B. Mozhaev: 'Iz zhizni Fedora Kuz'kina', *Novy mir*, No. 7 (1966).

Appendix

[1] A. Bergson: *The Real National Income of Soviet Russia since 1928*, pp. 228, 237.

[2] For an account of the different kinds of *obyedineniye* see A. Nove: *The Soviet Economic System* (Allen & Unwin, 1977).

[3] A. Katsenelinboigen and S. Shatalin, in *Voprosy ekonomiki*, No. 4 (1967), pp. 98–106.

[4] See Bibliography for references.

39 See A. Mozer, *Novye industrial'nye tsentry i mesta*, Publishing Corporation (Novemberr December 1972).

40 *Pravda*, 2 July 1973.

41 Particularly useful is O. Antonov, *Dlya vsekh i dlya kazhdogo* (Moscow, 1965).

42 *Ekonomicheskaya gazeta*, 10 November 1964, p. 8.

43 For a fully documented account of the revitalised academy, see A. Nove, *The industrial planning system: reforms in prospect*, *Soviet Studies* (July 1962).

44 L. Brezhnev's speech to the March 1965 plenum, and M. Makievich (Minister of Agriculture), *Pravda*, 9 November 1965 &[??]. To show that the author did not have himself the such evidence, see A. Ts ...

45 *Sotsialisticheskaya industriya*, ... *Pravda* (July 1965).

46 G. Shmelev, *Lgovye obozreniya Nauk* (1965).

47 *Ekonomicheskaya nauka*, No. 1, April 1967, p. 2.

Chapter 13

1 According to the Minister of Finance Garbuzov, the subsidy was to total 100 milliard roubles in the five years 1976–80 (*Pravda*, February 1977, p. 2). Another increase in prices in 1979 was to cost a further 9 milliards a year, according to *Pravda*, No. 4201, No. 70, 1978.

2 K. Chakhesov, *Plan khoz.*, No. 1, 1980, p. 19.

3 For guidance see, for instance, V. Dobrynin, *Vop. ekon.*, No. 11, 1974. And S. Kim etal in E.K.O., Nos. 5 and 6, 1980.

4 *Vop. ekon.*, 1970/56, 131.

5 Calculated from data in *Nar. khoz.*, 1979, p. 193.

6 For excellent biographies of some key managers of the economy, see R. W. Davies, 'Some Soviet economic controllers', *Soviet Studies* (January, April and July 1960).

7 For a very interesting biographical survey of the role of authority in Russia, see T. H. Rigby, *Stability and change in Russian politics*, *Survey* (July 1967).

8 O. Lewin, *Le Pain* (Paris, 1968).

9 R. Manevich in *Mezh i eksiora E.K.O.* (Baku), No. 7, 1969.

Appendix

1 A. Bergson, *The Real National Income of Soviet Russia since 1928*, pp. 222/227.

2 For an account of the different kinds of controversies, see A. Nove, *The Soviet Economic System* (Allen & Unwin, 1977).

3 A. K sterial'shchikov and S. Khoziaistvo, in *Voprosy ekonomiki* (2) (1969) pp. 98–106.

4 See Bibliography for relevance.

Glossary

The following are some Russian words used frequently in the text. They are first given in italic in the text, but some very frequently repeated terms, like kulak or kolkhoz, have been treated as English words and given English plurals.

Artel': Any cooperative work group, but particularly an agricultural collective.

Glavk (plural glavki): Chief department, e.g. in a ministry or other central institution.

Gosekonomkomissiya: State Economic Commission for current planning.

Gosplan: State Planning Commission (Committee), attached to government.

Kulak: Rich peasant.

Kolkhoz (*kollektivnoe khozyaistvo*): Collective farm.

Khozraschyot: Economic or commercial accounting.

MTS: Machine Tractor Station (abolished 1958).

Nachalstvo: Bosses.

Nar. khoz. (narodnoe khozyaistvo): Literally 'People's (or national) economy'. Often used as title of a statistical handbook or annual.

NEP: New Economic Policy (1921–9?).

Nepman: Beneficiary of NEP, private trader or manufacturer.

Obkom: Party committee of *oblast'*.

Oblast': Province.

Politotdel: Political department in some institution (e.g. MTS, railways, army).

Politbureau: Political bureau of central committee of the communist party; top authority in party (called presidium, 1952–66).

Prodrazverstka: Compulsory delivery to the government of 'surplus' foodstuffs by the peasants.

RSFSR: Russian Soviet Federative Socialist Republic (Russia proper).

Raikom: Party committee of a *raion*.

Raion: District.

Sovnarkhoz: Regional or republican council of national economy (except that a U.S.S.R. sovnarkhoz existed, 1962–5).

STO: Council of Labour and Defence.

Travopolye: Crop rotation system.

Trudoden' (plural Trudodni): Workday units, used for calculating remuneration in collective farms.

VSNKH (Vysshyi Sovet Narodnovo Khozyaistva): Supreme Council of National Economy (December 1917–January 1932, also 1963–5).

Index of Subjects

Index of Names